CLUSTERING

IEEE Press
445 Hoes Lane
Piscataway, NJ 08854

Books in the IEEE Press Series on Computational Intelligence

Introduction to Evolvable Hardware: A Practical Guide for Designing Self-Adaptive Systems
Garrison W. Greenwood and Andrew M. Tyrrell
2007 978-0471-71977-9

Evolutionary Computation: Toward a New Philosophy of Machine Intelligence, Third Edition
David B. Fogel
2006 978-0471-66951-7

Emergent Information Technologies and Enabling Policies for Counter-Terrorism
Edited by Robert L. Popp and John Yen
2006 978-0471-77615-4

Computationally Intelligent Hybrid Systems
Edited by Seppo J. Ovaska
2005 0-471-47668-4

Handbook of Learning and Appropriate Dynamic Programming
Edited by Jennie Si, Andrew G. Barto, Warren B. Powell, Donald Wunsch II
2004 0-471-66054-X

Computational Intelligence: The Experts Speak
Edited by David B. Fogel and Charles J. Robinson
2003 0-471-27454-2

Computational Intelligence in Bioinformatics
Edited by Gary B. Fogel, David W. Corne, Yi Pan
2008 978-0470-10526-9

Computational Intelligence and Feature Selection: Rough and Fuzzy Approaches
Richard Jensen and Qiang Shen
2008 978-0470-22975-0

CLUSTERING

RUI XU

DONALD C. WUNSCH, II

IEEE Computational Intelligence Society, *Sponsor*

IEEE Press Series on Computational Intelligence
David B. Fogel, *Series Editor*

IEEE PRESS

A JOHN WILEY & SONS, INC., PUBLICATION

Published by John Wiley & Sons, Inc., Hoboken, New Jersey.
Published simultaneously in Canada.

For general information on our other products and services or for technical support, please contact our Customer Care Department within the United States at (800) 762-2974, outside the United States at (317) 572-3993 or fax (317) 572-4002.

Wiley also publishes its books in a variety of electronic formats. Some content that appears in print may not be available in electronic formats. For more information about Wiley products, visit our web site at www.wiley.com.

Library of Congress Cataloging-in-Publication Data is available.

ISBN: 978-0-470-27680-8

Printed in the United States of America.

10 9 8 7 6 5 4 3 2 1

CONTENTS

PREFACE

Clustering has become an increasingly important topic in recent years, caused by the glut of data from a wide variety of disciplines. However, due to the lack of good communication among these communities, similar theories or algorithms are redeveloped many times, causing unnecessary waste of time and resources. Furthermore, different terminologies confuse practitioners, especially those new to cluster analysis. Clear and comprehensive information in this field is needed. This need, among others, has encouraged us to produce this book, seeking to provide a comprehensive and systematic description of the important clustering algorithms rooted in statistics, computer science, computational intelligence, and machine learning, with an emphasis on the new advances in recent years. The book consists of 11 chapters, ranging from the basic concept of cluster analysis, proximity measures, and cluster validation, to a wide variety of clustering algorithms, including hierarchical clustering, partitional clustering, neural network-based clustering, kernel-based clustering, sequential data clustering, large-scale data clustering, and high dimensional data clustering. It also includes rich references and illustrates examples in recent applications, such as bioinformatics and web document clustering. Exercises are provided at the end of the chapters to help readers understand the corresponding topics.

The book is intended as a professional reference and also as a course textbook for graduate students in math, science, or engineering. We expect it to be particularly interesting to computer scientists and applied mathematicians applying it to data-intensive applications like bioinformatics, data mining, sensor networks, and computer security, among many other fields. It is a natural fit for computational intelligence researchers, who often must use clustering for feature selection or data reduction. The book will not have extensive assumptions of prerequisite background but will provide enough detail to allow the reader to select the method that best fits his or her application.

We have been working on cluster analysis for many years. Support from the National Science Foundation, Sandia Laboratories, and the M.K. Finley Missouri endowment is gratefully acknowledged.

We are grateful to the thousands of researchers who have contributed to this field, many of whom are our current and past collaborators, mentors, role models, and friends. It is not possible to reference all of the countless publications in this area, but we are always interested in finding outstanding ones we may have overlooked, perhaps to cover in a future edition. We thank the anonymous Associate Editor of our 2005 paper in *IEEE Transactions on Neural Networks** for the part on classification and clustering. We wish to thank the reviewers for their helpful comments. We are grateful to Bart Kosko for encouraging us to write this book after the success of the journal article. The manuscript of the book has been used in a course at Missouri University of Science and Technology. Many thanks to the graduate students Soumya De, Tae-hyung Kim, Ryan Meuth, Paul Robinette, John Seiffertt IV, and Hanzheng Wang for their valuable feedback and help in solving the homework problems. We also wish to thank Ms. Barbie Kuntemeier for her proofreading assistance.

Finally, Rui Xu would like to thank his family: Xiaomin, Benlin, Shuifeng, Wei, Jie, and Qiong; and Don Wunsch would like to thank Hong and Donnie. Without their encouragement, understanding, and patience, this book would not exist.

* Reference (Xu and Wunsch, 2005), which remained on the IEEE Explore top 100 list for over a year.

CHAPTER 1

CLUSTER ANALYSIS

1.1. CLASSIFICATION AND CLUSTERING

We are living in a world full of data. Every day, people deal with different types of data coming from all types of measurements and observations. Data describe the characteristics of a living species, depict the properties of a natural phenomenon, summarize the results of a scientific experiment, and record the dynamics of a running machinery system. More importantly, data provide a basis for further analysis, reasoning, decisions, and ultimately, for the understanding of all kinds of objects and phenomena. One of the most important of the myriad of data analysis activities is to classify or group data into a set of categories or clusters. Data objects that are classified in the same group should display similar properties based on some criteria. Actually, as one of the most primitive activities of human beings (Anderberg, 1973; Everitt et al., 2001), classification plays an important and indispensable role in the long history of human development. In order to learn a new object or understand a new phenomenon, people always try to identify descriptive features and further compare these features with those of known objects or phenomena, based on their similarity or dissimilarity, generalized as proximity, according to some certain standards or rules. As an example, all natural objects are basically classified into three groups: animal, plant, and mineral. According to the biological taxonomy, all animals are further classified into categories of kingdom, phylum, class, order, family, genus, and species, from general to specific. Thus, we have animals named tigers, lions, wolves, dogs, horses,

Clustering, by Rui Xu and Donald C. Wunsch, II

sheep, cats, mice, and so on. Actually, naming and classifying are essentially synonymous, according to Everitt et al. (2001). With such classification information at hand, we can infer the properties of a specific object based on the category to which it belongs. For instance, when we see a seal lying easily on the ground, we know immediately that it is a good swimmer without really seeing it swim.

Basically, classification systems are either supervised or unsupervised, depending on whether they assign new data objects to one of a finite number of discrete supervised classes or unsupervised categories, respectively (Bishop, 1995; Cherkassky and Mulier, 1998; Duda et al., 2001). In supervised classification, the mapping from a set of input data vectors, denoted as $\mathbf{x} \in \Re^d$, where d is the input space dimensionality, to a finite set of discrete class labels, represented as $y \in 1, \ldots, C$, where C is the total number of class types, is modeled in terms of some mathematical function $y = y(\mathbf{x},\mathbf{w})$, where $\mathbf{w}$ is a vector of adjustable parameters. The values of these parameters are determined (optimized) by an inductive learning algorithm (also termed inducer), whose aim is to minimize an empirical risk functional (related to an inductive principle) on a finite data set of input-output examples, $(\mathbf{x}_i, y_i)$, $i = 1, \ldots, N$, where N is the finite cardinality of the available representative data set (Bishop, 1995; Cherkassky and Mulier, 1998; Kohavi, 1995). When the inducer reaches convergence or terminates, an induced classifier is generated (Kohavi, 1995).

In unsupervised classification, also called clustering or exploratory data analysis, no labeled data are available (Everitt et al., 2001; Jain and Dubes, 1988). The goal of clustering is to separate a finite, unlabeled data set into a finite and discrete set of "natural," hidden data structures, rather than to provide an accurate characterization of unobserved samples generated from the same probability distribution (Baraldi and Alpaydin, 2002; Cherkassky and Mulier, 1998). This can make the task of clustering fall outside of the framework of unsupervised predictive learning problems, such as vector quantization (Cherkassky and Mulier, 1998) (see Chapter 4), probability density function estimation (Bishop, 1995) (see Chapter 4), and entropy maximization (Fritzke, 1997). It is noteworthy that clustering differs from multidimensional scaling (perceptual maps), whose goal is to depict all the evaluated objects in a way that minimizes topographical distortion while using as few dimensions as possible. Also note that, in practice, many (predictive) vector quantizers are also used for (non-predictive) clustering analysis (Cherkassky and Mulier, 1998).

It is clear from the above discussion that a direct reason for unsupervised clustering comes from the requirement of exploring the unknown natures of the data that are integrated with little or no prior information. Consider, for example, disease diagnosis and treatment in clinics. For a particular type of disease, there may exist several unknown subtypes that exhibit similar morphological appearances while responding differently to the same therapy. In this context, cluster analysis with gene expression data that measure the activities of genes provides a promising method to uncover the subtypes and thereby determine the corresponding therapies. Sometimes, the process of labeling data samples may become extremely expensive and time consuming, which also makes clustering a good choice considering the great savings in both cost and time. In addition, cluster analysis provides a compressed representation of the data and is useful in large-scale data analysis. Aldenderfer and Blashfield (1984) summarized the goals of cluster analysis in the following four major aspects:

- Development of a classification;
- Investigation of useful conceptual schemes for grouping entities;
- Hypothesis generation through data exploration;
- Hypothesis testing or the attempt to determine if types defined through other procedures are in fact present in a data set.

Nonpredictive clustering is a subjective process in nature that precludes an absolute judgment as to the relative efficacy of all clustering techniques (Baraldi and Alpaydin, 2002; Jain et al., 1999). As pointed out by Backer and Jain (1981), "in cluster analysis a group of objects is split up into a number of more or less homogeneous subgroups on the basis of an often subjectively chosen measure of similarity (i.e., chosen subjectively based on its ability to create "interesting" clusters), such that the similarity between objects within a subgroup is larger than the similarity between objects belonging to different subgroups." Moreover, a different clustering criterion or clustering algorithm, even for the same algorithm but with different selection of parameters, may cause completely different clustering results. For instance, human beings may be classified based on their ethnicity, region, age, socioeconomic status, education, career, hobby, weight and height, favorite food, dressing style, and so on. Apparently, different clustering criteria may assign a specific individual to very different groups and therefore produce different partitions. However, there is absolutely no way to determine which criterion is the best in general. As a matter of fact, each criterion has its own appropriate use corresponding to particular occasions, although some of them may be applied to wider situations than others. Figure 1.1 illustrates another example of the effect of subjectivity on the resulting clusters. A coarse partition divides the regions into four major clusters, while a finer one suggests that the data consist of nine clusters. Whether we adopt a coarse or fine scheme depends on the requirement of the specific problem, and in this sense, we would not say which clustering results are better, in general.

1.2. DEFINITION OF CLUSTERS

Clustering algorithms partition data objects (patterns, entities, instances, observances, units) into a certain number of clusters (groups, subsets, or

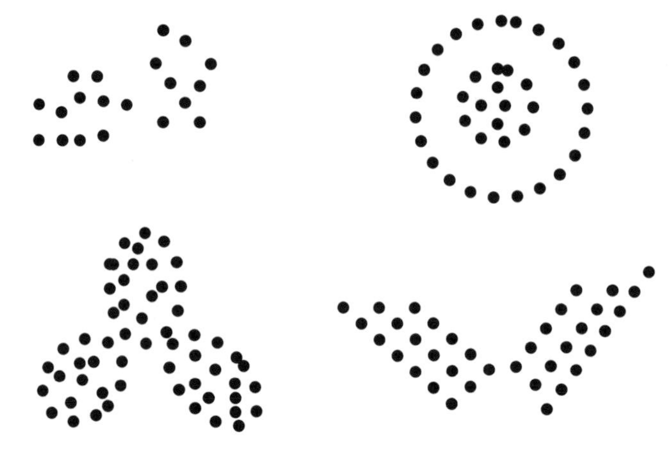

Fig. 1.1. Illustration of subjectivity of cluster analysis. Clustering at a coarse level produces four major clusters, while a finer clustering leads to nine clusters.

categories). However, there is no universally agreed upon and precise definition of the term cluster. Everitt et al. (2001) indicate that "formal definition (of cluster) is not only difficult but may even be misplaced." In spite of this difficulty, several operational definitions are still available, as summarized by Everitt (1980) and illustrated as follows:

"A cluster is a set of entities which are alike, and entities from different clusters are not alike."

A cluster is "an aggregate of points in the test space such that the distance between any two points in the cluster is less than the distance between any point in the cluster and any point not in it."

"Clusters may be described as continuous regions of this space (d-dimensional feature space) containing a relatively high density of points, separated from other such regions by regions containing a relatively low density of points."

Clearly, a cluster in these definitions is described in terms of internal homogeneity and external separation (Gordon, 1999; Hansen and Jaumard, 1997; Jain and R. Dubes, 1988), i.e., data objects in the same cluster should be similar to each other, while data objects in different clusters should be dissimilar from one another. Both the similarity and the dissimilarity should be elucidated in a clear and meaningful way. Here, we give some simple mathematical descriptions of two types of clustering, known as partitional and hierarchical clustering, based on the discussion in Hansen and Jaumard (1997).

Given a set of input patterns $\mathbf{X} = \{\mathbf{x}_1, \ldots, \mathbf{x}_j, \ldots, \mathbf{x}_N\}$, where $\mathbf{x}_j = (x_{j1}, x_{j2}, \ldots, x_{jd}) \in \Re^d$, with each measure x_{ji} called a feature (attribute, dimension, or variable):

1. Hard partitional clustering attempts to seek a K-partition of $\mathbf{X}$, $C = \{C_1, \ldots, C_K\}$ $(K \le N)$, such that

- $C_i \neq \phi, i = 1, \cdots, K;$ (1.1)
- $\bigcup_{i=1}^{K} C_i = \mathbf{X};$ (1.2)
- $C_i \cap C_j = \phi, i, j = 1, \cdots, K$ and $i \neq j$. (1.3)

2. Hierarchical clustering attempts to construct a tree-like, nested structure partition of $\mathbf{X}$, $H = \{H_1, \ldots, H_Q\}$ $(Q \leq N)$, such that $C_i \in H_m$, $C_j \in H_l$, and $m > l$ imply $C_i \subset C_j$ or $C_i \cap C_j = \phi$ for all $i, j \neq i, m, l = 1, \ldots, Q$.

For hard partitional clustering, each data object is exclusively associated with a single cluster. It may also be possible that an object is allowed to belong to all K clusters with a degree of membership, $u_{i,j} \in [0,1]$, which represents the membership coefficient of the jth object in the ith cluster and satisfies the following two constraints:

$$\sum_{i=1}^{K} u_{i,j} = 1, \forall j, \tag{1.4}$$

and

$$\sum_{j=1}^{N} u_{i,j} < N, \forall i, \tag{1.5}$$

as introduced in fuzzy set theory (Zadeh, 1965). This is known as fuzzy clustering and will be discussed in Chapter 4.

Figure 1.2 depicts the procedure of cluster analysis with the following four basic steps:

1. *Feature selection or extraction.* As pointed out by Jain et al. (1999, 2000) and Bishop (1995), feature selection chooses distinguishing features from a set of candidates, while feature extraction utilizes some transformations to generate useful and novel features from the original ones. Clearly, feature extraction is potentially capable of producing features that could be of better use in uncovering the data structure. However, feature extraction may generate features that are not physically interpretable, while feature selection assures the retention of the original physical meaning of the selected features. In the literature, these two terms sometimes are used interchangeably without further identifying the difference. Both feature selection and feature extraction are very important to the effectiveness of clustering applications. Elegant selection or generation of salient features can greatly decrease the storage requirement and measurement cost, simplify the subsequent design process, and facilitate the understanding of the data. Generally, ideal features should be of use in distinguishing patterns belonging to different clusters, immune to noise, and easy to obtain and interpret. We elaborate

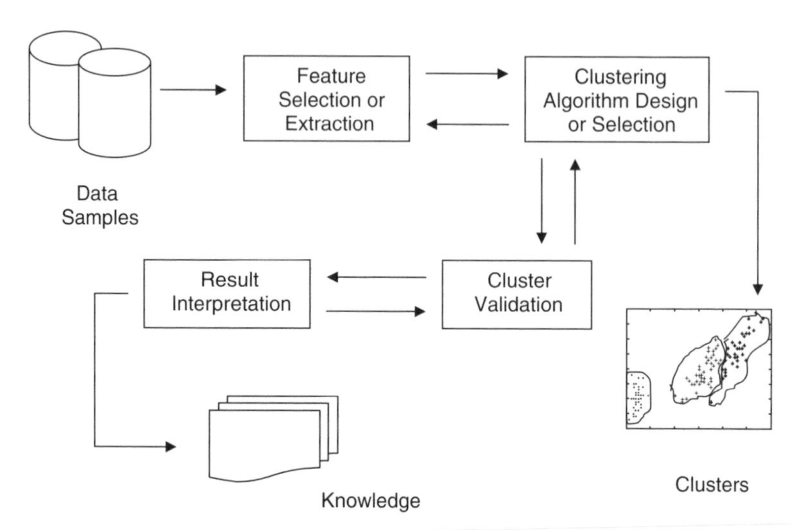

Fig. 1.2. Clustering procedure. The basic process of cluster analysis consists of four steps with a feedback pathway. These steps are closely related to each other and determine the derived clusters.

on the discussion of feature extraction in Chapter 9 in the context of data visualization and dimensionality reduction. Feature selection is more often used in the context of supervised classification with class labels available (Jain et al., 2000; Sklansky and Siedlecki, 1993). Jain et al. (2000), Liu and Yu (2005), and Theodoridis and Koutroumbas (2006) all provided good reviews of the feature selection techniques for supervised learning. A method of simultaneous feature selection and clustering, under the framework of finite mixture models, was proposed in Law et al. (2004). Kim et al. (2000) employed the genetic algorithm for feature selection in a K-means algorithm. Mitra et al. (2002) introduced a maximum information compression index to measure feature similarity and examine feature redundancy. More discussions on feature selection in clustering were given in Dy and Brodley (2000), Roth and Lange (2004), and Talavera (2000).

2. *Clustering algorithm design or selection.* This step usually consists of determining an appropriate proximity measure and constructing a criterion function. Intuitively, data objects are grouped into different clusters according to whether they resemble one another or not. Almost all clustering algorithms are explicitly or implicitly connected to some particular definition of proximity measure. Some algorithms even work directly on the proximity matrix, as defined in Chapter 2. Once a proximity measure is determined, clustering could be construed as an optimization problem with a specific criterion function. Again, the obtained clusters are dependent on the selection of the criterion function. The subjectivity of cluster analysis is thus inescapable.

Clustering is ubiquitous, and a wealth of clustering algorithms has been developed to solve different problems from a wide variety of fields. However, there is no universal clustering algorithm to solve all problems. "It has been very difficult to develop a unified framework for reasoning about it (clustering) at a technical level, and profoundly diverse approaches to clustering" (Kleinberg, 2002). Therefore, it is important to carefully investigate the characteristics of a problem in order to select or design an appropriate clustering strategy. Clustering algorithms that are developed to solve a particular problem in a specialized field usually make assumptions in favor of the application of interest. For example, the K-means algorithm is based on the Euclidean measure and hence tends to generate hyperspherical clusters. However, if the real clusters are in other geometric forms, K-means may no longer be effective, and we need to resort to other schemes. Similar considerations must be kept in mind for mixture-model clustering, in which data are assumed to come from some specific models that are already known in advance.

3. *Cluster validation*. Given a data set, each clustering algorithm can always produce a partition whether or not there really exists a particular structure in the data. Moreover, different clustering approaches usually lead to different clusters of data, and even for the same algorithm, the selection of a parameter or the presentation order of input patterns may affect the final results. Therefore, effective evaluation standards and criteria are critically important to provide users with a degree of confidence for the clustering results. These assessments should be objective and have no preferences to any algorithm. Also, they should be able to provide meaningful insights in answering questions like how many clusters are hidden in the data, whether the clusters obtained are meaningful from a practical point of view or just artifacts of the algorithms, or why we choose one algorithm instead of another. Generally, there are three categories of testing criteria: external indices, internal indices, and relative indices. They are defined on three types of clustering structures, known as partitional clustering, hierarchical clustering, and individual clusters (Gordon, 1998; Halkidi et al., 2002; Jain and Dubes, 1988). Tests for situations in which no clustering structure exists in the data are also considered (Gordon, 1998) but seldom used because users are usually confident of the presence of clusters in the data of interest. External indices are based on some prespecified structure, which is the reflection of prior information on the data and is used as a standard to validate the clustering solutions. Internal tests are not dependent on external information (prior knowledge). Instead, they examine the clustering structure directly from the original data. Relative criteria emphasize the comparison of different clustering structures in order to provide a reference to decide which one may best reveal the characteristics of the objects. Cluster validation will be discussed in Chapter 10, with a focus on the methods for estimating the number of clusters.

4. *Result interpretation.* The ultimate goal of clustering is to provide users with meaningful insights from the original data so that they can develop a clear understanding of the data and therefore effectively solve the problems encountered. Anderberg (1973) saw cluster analysis as "a device for suggesting hypotheses." He also suggested that "a set of clusters is not itself a finished result but only a possible outline." Experts in the relevant fields are encouraged to interpret the data partition, integrating other experimental evidence and domain information without restricting their observations and analyses to any specific clustering result. Consequently, further analyses and experiments may be required.

It is interesting to observe that the flow chart in Fig. 1.2 also includes a feedback pathway. Cluster analysis is not a one-shot process. In many circumstances, clustering requires a series of trials and repetitions. Moreover, there are no universally effective criteria to guide the selection of features and clustering schemes. Validation criteria provide some insights into the quality of clustering solutions, but even choosing an appropriate criterion is a demanding problem.

1.3. CLUSTERING APPLICATIONS

Clustering has been applied in a wide variety of fields, as illustrated below with a number of typical applications (Anderberg, 1973; Everitt et al., 2001; Hartigan, 1975).

1. Engineering (computational intelligence, machine learning, pattern recognition, mechanical engineering, electrical engineering). Typical applications of clustering in engineering range from biometric recognition and speech recognition, to radar signal analysis, information compression, and noise removal.
2. Computer sciences. We have seen more and more applications of clustering in web mining, spatial database analysis, information retrieval, textual document collection, and image segmentation.
3. Life and medical sciences (genetics, biology, microbiology, paleontology, psychiatry, clinic, phylogeny, pathology). These areas consist of the major applications of clustering in its early stage and will continue to be one of the main playing fields for clustering algorithms. Important applications include taxonomy definition, gene and protein function identification, disease diagnosis and treatment, and so on.
4. Astronomy and earth sciences (geography, geology, remote sensing). Clustering can be used to classify stars and planets, investigate land formations, partition regions and cities, and study river and mountain systems.

5. Social sciences (sociology, psychology, archeology, anthropology, education). Interesting applications can be found in behavior pattern analysis, relation identification among different cultures, construction of evolutionary history of languages, analysis of social networks, archeological finding and artifact classification, and the study of criminal psychology.
6. Economics (marketing, business). Applications in customer characteristics and purchasing pattern recognition, grouping of firms, and stock trend analysis all benefit from the use of cluster analysis.

1.4. LITERATURE OF CLUSTERING ALGORITHMS

Clustering has a long history (Hansen and Jaumard, 1997), dating back to Aristotle. There is a vast, ever-increasing literature on cluster analysis from a wide variety of disciplines. Web of Science ® (www.isinet.com/products/citation/wos/) alone reveals over 12,000 journal and conference papers using the phrase "cluster analysis" within the title, keywords, and abstract in the past decade. These papers come from over 3,000 journals published in more than 200 major subject categories. Detailed search results are depicted in Fig. 1.3. We can see a rapid increase in the past three years.

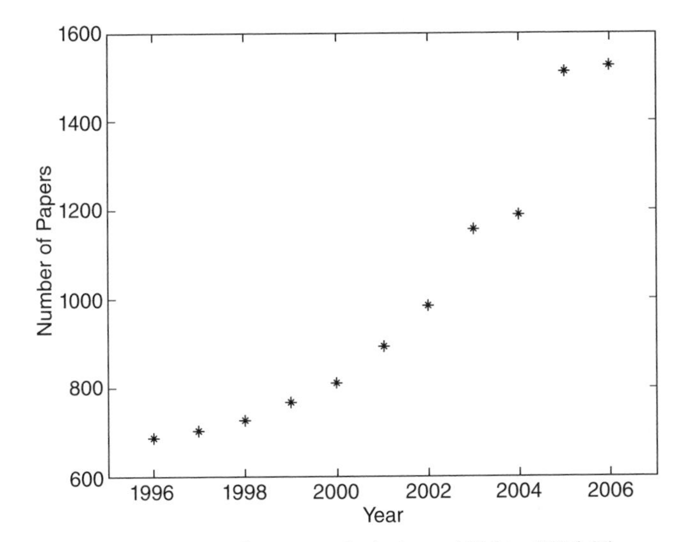

Fig. 1.3. Number of papers on cluster analysis from 1996 to 2006. The searches were performed using Web of Science ®, which includes three databases: the Science Citation Index Expanded™ (SCI_EXPANDED), the Social Sciences Citation Index ® (SSCI), and the Arts & Humanities Citation Index ® (A&HCI).[1]

[1] The Web of Science ®, the Social Sciences Citation Index ®, and the Arts & Humanities Citation Index ® are registered trademarks and the Science Citation Index Expanded™ is a trademark of The Thomson Corporation.

Clustering papers have been appearing in a large number of journals from different disciplines, as we have already seen in the above statistics. We illustrate the following major journals that contain publications on theories, algorithms, and applications of cluster analysis in alphabetical order, which, however, by no means is an exhaustive illustration of all the important journals on this topic:

1. *Applied and Environmental Microbiology*
2. *Bioinformatics*
3. *Biometrics*
4. *Biometrika*
5. *Computer Journal*
6. *Euphytica*
7. *Fuzzy Sets and Systems*
8. *IEEE Transactions on Fuzzy Systems*
9. *IEEE Transactions on Knowledge and Data Engineering*
10. *IEEE Transactions on Neural Networks*
11. *IEEE Transactions on Pattern Analysis and Machine Intelligence*
12. *IEEE Transactions on Systems, Man, and Cybernetics*
13. *Journal of Chemical Information and Computer Sciences*
14. *Journal of Classification*
15. *Journal of Computational Biology*
16. *Journal of the American Statistical Association*
17. *Journal of the Royal Statistical Society*
18. *Machine Learning*
19. *Multivariate Behavioral Research*
20. *Neural Computation*
21. *Neural Networks*
22. *Nucleic Acids Research*
23. *Pattern Recognition*
24. *Pattern Recognition Letters*
25. *Proceedings of the IEEE*
26. *Proceedings of the National Academy of Science of the United States of America*
27. *Psychometrika*
28. *Science*
29. *Theoretical and Applied Genetics*

Among these tens of thousands of papers on cluster analysis, there are a few clustering survey papers that deserve particular attention here. Starting from a statistical pattern recognition viewpoint, Jain et al. (1999) review a large number of clustering algorithms, together with their applications in image segmentation, object and character recognition, information retrieval, and data mining. Hansen and Jaumard (1997) investigate the clustering problems under a mathematical programming scheme. Kolatch (2001) and He (1999) discuss the applications of clustering algorithms in spatial database systems and information retrieval, respectively. Berkhin (2001) further expands the topic to the field of data mining. Murtagh (1983) reports the advances in hierarchical clustering algorithms. Baraldi and Blonda (1999) summarize clustering algorithms using the soft competitive paradigm. Liao (2005) focuses on the clustering of time series. Gordon (1998) and Halkidi et al. (2002) emphasize a specific topic in cluster validation. The most recent comprehensive review of clustering algorithms is offered by Xu and Wunsch (2005). Kettenring (2006) discusses the practical issues in cluster analysis. More survey papers can also be found in Baraldi and Schenato (1999), Bock (1996), Dubes (1993), Fasulo (1999), Jain et al. (2000), and Zhao and Karypis (2005). Moreover, comparative research on clustering algorithms has also been performed and reported in the literature. For example, Rauber et al. (2000) present empirical results for five typical clustering algorithms, including hierarchical agglomerative clustering, Bayesian clustering, self-organizing feature maps, growing hierarchical self-organizing maps, and generative topographic mapping. Wei et al. (2000) concentrate on comparing fast algorithms for large databases. Scheunders (1997) compares K-means, competitive learning clustering, hierarchical competitive learning clustering, and genetic K-means for color image quantization, with an emphasis on computational time and the possibility of obtaining global optima. Applications and evaluations of different clustering algorithms for the analysis of gene expression data from DNA microarray experiments are described in Jiang et al. (2004), Madeira and Oliveira (2004), Shamir and Sharan (2002), and Tibshirani et al. (1999). Experimental evaluations of document clustering algorithms, based on hierarchical and K-means clustering algorithms, are summarized by Steinbach et al. (2000).

In addition to the papers published in the literature, there are also some important books devoted to cluster analysis. Early comprehensive books include those of Aldenderfer and Blashfield (1984), Anderberg (1973), Duran and Odell (1974), Hartigan (1975), Romesburg (1984), and Späth (1980). Jain and Dubes (1988) provide an excellent introduction to and discussion of the major hierarchical and partitional clustering algorithms and other clustering related topics, such as data representation and cluster validation. Everitt et al. (2001) introduce cluster analysis in a readable way, combined with many practical applications. While describing the properties of the clustering algorithms, Kaufman and Rousseeuw (1990) also provide instructions on how to use the programs that implement these corresponding algorithms. Another book that comprehensively discusses cluster analysis is written by Gordon (1999). McLachlan and Peel (2000) discuss clustering using finite probabilistic models, while Höppner et al. (1999) explore fuzzy clustering. Duda et al. (2001)

contribute one chapter on cluster analysis in their excellent book on pattern recognition. More books that include cluster analysis related topics are Bishop (1995), Cherkassky and Mulier (1998), Haykin (1999), and Theodoridis and Koutroumbas (2006).

1.5. OUTLINE OF THE BOOK

As we have already seen, cluster analysis is a basic human mental activity and consists of research developed across a wide variety of communities. Accordingly, cluster analysis has many alternative names differing from one discipline to another. In biology and ecology, cluster analysis is more often known as *numerical taxonomy*. Researchers in computational intelligence and machine learning are more likely to use the terms *unsupervised learning* or *learning without a teacher*. In social science, *typological analysis* is preferred, while in graph theory, *partition* is usually employed. This diversity reflects the important position of clustering in scientific research. On the other hand, it causes confusion because of the differing terminologies and goals. Frequently, similar theories or algorithms are redeveloped several times in different disciplines due to the lack of good communication, which causes unnecessary burdens and wastes time. Thus, the main purpose of this book is to provide a comprehensive and systematic description of the influential and important clustering algorithms rooted in statistics, computer science, computational intelligence, and machine learning, with an emphasis on recent advances.

We begin our discussion in Chapter 2 on the proximity measure between data objects, which is the basis for most clustering algorithms. Considering the different feature types and measurement levels, we describe the similarity and dissimilarity measures for continuous, discrete, and mixed variables. From Chapter 3 to Chapter 9, we review a wide variety of clustering algorithms based on the natures of the formed clusters and the technologies and theories behind these algorithms. For each chapter, we also illustrate examples of some particular applications for the purpose of providing a better way of understanding the corresponding clustering algorithms.

In Chapter 3, we deal with hierarchical clustering algorithms, which group data objects with a sequence of nested partitions. We discuss the pros and cons of classical agglomerative or divisive hierarchical clustering and then describe the recent advances in addressing the disadvantages. In Chapter 4, we move to partitional clustering algorithms that directly divide data objects into some pre-specified number of clusters without the hierarchical structure. The well-known K-means algorithm, graph theory–based methods, and search technique–based methods are all included in this chapter. Further, we see how to generate a fuzzy clustering partition when an object no longer belongs to an exclusive cluster. We also assume that the clusters are generated from different probability distributions, and cluster analysis, in this sense, becomes a procedure of estimating a finite mixture probability density.

Since neural networks have been applied successfully to a wide range of problems, we discuss their use in cluster analysis in Chapter 5, based on a division of either a hard or soft competitive learning strategy used in these technologies. Chapter 6 concentrates on kernel-based clustering algorithms, where the input patterns are mapped into a high-dimensional feature space in order to achieve a more effective partition. We describe the nonlinear version of principal component analysis and two kernel clustering algorithms based on the different construction of the optimization problems in the feature space. Compared with graph theory and fuzzy set theory, which had already been widely used in cluster analysis before the 1980s, neural networks and kernel learning technologies have been finding their applications in clustering more recently. However, in spite of the short history, much progress has been achieved.

Considering the more frequent requirement of tackling sequential, large-scale, and high-dimensional data sets in many current applications, in the next three chapters, we focus on the clustering approaches related to these topics. In Chapter 7, we describe three different strategies for clustering sequential data, which usually have different characteristics from the feature vector-based data. Applications in this chapter are particularly focused on genomic and biological sequences. In Chapter 8, we address the problem of clustering with large-scale data, based on six different types of strategies, such as random sampling and divide and conquer. Then in Chapter 9, visualization of high-dimensional data is discussed in the framework of linear and nonlinear projection approaches. Strictly speaking, these methods are not clustering algorithms because they do not directly partition the data. However, they allow the representation of data in two or three dimensions, where the structure of the data may become clear and the clusters may be revealed.

Cluster validation is an indispensable step in cluster analysis because the resulting clusters are usually largely dependent on the choice of clustering algorithms. In Chapter 10, we describe three categories of testing criteria as external indices, internal indices, and relative indices, and we further address the problem of how to estimate an appropriate number of clusters, which is regarded as the most important problem in cluster validity. We conclude in Chapter 11 by summarizing the major challenges in cluster analysis and the important trends in their development.

CHAPTER 2

PROXIMITY MEASURES

2.1. INTRODUCTION

Clusters are considered as groups containing data objects that are similar to each other, while data objects in different clusters are not. Thus, it is natural to ask what kind of standards we should use to determine the closeness, or how to measure the distance (dissimilarity) or similarity between a pair of objects, an object and a cluster, or a pair of clusters. In Chapter 3 on hierarchical clustering, we will discuss linkage metrics for measuring proximity between clusters. Usually, a prototype is used to represent a cluster so that it can be further processed like other individual objects. Here, we focus on reviewing measurement approaches between individuals.

The chapter begins with an introduction of the characteristics of data themselves, including feature types and measurement levels. The following section provides the basic definition of distance and similarity measurement, as well as metric. Sections 2.4 through 2.6 discuss the proximity measures for data with continuous, discrete (including binary), and mixed variables, respectively.

2.2. FEATURE TYPES AND MEASUREMENT LEVELS

A data object is described by a set of features or variables, usually represented as a multidimensional vector. For N data objects with d features, an $N \times d$

Clustering, by Rui Xu and Donald C. Wunsch, II

pattern matrix is built from the corresponding vectors. Each row in the matrix denotes an object while each column represents a feature. Table 2.1 illustrates the first 25 records of the Horse Colic data set from the Repository of Machine Learning databases (Asuncion and Newman, 2007), and Table 2.2 further shows the possible values for each of the 28 features. We use this data as an example for the following discussion.

A feature can be classified as continuous, discrete, or binary (Anderberg, 1973; Jain and Dubes, 1988). A continuous feature takes values from an uncountably infinite range set, such as the packed cell volume and total protein in Table 2.2. The length or width of petals, the expression level of a gene at a certain state, and the weight or height of a human being are all real examples of continuous features. In comparison, discrete features only have finite, or at most, a countably infinite number of values. Many features in Table 2.2 belong to this category, such as peripheral pulse, peristalsis, abdomen, and nasogastric reflux. More examples include the color and model of cars, the taste of ice cream, the rating of movies, and so on. Binary or dichotomous features are a special case of discrete features when they have exactly two values, like the features of surgical lesion and cp_data in the Horse Colic data, which have only yes or no values.

Another property of features is the measurement level, which reflects the relative significance of numbers (Jain and Dubes, 1988). The levels of measurement consist of four scales, in order from lowest to highest, recognized as nominal, ordinal, interval, and ratio:

1. Nominal. Features at this level are represented with labels, states, or names. The numbers are meaningless in any mathematical sense and no mathematical calculation can be made. The features of peristalsis and outcome in Table 2.2 belong to this level. The feature peristalsis uses 1 to 4 to represent 4 possible values, "hypermotile," "normal," "hypomotile," and "absent," and the outcome feature has three possible values, 1 for "lived," 2 for "died," and 3 for "was euthanized."
2. Ordinal. Features at this level are also names, but with a certain order implied. However, the difference between the values is again meaningless. A simple example is the feature of abdominal distension, which takes values in a meaningful order as "none," "slight," "moderate," and "severe."
3. Interval. Features at this level offer a meaningful interpretation of the difference between two values. However, there exists no true zero and the ratio between two values is meaningless. The feature of rectal temperature, which is measured in degrees Celsius, provides such an example. Obviously, a temperature of 0 °C does not mean there is no heat at all. The ratio of 40 °C to 20 °C is not, in any meaningful sense, a factor of two. Contrast this with degrees Kelvin, where such a ratio does hold true, because it is a true ratio scale, defined next.

TABLE 2.1. The first 25 records of the Horse Colic data set.

Surgery?	Age	Hospital Number	Rectal Temperature	Pulse	Respiratory Rate	Temperature of Extremities	Peripheral Pulse	Mucous Membranes	Capillary Refill Time	Pain	Peristalsis	Abdominal Distension	Nasogastric Tube
2	1	530101	38.50	66	28	3	3	?	2	5	4	4	?
1	1	534817	39.2	88	20	?	?	4	1	3	4	2	?
2	1	530334	38.30	40	24	1	1	3	1	3	3	1	?
1	9	5290409	39.10	164	84	4	1	6	2	2	4	4	1
2	1	530255	37.30	104	35	?	?	6	2	?	?	?	?
2	1	528355	?	?	?	2	1	3	1	2	3	2	2
1	1	526802	37.90	48	16	1	1	1	1	3	3	3	1
1	1	529607	?	60	?	3	?	?	1	?	4	2	2
2	1	530051	?	80	36	3	4	3	1	4	4	4	2
2	9	5299629	38.30	90	?	1	?	1	1	5	3	1	2
1	1	528548	38.10	66	12	3	3	5	1	3	3	1	2
2	1	527927	39.10	72	52	2	?	2	1	2	1	2	1
1	1	528031	37.20	42	12	2	1	1	1	3	3	3	3
2	9	5291329	38.00	92	28	1	1	2	1	1	3	2	3
1	1	534917	38.2	76	28	3	1	1	1	3	4	1	2
1	1	530233	37.60	96	48	3	1	4	1	5	3	3	2
1	9	5301219	?	128	36	3	3	4	2	4	4	3	3
2	1	526639	37.50	48	24	?	?	?	?	?	?	?	?
1	1	5290481	37.60	64	21	1	1	2	1	2	3	1	1
2	1	532110	39.4	110	35	4	3	6	?	?	3	3	?
1	1	530157	39.90	72	60	1	1	5	2	5	4	4	3
2	1	529340	38.40	48	16	1	?	1	1	1	3	1	2
1	1	521681	38.60	42	34	2	1	4	?	2	3	1	?
1	9	534998	38.3	130	60	?	3	?	1	2	4	?	?
1	1	533692	38.1	60	12	3	3	3	1	?	4	3	3

TABLE 2.1. Continued.

Nasogastric Reflux	Nasogastric Reflux PH	Rectal Examination—Feces	Abdomen	Packed Cell Volume	Total Protein	Abdominocentesis Appearance	Abdomcentesis Total Protein	Outcome	Surgical Lesion?	Type of Lesion	Type of Lesion	Type of Lesion	cp_data
?	?	3	5	45.00	8.40	?	?	2	2	11300	00000	00000	2
?	?	4	2	50	85	2	2	3	2	02208	00000	00000	2
?	?	1	1	33.00	6.70	?	?	1	2	00000	00000	00000	1
2	5.00	3	?	48.00	7.20	3	5.30	2	1	02208	00000	00000	1
?	?	?	?	74.00	7.40	?	?	2	2	04300	00000	00000	2
1	?	3	3	?	?	?	?	1	2	00000	00000	00000	2
1	?	3	5	37.00	7.00	?	?	1	1	03124	00000	00000	2
1	?	3	4	44.00	8.30	?	?	2	1	02208	00000	00000	2
1	?	3	5	38.00	6.20	?	?	3	1	03205	00000	00000	2
1	?	3	?	40.00	6.20	1	2.20	1	2	00000	00000	00000	1
1	3.00	2	5	44.00	6.00	2	3.60	1	1	02124	00000	00000	1
1	?	4	4	50.00	7.80	?	?	1	1	02111	00000	00000	2
1	?	4	5	?	7.00	?	?	1	2	04124	00000	00000	2
?	7.20	1	1	37.00	6.10	1	?	2	2	00000	00000	00000	1
2	?	4	4	46	81	1	2	1	1	02112	00000	00000	2
3	4.50	4	?	45.00	6.80	?	?	2	1	03207	00000	00000	2
?	?	4	5	53.00	7.80	3	4.70	2	2	01400	00000	00000	1
?	?	?	?	?	?	?	?	1	2	00000	00000	00000	2
1	?	2	5	40.00	7.00	1	?	1	1	04205	00000	00000	1
?	?	?	?	55	8.7	?	?	1	2	00000	00000	00000	2
1	?	4	4	46.00	6.10	2	?	1	1	02111	00000	00000	2
3	5.50	4	3	49.00	6.80	?	?	1	2	00000	00000	00000	2
?	?	1	?	48.00	7.20	?	?	1	1	03111	00000	00000	2
?	?	?	?	50	70	?	?	1	1	03111	00000	00000	2
2	2	?	?	51	65	?	?	1	1	03111	00000	00000	2

The symbol ? represents that the value for the feature is missing.

TABLE 2.2. Features of the Horse Colic data set.

1	Surgery?:	1 = Yes, it had surgery; 2 = It was treated without surgery
2	Age:	1 = Adult horse; 2 = Young horse (<6 months)
3	Hospital Number:	the case number assigned to the horse
4	Rectal Temperature:	in degrees Celsius, linear
5	Pulse:	the heart rate in beats per minute, linear
6	Respiratory Rate:	linear
7	Temperature of Extremities:	1 = normal; 2 = warm; 3 = cool; 4 = cold
8	Peripheral Pulse:	1 = normal; 2 = increased; 3 = reduced; 4 = absent
9	Mucous Membranes:	1 = normal pink; 2 = bright pink; 3 = pale pink; 4 = pale cyanotic; 5 = bright red / injected; 6 = dark cyanotic
10	Capillary Refill Time:	1 = <3 seconds; 2 = >= 3 seconds
11	Pain:	1 = alert; no pain; 2 = depressed; 3 = intermittent mild pain; 4 = intermittent severe pain; 5 = continuous severe pain
12	Peristalsis:	1 = hypermotile; 2 = normal; 3 = hypomotile; 4 = absent
13	Abdominal Distension:	1 = none; 2 = slight; 3 = moderate; 4 = severe
14	Nasogastric Tube:	1 = none; 2 = slight; 3 = significant
15	Nasogastric Reflux:	1 = none; 2 = >1 liter; 3 = <1 liter
16	Nasogastric Reflux pH:	scale is from 0 to 14 with 7 being neutral, linear
17	Rectal Examination—Feces:	1 = normal; 2 = increased; 3 = decreased; 4 = absent
18	Abdomen:	1 = normal; 2 = other; 3 = firm feces in the large intestine; 4 = distended small intestine; 5 = distended large intestine
19	Packed Cell Volume:	the # of red cells by volume in the blood, linear
20	Total Protein:	linear
21	Abdominocentesis Appearance:	1 = clear; 2 = cloudy; 3 = serosanguinous
22	Abdomcentesis Total Protein:	linear
23	Outcome:	1 = lived; 2 = died; 3 = was euthanized
24	Surgical Lesion?:	1 = Yes; 2 = No
25	Type of Lesion:	First number is site of lesion: 1 = gastric; 2 = sm intestine; 3 = lg colon; 4 = lg colon and cecum; 5 = cecum; 6 = transverse colon; 7 = rectum/descending colon; 8 = uterus; 9 = bladder; 11 = all intestinal sites; 00 = none. Second number is type: 1 = simple; 2 = strangulation; 3 = inflammation; 4 = other. Third number is subtype: 1 = mechanical; 2 = paralytic; 0 = n/a. Fourth number is specific code: 1 = obturation; 2 = intrinsic; 3 = extrinsic; 4 = adynamic; 5 = volvulus/torsion; 6 = intussuption; 7 = thromboembolic; 8 = hernia; 9 = lipoma/slenic incarceration; 10 = displacement; 0 = n/a
26	Type of Lesion:	similar to 25
27	Type of Lesion:	similar to 25
28	cp_data:	1 = Yes; 2 = No

4. Ratio. Features at this level possess all the properties of the above levels but also have an absolute zero (such as exists with degrees Kelvin), so that the ratio between two values is meaningful. The feature of pulse is also regarded as a ratio variable because it is meaningful to say that a pulse of 120 is twice as fast as one with a pulse of 60.

More often, nominal and ordinal features are called qualitative or categorical features while interval and ratio features are referred to as quantitative features (Anderberg, 1973; Jain and Dubes, 1988). Anderberg (1973) further elaborated the discussion of the conversions among these scales.

From the Horse Colic data set, it is also noticed that there exist missing feature values for most records, represented as?, which is quite common for real data sets due to all kinds of limitations and uncertainties. The first thought of how to deal with missing data may be to discard the records that contain missing features. However, this approach can only work when the number of objects that have missing features is much smaller than the total number of objects in the data set. For the Horse Colic data set with almost all objects having missing features, this strategy obviously cannot be used. Now we consider calculating the proximity by only using the feature values that are available. In other words, we simply treat the contributions from the missing features to the total proximity as zero. Therefore, the proximity for a pair of d-dimensional objects $\mathbf{x}_i$ and $\mathbf{x}_j$, is defined as

$$D(\mathbf{x}_i, \mathbf{x}_j) = \frac{d}{d - \sum_{l=1}^{d} \delta_{ijl}} \sum_{\substack{\text{all } l \text{ and} \\ \delta_l = 0}} d_l(x_{il}, x_{jl}), \tag{2.1}$$

where d_l (x_{il}, x_{jl}) is the distance between each pair of components of the two objects and

$$\delta_{ijl} = \begin{cases} 1, & \text{if } x_{il} \text{ or } x_{jl} \text{ is missing} \\ 0, & \text{otherwise} \end{cases}. \tag{2.2}$$

An alternate strategy is based on the direct estimation of the missing values or the distance using the available information. Specifically, for the lth feature, we can replace the missing values with the mean value obtained from all objects or some certain number of neighbors (this requires extra computation and may not be available) whose values in this feature are available. On the other hand, we can directly calculate the average distances between all pairs of data objects along all the features and use that to estimate distances for the missing features. The average distance for the lth feature is given as

$$\bar{d}_l = \frac{2}{N(N-1)} \sum_{i=2}^{N} \sum_{j=1}^{i-1} d(x_{il}, x_{jl}), \; x_{il} \text{ and } x_{jl} \text{ are available;} \tag{2.3}$$

therefore, the distance for the missing feature is obtained as

$$d_l(x_{il}, x_{jl}) = \bar{d}_l, \text{ if } x_{il} \text{ or } x_{jl} \text{ is missing.} \tag{2.4}$$

Dixon (1979) investigated the performance of the above methods and suggested that the method based on Eq. 2.1 may be the best selection overall. Not surprisingly, this can be expected to vary with the applications. More details on the topic are given in Little and Rubin (1987).

2.3. DEFINITION OF PROXIMITY MEASURES

Proximity is the generalization of both dissimilarity and similarity. A dissimilarity or distance function on a data set **X** is defined to satisfy the following conditions:

1. Symmetry,

$$D(\mathbf{x}_i, \mathbf{x}_j) = D(\mathbf{x}_j, \mathbf{x}_i); \tag{2.5}$$

2. Positivity,

$$D(\mathbf{x}_i, \mathbf{x}_j) \geq 0 \quad \text{for all } \mathbf{x}_i \text{ and } \mathbf{x}_j. \tag{2.6}$$

If the conditions

3. Triangle inequality,

$$D(\mathbf{x}_i, \mathbf{x}_j) \leq D(\mathbf{x}_i, \mathbf{x}_k) + D(\mathbf{x}_k, \mathbf{x}_j) \quad \text{for all } \mathbf{x}_i, \mathbf{x}_j \text{ and } \mathbf{x}_k \tag{2.7}$$

and

4. Reflexivity,

$$D(\mathbf{x}_i, \mathbf{x}_j) = 0 \quad \text{iff } \mathbf{x}_i = \mathbf{x}_j \tag{2.8}$$

also hold, it is called a metric. If only the triangle inequality is not satisfied, the function is called a semimetric. A metric is referred to as ultrametric (Johnson, 1967) if it satisfies a stronger condition,

$$D(\mathbf{x}_i, \mathbf{x}_j) \leq \max(D(\mathbf{x}_i, \mathbf{x}_k), D(\mathbf{x}_j, \mathbf{x}_k)) \quad \text{for all } \mathbf{x}_i, \mathbf{x}_j \text{ and } \mathbf{x}_k. \tag{2.9}$$

Likewise, a similarity function is defined to satisfy the conditions below:

1. Symmetry,

$$S(\mathbf{x}_i, \mathbf{x}_j) = S(\mathbf{x}_j, \mathbf{x}_i); \tag{2.10}$$

2. Positivity,

$$0 \le S(\mathbf{x}_i, \mathbf{x}_j) \le 1, \quad \text{for all } \mathbf{x}_i \text{ and } \mathbf{x}_j. \tag{2.11}$$

If it also satisfies the following additional conditions

3. For all $\mathbf{x}_i$, $\mathbf{x}_j$, and $\mathbf{x}_k$,

$$S(\mathbf{x}_i, \mathbf{x}_j)S(\mathbf{x}_j, \mathbf{x}_k) \le [S(\mathbf{x}_i, \mathbf{x}_j) + S(\mathbf{x}_j, \mathbf{x}_k)]S(\mathbf{x}_i, \mathbf{x}_k); \tag{2.12}$$

4.

$$S(\mathbf{x}_i, \mathbf{x}_j) = 1 \quad \text{iff } \mathbf{x}_i = \mathbf{x}_j, \tag{2.13}$$

it is called a similarity metric.

For a data set with N data objects, we can define an $N \times N$ symmetric matrix, called a proximity matrix, whose $(i,j)^{\text{th}}$ element represents the similarity or dissimilarity measure for the ith and j^{th} objects ($i, j = 1, \ldots, N$). The proximity matrix is called one-mode, since both its rows and columns indices have the same meaning. Correspondingly, the original $N \times d$ data matrix, where d is the dimension of the data, is designated two-mode.

2.4. PROXIMITY MEASURES FOR CONTINUOUS VARIABLES

Perhaps the most commonly used distance measure is the Euclidean distance, also known as L_2 norm, represented as,

$$D(\mathbf{x}_i, \mathbf{x}_j) = \left(\sum_{l=1}^{d} |x_{il} - x_{jl}|^{1/2} \right)^2, \tag{2.14}$$

where $\mathbf{x}_i$ and $\mathbf{x}_j$ are d-dimensional data objects. It is not difficult to see that the Euclidean distance satisfies all the conditions in Eq. 2.5– 2.8 and therefore is a metric. Further investigation shows that the Euclidean distance tends to form hyperspherical clusters. Also, clusters formed with the Euclidean distance are invariant to translations and rotations in the feature space (Duda et al., 2001). However, if the features are measured with units that are quite different, features with large values and variances will tend to dominate over other features. Linear or other transformations can also cause the distortion of the distance relations, as pointed out in Duda et al. (2001).

One possible way to deal with this problem is to normalize the data to make each feature contribute equally to the distance. A commonly used method is data standardization, i.e., each feature has zero mean and unit variance,

$$x_{il} = \frac{x_{il}^* - m_l}{s_l}, i = 1, \ldots, N, l = 1, \ldots, d, \quad (2.15)$$

where x_{il}^* represents the raw data, and sample mean m_l and sample standard deviation s_l are defined as

$$m_l = \frac{1}{N}\sum_{i=1}^{N} x_{il}^*, \quad (2.16)$$

and

$$s_l = \sqrt{\frac{1}{N-1}\sum_{i=1}^{N}(x_{il}^* - m_l)^2}, \quad (2.17)$$

respectively. This is also known as z-score or standard score in statistical analysis, which measures the position of a given data point in the entire data set, in terms of number of standard deviations the data point's value deviating from the mean (Hogg and Tanis, 2005).

Another normalization approach is based on the maximum and minimum of the data and all features in the range of [0,1],

$$x_{il} = \frac{x_{il}^* - \min(x_{il}^*)}{\max(x_{il}^*) - \min(x_{il}^*)}. \quad (2.18)$$

The Euclidean distance can be generalized as a special case of a family of metrics, called Minkowski distance or L_p norm, defined as,

$$D(\mathbf{x}_i, \mathbf{x}_j) = \left(\sum_{l=1}^{d} |x_{il} - x_{jl}|^{1/p}\right)^p. \quad (2.19)$$

Note that when $p = 2$, the distance becomes the Euclidean distance. By letting $p = 1$ and $p \to \infty$, we can obtain another two common special cases of Minkowski distance: the city-block, also called Manhattan distance, or L_1 norm,

$$D(\mathbf{x}_i, \mathbf{x}_j) = \sum_{l=1}^{d} |x_{il} - x_{jl}|, \quad (2.20)$$

and the sup distance or L_∞ norm,

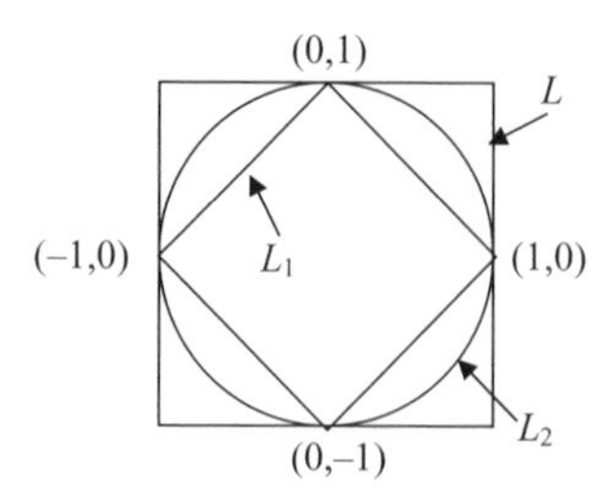

Fig. 2.1. The surface consists of points that have a unit distance from the origin, measured by the L_1, L_2, and L_∞ metrics under two dimensions. A circle is obtained when L_2 is used, while squares are generated for L_1 and L_∞ metrics. For L_1, the vertices of the square are on the axes, and for L_∞, the square's sides are parallel to the axes.

$$D(\mathbf{x}_i, \mathbf{x}_j) = \max_{1 \le l \le d} |x_{il} - x_{jl}|. \tag{2.21}$$

Note that the invariance to translations and rotations is no longer valid for other cases of the Minkowski metric (Jain and Dubes, 1988).

Now let us consider a set of points satisfying $D_p(\mathbf{x}_i, \mathbf{x}_0) = 1$, where $\mathbf{x}_0$ represents the origin. Here, we use the subscript p to represent that the distance is measured with the Minkowski metric. In Fig. 2.1, the shape of the surface formed by a set of points that has a unit distance from the origin are shown for the city block, Euclidean, and sup distances under two dimensions. It is also observed that the shapes for Minkowski metrics with $1 < p < 2$ or $2 < p < \infty$ are convex curves lying between the unit balls for L_1 and L_2 metrics or L_2 and L_∞ metrics, respectively (Anderberg, 1973).

The squared Mahalanobis distance is also a metric and defined as,

$$D(\mathbf{x}_i, \mathbf{x}_j) = (\mathbf{x}_i - \mathbf{x}_j)^T \mathbf{S}^{-1} (\mathbf{x}_i - \mathbf{x}_j), \tag{2.22}$$

where $\mathbf{S}$ is the within-class covariance matrix defined as $\mathbf{S} = E[(\mathbf{x} - \boldsymbol{\mu})(\mathbf{x} - \boldsymbol{\mu})^T]$, where $\boldsymbol{\mu}$ is the mean vector and $E[\cdot]$ calculates the expected value of a random variable. Mahalanobis distance tends to form hyperellipsoidal clusters, which are invariant to any nonsingular linear transformation. However, the calculation of the inverse of $\mathbf{S}$ may cause some computational burden for large-scale data. When features are not correlated, which leads $\mathbf{S}$ to an identity matrix, the squared Mahalanobis distance is equivalent to the squared Euclidean distance (Jain and Dubes, 1988; Mao and Jain, 1996).

The point symmetry distance is based on the symmetry assumption in the cluster's structure and is defined as,

$$D(\mathbf{x}_i, \mathbf{x}_r) = \min_{\substack{j=1,\ldots,N \\ \text{and } j \neq i}} \frac{\|(\mathbf{x}_i - \mathbf{x}_r) + (\mathbf{x}_j - \mathbf{x}_r)\|}{\|(\mathbf{x}_i - \mathbf{x}_r)\| + \|(\mathbf{x}_j - \mathbf{x}_r)\|}, \tag{2.23}$$

where $\mathbf{x}_r$ is a reference point, such as the centroid of the cluster, and $\|\cdot\|$ represents the Euclidean norm. The point symmetry distance calculates the distance between an object $\mathbf{x}_i$ and the reference point $\mathbf{x}_r$ given other $N-1$ objects and is minimized when a symmetric pattern exists.

The distance measure can also be derived from the correlation coefficient, such as the Pearson correlation coefficient (Kaufman and Rousseeuw, 1990), defined as,

$$r_{ij} = \frac{\sum_{l=1}^{d}(x_{il}-\bar{x}_i)(x_{jl}-\bar{x}_j)}{\sqrt{\sum_{l=1}^{d}(x_{il}-\bar{x}_i)^2\sum_{l=1}^{d}(x_{jl}-\bar{x}_j)^2}}, \tag{2.24}$$

where $\bar{x}_i = \frac{1}{d}\sum_{l=1}^{d} x_{il}$. Noticing that the correlation coefficient is in the range of [−1, 1], with 1 and −1 indicating the strongest positive and negative correlation respectively, we can define the distance measure as

$$D(\mathbf{x}_i, \mathbf{x}_j) = (1 - r_{ij})/2, \tag{2.25}$$

which falls into the range of [0, 1]. When correlation coefficients are used for distance measures, it is important to keep in mind that they tend to disclose the difference in shapes rather than to detect the magnitude of differences between two objects. Furthermore, when the features are not measured on the same scales, the calculation of mean or variance in Eq. 2.24 has no meaning at all, as pointed out by Everitt et al. (2001).

Finally, we consider the direct application of similarity measures to compare a pair of data objects with continuous variables. One such example is the cosine similarity, given as

$$S(\mathbf{x}_i, \mathbf{x}_j) = \cos\alpha = \frac{\mathbf{x}_i^T \mathbf{x}_j}{\|\mathbf{x}_i\|\|\mathbf{x}_j\|}, \tag{2.26}$$

which can also be regarded as a normalized inner product. The more similar the two objects, the more parallel they are in the feature space, and the greater the cosine value. Note that this can also be constructed as a distance measure by simply using $D(\mathbf{x}_i, \mathbf{x}_j) = 1 - S(\mathbf{x}_i, \mathbf{x}_j)$. Similar to the Pearson correlation coefficient, the cosine similarity is unable to provide information on the magnitude of differences.

We illustrate the examples and applications of the above dissimilarity and similarity measures in Table 2.3.

TABLE 2.3. Proximity measures and their applications.

Measure	Metric	Examples and Applications
Minkowski distance	Yes	Fuzzy c-means with measures based on Minkowski family (Hathaway et al., 2000)
City block distance	Yes	Fuzzy ART (Carpenter et al., 1991)
Euclidean distance	Yes	K-means with its variants (Ball and Hall, 1967; Forgy, 1965; MacQueen, 1967)
Sup distance	Yes	Fuzzy c-means with sup norm (Bobrowski and Bezdek, 1991)
Mahalanobis	Yes	Ellipsoidal ART (Anagnostopoulos and M. Georgiopoulos, 2001); Hyperellipsoidal clustering algorithm (Mao and Jain, 1996)
Point symmetry distance	No	Symmetry-based K-means (Su and Chou, 2001)
Pearson correlation	No	Widely used as the measure for microarray gene expression data analysis (Eisen et al., 1998)
Cosine similarity	No	The most commonly used measure in document clustering (Steinbach et al., 2000)

TABLE 2.4. Contingency table for a pair of binary objects. One indicates the presence of features in the objects, while zero represents the absence of the features.

		Object $\mathbf{x}_i$		
		1	0	Totals
	1	n_{11}	n_{10}	$n_{11} + n_{10}$
Object $\mathbf{x}_j$	0	n_{01}	n_{00}	$n_{01} + n_{00}$
	Totals	$n_{11} + n_{01}$	$n_{10} + n_{00}$	d

2.5. PROXIMITY MEASURES FOR DISCRETE VARIABLES

2.5.1 Similarity Measures for Binary Variables

Similarity measures are most commonly used for discrete features (Anderberg, 1973; Everitt et al., 2001). Particularly, there exist many similarity measures when the features take only two values.

Suppose we use two binary subscripts to count features in two d-dimensional objects, $\mathbf{x}_i$ and $\mathbf{x}_j$. n_{00} and n_{11} represent the number of the simultaneous absence or presence of features in these objects, and n_{01} and n_{10} count the features present only in one object. These can be depicted in a contingency table as shown in Table 2.4.

According to Kaufman and Rousseeuw (1990), the binary features can be classified as symmetric or asymmetric, based on whether the two values are equally important or not. A symmetric feature considers both values equally significant, while an asymmetric feature treats one value, usually denoted with

1, as more important than the other, usually denoted with 0. For example, the feature of gender is symmetric. The value of "female" or "male" can be encoded as 1 or 0 without affecting the evaluation of the similarity. However, if a feature contains information related to the presence or absence of a property, such as some rare form of a certain gene, we cannot justify saying that the absence of the special form helps to indicate some similarity between the two objects. Based on this consideration, we have invariant similarities for symmetric binary variables and noninvaraint similarities for asymmetric binary variables, as illustrated below:

$$S(\mathbf{x}_i, \mathbf{x}_j) = \frac{n_{11} + n_{00}}{n_{11} + n_{00} + w(n_{10} + n_{01})}, \quad \begin{array}{l} w = 1 \text{ simple matching coefficient} \\ \text{(Zubin, 1938)} \\ w = 2 \text{ Rogers and Tanimoto} \\ \text{(1960)} \\ w = 1/2 \text{ Sokal and Sneath} \\ \text{(1963)} \end{array} . \tag{2.27}$$

These invariant similarity measures regard the 1-1 match and 0-0 match as equally important. Unmatched pairs are weighted based on their contribution to the similarity. For the simple matching coefficient, the corresponding dissimilarity measure from $D(\mathbf{x}_i, \mathbf{x}_j) = 1 - S(\mathbf{x}_i, \mathbf{x}_j)$ is known as the Hamming distance (Gersho and Gray, 1992).

$$S(\mathbf{x}_i, \mathbf{x}_j) = \frac{n_{11}}{n_{11} + w(n_{10} + n_{01})}, \quad \begin{array}{l} w = 1, \text{ Jaccard coefficient} \\ \text{(Jaccard, 1908)} \\ w = 2, \text{ Sokal and Sneath (1963)} \\ w = 1/2, \text{ Dice (1945)} \end{array} . \tag{2.28}$$

These non-invariant similarity measures focus on the 1-1 match features while ignoring the effect of the 0-0 match, which is considered uninformative. Again, the unmatched pairs are weighted depending on their importance.

2.5.2 Similarity Measures for Discrete Variables with More than Two Values

For discrete features that have more than two states, a simple and direct strategy is to map them into a larger number of new binary features (Anderberg, 1973; Kaufman and Rousseeuw, 1990). For example, the feature of abdominocentesis appearance in Table 2.2, which has three values, clear, cloudy, or serosanguinous, can be coded with three new binary features denoted as abdominocentesis appearance—clear, abdominocentesis appearance—cloudy, and abdominocentesis appearance—serosanguinous. If an object is in the state of "clear" for this feature, the new feature, abdominocentesis appearance—

clear, will have a value of 1 indicating its occurrence, while the values for the other two derived features will be 0. Note that these new binary features are asymmetric. The disadvantage of this method is the possibility of introducing too many binary variables.

A more effective and commonly used method is based on the simple matching criterion (Everitt et al., 2001; Kaufman and Rousseeuw, 1990). For a pair of d-dimensional objects $\mathbf{x}_i$ and $\mathbf{x}_j$, the similarity using the simple matching criterion is obtained as

$$S(\mathbf{x}_i, \mathbf{x}_j) = \frac{1}{d}\sum_{l=1}^{d} S_{ijl}, \tag{2.29}$$

where

$$S_{ijl} = \begin{cases} 0 & \text{if } \mathbf{x}_i \text{ and } \mathbf{x}_j \text{ do not match in the } l^{th} \text{ feature} \\ w & \text{if } \mathbf{x}_i \text{ and } \mathbf{x}_j \text{ match in the } l^{th} \text{ feature} \end{cases}. \tag{2.30}$$

Usually, the score for the matches is set as 1. However, similar to the similarity measures used for binary variables, the number of matches could be weighted based on the properties and requirements of the data. For example, if there exists a large number of possible values for a feature, it is reasonable to give more weight to the match in this feature. In this case, w will be larger than 1.

Sometimes, the categorical features display certain orders, known as the ordinal features discussed in Section 2.2. In this case, the codes from 1 to M_l, where M_l is the number of levels for the feature l or the highest level, are no longer meaningless in similarity measures. Rather, the closer the two levels are, the more similar the two objects in this certain feature. For example, the feature of abdominal distension in Table 2.2 is modeled as four levels: 1 = none, 2 = slight, 3 = moderate, 4 = severe. Obviously, a horse having a slight abdominal distension (recorded as 1) is more similar to the one with moderate distension (2) than to the one with severe distension (3). Thus, objects with this type of feature can be compared just by using the continuous dissimilarity measures, such as city-block or Euclidean distance (Kaufman and Rousseeuw, 1990). Since the number of possible levels varies with the different features, the original ranks r_{il}^* for the ith object in the lth feature are usually converted into the new ranks r_{il} in the range of [0,1], using the following method:

$$r_{il} = \frac{r_{il}^* - 1}{M_l - 1}. \tag{2.31}$$

Considering the restrictions of the Jaccard coefficient and other similarity functions in explaining the global relations among data points in a neighborhood, Guha et al. (2000) suggested using the links to measure the similarity of categorical variables with two or more levels. The link between a pair of data points is defined as the number of neighbors shared by these two points. An object $\mathbf{x}_i$ is said to be a neighbor of another object $\mathbf{x}_j$ if $S_{ij} \geq \delta$, where S_{ij} could be any appropriate similarity function, and δ is a threshold parameter provided by the users. Therefore, data objects with more common neighbors are more likely to belong to the same cluster. The measure is applied in the agglomerative hierarchical clustering algorithm ROCK (Guha et al., 2000), described in Chapter 3.

2.6. PROXIMITY MEASURES FOR MIXED VARIABLES

For real data sets, it is more common to see both continuous and categorical features at the same time. In other words, the data are in a mixed mode. For example, the Horse Colic data set in Table 2.1 contains 7 continuous variables and 21 discrete variables, including 5 binary variables. Generally, we can map data consisting of mixed variables into the interval [0,1] and use measures like the Euclidean distance. However, for categorical variables whose numbers are just names without any meaning, this method is inappropriate. Alternatively, we can transform all features into binary variables and only use binary similarity functions. The drawback of this method is the information loss.

A more powerful method was proposed by Gower (1971) and extended by Kaufman and Rousseeuw (1990). The similarity measure for a pair of d-dimensional mixed data objects $\mathbf{x}_i$ and $\mathbf{x}_j$ is defined as

$$S(\mathbf{x}_i, \mathbf{x}_j) = \frac{\sum_{l=1}^{d} \delta_{ijl} S_{ijl}}{\sum_{l=1}^{d} \delta_{ijl}}, \tag{2.32}$$

where S_{ijl} indicates the similarity for the lth feature between the two objects, and δ_{ijl} is a 0-1 coefficient based on whether the measure of the two objects is missing, as defined in Eq. 2.2. Correspondingly, the dissimilarity measure can be obtained by simply using $D(\mathbf{x}_i, \mathbf{x}_j) = 1 - S(\mathbf{x}_i, \mathbf{x}_j)$. For discrete variables (including binary variables), the component similarity is obtained as

$$S_{ijl} = \begin{cases} 1 & \text{if } x_{il} = x_{jl} \\ 0 & \text{if } x_{il} \neq x_{jl} \end{cases}. \tag{2.33}$$

For continuous variables, S_{ijl} is then given as

$$S_{ijl} = 1 - \frac{|x_{il} - x_{jl}|}{R_l}, \tag{2.34}$$

where R_l is the range of the lth variable over all objects, written as

$$R_l = \max_m x_{ml} - \min_m x_{ml}. \quad (2.35)$$

Thus, Eq. 2.34 uses the city-block distance after applying a standardization process to the variables. This equation can also be employed to calculate the similarity for ordinal variables.

2.7. SUMMARY

In this chapter, we summarize many typical proximity measures for continuous, discrete, and mixed data. More details are in Anderberg (1973), Everitt et al. (2001), Gower and Legendre (1986), Kaufman and Rousseeuw (1990), and Sneath and Sokal (1973). A discussion of the edit distance for alphabetic sequences is deferred until Chapter 7 and further elaborated in Gusfield (1997) and Sankoff and Kruskal (1999).

Different proximity measures will affect the formations of the resulting clusters, so the selection of an appropriate proximity function is important. How can effective measures be selected? How can data objects with features that have quite different physical meanings be compared? Furthermore, what kinds of standards can be used to weight the features or to select the features that are important to the clustering? Unfortunately, we cannot answer these questions in a conclusive and absolute way. Cluster analysis is a subjective and problem-dependent process. As an indispensable part of this exploration, a proximity measure, together with the features selected, ultimately has to be put into the framework of the entire data set and determined based on the understanding and judgment of the investigators.

CHAPTER 3

HIERARCHICAL CLUSTERING

3.1. INTRODUCTION

Clustering techniques are generally classified as partitional clustering and hierarchical clustering, based on the properties of the generated clusters (Everitt et al., 2001; Hansen and Jaumard, 1997; Jain et al., 1999; Jain and Dubes, 1988). Partitional clustering directly divides data points into some prespecified number of clusters without the hierarchical structure, while hierarchical clustering groups data with a sequence of nested partitions, either from singleton clusters to a cluster including all individuals or vice versa. The former is known as agglomerative hierarchical clustering, and the latter is called divisive hierarchical clustering.

Both agglomerative and divisive clustering methods organize data into the hierarchical structure based on the proximity matrix. The results of hierarchical clustering are usually depicted by a binary tree or dendrogram, as depicted in Fig. 3.1. The root node of the dendrogram represents the whole data set, and each leaf node is regarded as a data point. The intermediate nodes thus describe the extent to which the objects are proximal to each other; and the height of the dendrogram usually expresses the distance between each pair of data points or clusters, or a data point and a cluster. The ultimate clustering results can be obtained by cutting the dendrogram at different levels (the dashed line in Fig. 3.1). This representation provides very informative descriptions and a visualization of the potential data clustering structures, especially when real hierarchical relations exist in the data, such as the data from evolu-

Clustering, by Rui Xu and Donald C. Wunsch, II

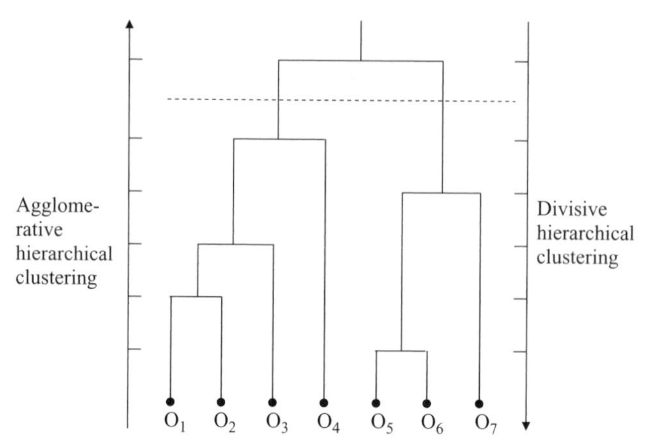

Fig. 3.1. Example of a dendrogram from hierarchical clustering. The clustering direction for the divisive hierarchical clustering is opposite to that of the agglomerative hierarchical clustering. Two clusters are obtained by cutting the dendrogram at an appropriate level.

tionary research on different species of organisms, or other applications in medicine, biology, and archaeology (Everitt et al., 2001; Theodoridis and Koutroumbas, 2006).

Compared with agglomerative methods, divisive methods need to consider $2^{N-1} - 1$ possible two-subset divisions for a cluster with N data points, which is very computationally intensive. Therefore, agglomerative methods are more widely used. We focus on agglomerative clustering in the following section, and some divisive clustering algorithms are described briefly in Section 3.3. The common criticism of classical hierarchical clustering algorithms is high computational complexity, which is at least $O(N^2)$. This high computational burden limits their application in large-scale data sets. In order to address this problem and other disadvantages, some new hierarchical clustering algorithms have been proposed, such as BIRCH (Balanced Iterative Reducing and Clustering using Hierarchies) (Zhang et al., 1996) and CURE (Clustering Using Representatives) (Guha et al., 1998). These algorithms are described in Section 3.4.

3.2. AGGLOMERATIVE HIERARCHICAL CLUSTERING

3.2.1 General Agglomerative Hierarchical Clustering

Agglomerative clustering starts with N clusters, each of which includes exactly one data point. A series of merge operations is then followed that eventually forces all objects into the same group. The general agglomerative clustering can be summarized by the following procedure, which is also summarized in Fig. 3.2.

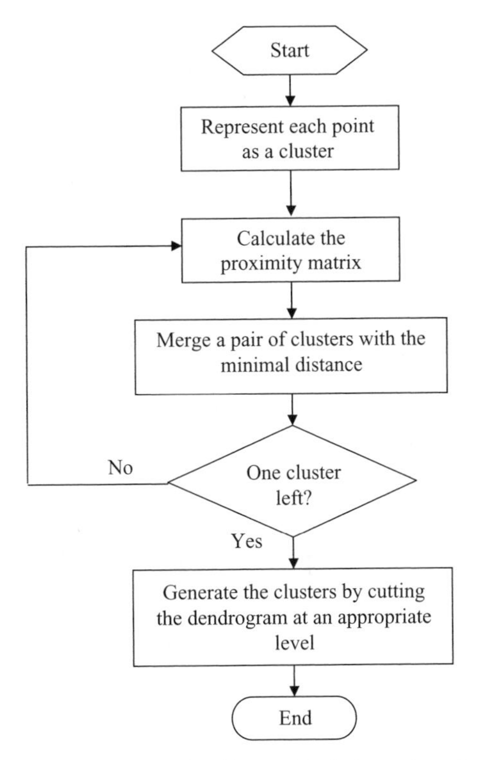

Fig. 3.2. Flowchart of the agglomerative hierarchical clustering algorithm. Agglomerative clustering considers each data point as a cluster in the beginning. Two clusters are then merged in each step until all objects are forced into the same group.

1. Start with N singleton clusters. Calculate the proximity matrix (usually based on the distance function) for the N clusters;
2. In the proximity matrix, search the minimal distance $D(C_i, C_j) = \min\limits_{\substack{1 \le m, l \le N \\ m \ne l}} D(C_m, C_l)$, where $D(\cdot,\cdot)$ is the distance function discussed later in the section, and combine cluster C_i and C_j to form a new cluster C_{ij};
3. Update the proximity matrix by computing the distances between the cluster C_{ij} and the other clusters;
4. Repeat steps 2 and 3 until only one cluster remains.

3.2.2 Clustering Linkage

Obviously, the merge of a pair of clusters or the formation of a new cluster is dependent on the definition of the distance function between two clusters. There exist a large number of distance definitions between a cluster C_l and a new cluster $C_{(ij)}$ formed by the merge of two clusters C_i and C_j, which can be generalized by the recurrence formula proposed by Lance and Williams (1967) as

TABLE 3.1. Lance and Williams' parameters for agglomerative hierarchical clustering. α_i, α_j, β, and γ are parameters defined in Eq. 3.1. n_i, n_j, and n_l are the number of data points in cluster C_i, C_j, and C_l, respectively.

Clustering algorithms	α_i	α_j	β	γ
Single linkage (nearest neighbor)	1/2	1/2	0	−1/2
Complete linkage (farthest neighbor)	1/2	1/2	0	1/2
Group average linkage (UPGMA)	$\frac{n_i}{n_i+n_j}$	$\frac{n_j}{n_i+n_j}$	0	0
Weighted average linkage (WPGMA)	1/2	1/2	0	0
Median linkage (WPGMC)	1/2	1/2	−1/4	0
Centroid linkage (UPGMC)	$\frac{n_i}{n_i+n_j}$	$\frac{n_j}{n_i+n_j}$	$\frac{-n_in_j}{(n_i+n_j)^2}$	0
Ward's method	$\frac{n_i+n_l}{n_i+n_j+n_l}$	$\frac{n_j+n_l}{n_i+n_j+n_l}$	$\frac{-n_l}{n_i+n_j+n_l}$	0

U: unweighted; W: weighted; PGM: pair group method; A: average; C: centroid.

$$D(C_l,(C_i,C_j)) = \alpha_i D(C_l,C_i) + \alpha_j D(C_l,C_j) + \beta D(C_i,C_j) + \gamma\left|D(C_l,C_i) - D(C_l,C_j)\right|, \tag{3.1}$$

where $D(\cdot,\cdot)$ is the distance function and α_i, α_j, β, and γ are coefficients that take values dependent on the scheme used. The parameter values for the commonly used algorithms are summarized in Table 3.1, which are also given in Everitt et al. (2001), Jain and Dubes (1988), and Murtagh (1983).

- The single linkage algorithm (Everitt et al., 2001; Johnson, 1967; Jain and Dubes, 1988; Sneath, 1957). For single linkage, the distance between a pair of clusters is determined by the two closest objects to the different clusters. So single linkage clustering is also called the nearest neighbor method. Following the parameters identified in Table 3.1, Eq. 3.1 becomes

$$D(C_l,(C_i,C_j)) = \min(D(C_l,C_i), D(C_l,C_j)). \tag{3.2}$$

  Therefore, the distance between the newly generated cluster and the old one is dependent on the minimal distance of $D(C_l, C_i)$ and $D(C_l, C_j)$. Single linkage clustering tends to generate elongated clusters, which causes the chaining effect (Everitt et al., 2001). As a result, two clusters with quite different properties may be connected due to the existence of noise. However, if the clusters are separated far from each other, the single linkage method works well.
- The complete linkage algorithm (Everitt et al., 2001; Jain and Dubes, 1988; Sorensen, 1948). In contrast to single linkage clustering, the com-

plete linkage method uses the farthest distance of a pair of objects to define inter-cluster distance. In this case, Eq. 3.1 becomes

$$D(C_l,(C_i,C_j)) = \max(D(C_l,C_i), D(C_l,C_j)). \quad (3.3)$$

It is effective in uncovering small and compact clusters.

- The group average linkage algorithm, also known as the unweighted pair group method average (UPGMA) (Everitt et al., 2001; Jain and Dubes, 1988; Sokal and Michener, 1958). The distance between two clusters is defined as the average of the distance between all pairs of data points, each of which comes from a different group. Eq. 3.1 is written as

$$D(C_l,(C_i,C_j)) = \frac{1}{2}(D(C_l,C_i) + D(C_l,C_j)). \quad (3.4)$$

The distance between the new cluster and the old one is the average of the distances of $D(C_l, C_i)$ and $D(C_l, C_j)$.

- The weighted average linkage algorithm, also known as the weighted pair group method average (WPGMA) (Jain and Dubes, 1988; McQuitty, 1966). Similar to UPGMA, the average linkage is also used to calculate the distance between two clusters. The difference is that the distances between the newly formed cluster and the rest are weighted based on the number of data points in each cluster. In this case, Eq. 3.1 is written as

$$D(C_l,(C_i,C_j)) = \frac{n_i}{n_i+n_j}D(C_l,C_i) + \frac{n_j}{n_i+n_j}D(C_l,C_j). \quad (3.5)$$

- The centroid linkage algorithm, also known as the unweighted pair group method centroid (UPGMC) (Everitt et al., 2001; Jain and Dubes, 1988; Sokal and Michener, 1958). Two clusters are merged based on the distance of their centroids (means), defined as

$$\mathbf{m}_i = \frac{1}{n_i}\sum_{\mathbf{x}\in C_i}\mathbf{x}, \quad (3.6)$$

where n_i is the number of data points belonging to the cluster. Eq. 3.1 now is written as

$$D(C_l,(C_i,C_j)) = \frac{n_i}{n_i+n_j}D(C_l,C_i) + \frac{n_j}{n_i+n_j}D(C_l,C_j) - \frac{n_i n_j}{(n_i+n_j)^2}D(C_i,C_j). \quad (3.7)$$

This definition is equivalent to the calculation of the squared Euclidean distance between the centroids of the two clusters,

$$D(C_l,(C_i,C_j)) = \|\mathbf{m}_l - \mathbf{m}_{(ij)}\|^2. \quad (3.8)$$

- The median linkage algorithm, also known as the weighted pair group method centroid (WPGMC) (Everitt et al., 2001; Gower, 1967; Jain and Dubes, 1988). The median linkage is similar to the centroid linkage, except that equal weight is given to the clusters to be merged. Eq. 3.1 is written as

$$D(C_l,(C_i,C_j)) = \frac{1}{2}D(C_l,C_i) + \frac{1}{2}D(C_l,C_j) - \frac{1}{4}D(C_i,C_j) \quad (3.9)$$

  This is a special case when the number of data points in the two merging clusters is the same.
- Ward's method, also known as the minimum variance method (Everitt et al., 2001; Jain and Dubes, 1988; Ward, 1963). The object of Ward's method is to minimize the increase of the within-class sum of the squared errors,

$$E = \sum_{k=1}^{K} \sum_{\mathbf{x}_i \in C_k} \|\mathbf{x}_i - \mathbf{m}_k\|^2, \quad (3.10)$$

  where K is the number of clusters and $\mathbf{m}_k$ is the centroid cluster C_k as defined in Eq. 3.6, caused by the merge of two clusters. This change is only computed on the formed cluster and the two clusters to be merged, and can be represented as

$$\Delta E_{ij} = \frac{n_i n_j}{n_i + n_j}\|\mathbf{m}_i - \mathbf{m}_j\|^2, \quad (3.11)$$

  Eq. 3.1 is then written as

$$D(C_l,(C_i,C_j)) = \frac{n_i + n_l}{n_i + n_j + n_l}D(C_l,C_i) + \frac{n_j + n_l}{n_i + n_j + n_l}D(C_l,C_j) - \frac{n_l}{(n_i + n_j)^2}D(C_i,C_j). \quad (3.12)$$

Single linkage, complete linkage, and average linkage consider all points of a pair of clusters when calculating their inter-cluster distance, and they are also called graph methods. The others are called geometric methods because they use geometric centers to represent clusters and determine their distances. Everitt et al. (2001) summarize the important features and properties of these methods, together with a brief review of experimental comparative studies. Yager (2000) discusses a family of inter-cluster distance measures, based on the generalized mean operators, with their possible effect on the hierarchical clustering process.

3.2.3 Relation with Graph Theory

The concepts and properties of graph theory (Harary, 1969) make it very convenient to describe clustering problems by means of graphs. Nodes $V = \{v_i, i = 1, \ldots, N\}$ of a weighted graph G correspond to N data points in the pattern space, and edges $E = \{e_{ij}, i, j \in V, i \neq j\}$ reflect the proximities between each pair of data points. If the dissimilarity matrix is defined as

$$\mathbf{D}_{ij} = \begin{cases} 1 & \text{if } d(\mathbf{x}_i, \mathbf{x}_j) < d_0 \\ 0 & \text{otherwise} \end{cases}, \tag{3.13}$$

where d_0 is a threshold value, the graph is simplified to an unweighted threshold graph. Both the single linkage method and the complete linkage method can be described on the basis of the threshold graph. Single linkage clustering is equivalent to seeking maximally connected sub-graphs (components), while complete linkage clustering corresponds to finding maximally complete subgraphs (cliques). Jain and Dubes (1988) provide a detailed description and discussion of hierarchical clustering from the point of view of graph theory. More discussion of graph theory in clustering can be found in Theodoridis and Koutroumbas (2003). Section 3.4.3 introduces a k-nearest-neighbor graph-based algorithm, Chameleon (Karypis et al., 1999).

3.3. DIVISIVE HIERARCHICAL CLUSTERING

Compared to agglomerative hierarchical clustering, divisive clustering proceeds in the opposite way. In the beginning, the entire data set belongs to a cluster, and a procedure successively divides it until all clusters are singletons. For a data set with N objects, a divisive hierarchical algorithm would start by considering $2^{N-1} - 1$ possible divisions of the data into two nonempty subsets, which is computationally expensive even for small-scale data sets. Therefore, divisive clustering is not a common choice in practice. However, the divisive clustering algorithms do provide clearer insights of the main structure of the data since the larger clusters are generated at the early stage of the clustering process and are less likely to suffer from the accumulated erroneous decisions, which cannot be corrected by the successive process (Kaufman and

Rousseeuw, 1990). Heuristic methods have been proposed, such as the algorithm DIANA (Divisive ANAlysis) (Kaufman and Rousseeuw, 1990), based on the earlier work of Macnaughton-Smith et al. (1964), which consider only a part of all possible divisions.

According to Kaufman and Rousseeuw (1990), at each stage, DIANA consists of a series of iterative steps in order to move the closer objects into the splinter group, which is seeded with the object that is farthest from the others in the cluster to be divided. The cluster with the largest diameter, defined as the largest distance between any pair of objects, is selected for further division. Supposing that the cluster C_l is going to split into cluster C_i and C_j, the iterative steps of DIANA can be described as follows, and the algorithm is summarized in Fig. 3.3.

1. Start with C_i equal to C_l and C_j as an empty cluster;
2. For each data object $\mathbf{x}_m \in C_i$,
   a) For the first iteration, compute its average distance to all other objects:

$$d(\mathbf{x}_m, C_i \setminus \{\mathbf{x}_m\}) = \frac{1}{N_{c_i} - 1} \sum_{\substack{\mathbf{x}_p \in C_i \\ p \neq m}} d(\mathbf{x}_m, \mathbf{x}_p); \tag{3.14}$$

   b) For the remaining iterations, compute the difference between the average distance to C_i and the average distance to C_j:

$$d(\mathbf{x}_m, C_i \setminus \{\mathbf{x}_m\}) - d(\mathbf{x}_m, C_j) = \frac{1}{N_{c_i} - 1} \sum_{\substack{\mathbf{x}_p \in C_i \\ p \neq m}} d(\mathbf{x}_m, \mathbf{x}_p) - \frac{1}{N_{c_j}} \sum_{\mathbf{x}_q \in C_j} d(\mathbf{x}_m, \mathbf{x}_q); \tag{3.15}$$

3. a) For the first iteration, move the data object with the maximum value into C_j;
   b) For the remaining iterations, if the maximum value of Eq. 3.15 is greater than 0, move the data object with the maximum difference into C_j. Repeat step 2 b) and 3 b). Otherwise, stop.

During each division of DIANA, all features are used; hence the divisive algorithm is called polythetic. In fact, most of the algorithms introduced in this book are polythetic. On the other hand, if the division is on a one-feature–one-step basis, the corresponding algorithm is said to be monothetic. One such algorithm is called MONA (Monothetic Analysis) and is used for data objects with binary features (Kaufman and Rousseeuw, 1990). The criterion for selecting a feature to divide the data is based on its similarity with other features, through the measures of association. Given the contingency table of two binary features f_i and f_j, in Table 3.2, the following formulae can be used to calculate the association:

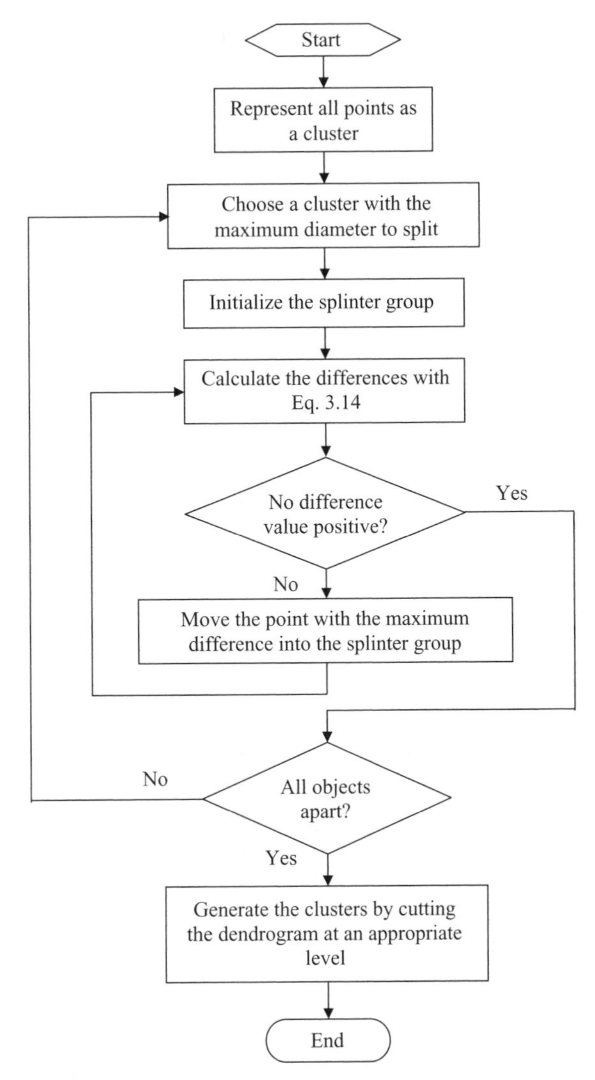

Fig. 3.3. Flowchart of the DIANA algorithm. Contrary to agglomerative hierarchical clustering, DIANA is a divisive hierarchical clustering algorithm that starts with a cluster consisting of all data points, and successively divides the data set until all clusters are singletons.

TABLE 3.2. The contingency table of two features with values of 0 and 1.

f_i \ f_j	0	1
0	n_{00}	n_{01}
1	n_{10}	n_{11}

$$a_1 = |n_{11}n_{00} - n_{10}n_{01}|, \tag{3.16}$$

$$a_2 = (n_{11}n_{00} - n_{10}n_{01})^2, \tag{3.17}$$

$$a_3 = \frac{(n_{11}n_{00} - n_{10}n_{01})^2 N}{(n_{11}+n_{01})(n_{11}+n_{10})(n_{00}+n_{01})(n_{00}+n_{10})}, \tag{3.18}$$

As an example, a_3 is a statistic that has approximately a chi-squared distribution with 1 degree of freedom (Kaufman and Rousseeuw, 1990). A large value of a_3 indicates that the two variables are highly related and the distance between them is small. More discussion and references pertaining to monothetic divisive algorithms can be found in Kaufman and Rousseeuw (1990) and Everitt et al. (2001).

3.4. RECENT ADVANCES

Common criticisms of classical hierarchical clustering algorithms focus on their lack of robustness and their sensitivity to noise and outliers. Once an object is assigned to a cluster, it will not be considered again, which means that hierarchical clustering algorithms are not capable of correcting possible previous misclassification. Another disadvantage of these algorithms is their computational complexity, which is at least $O(N^2)$ and cannot meet the requirement for dealing with large-scale data sets in data mining and other tasks in recent years. As a result of such requirements, many new clustering methods, with hierarchical cluster results, have appeared and have greatly improved the clustering performance. Typical examples include BIRCH (Balanced Iterative Reducing and Clustering using Hierarchies) (Zhang, et al., 1996), CURE (Clustering Using REpresentatives) (Guha et al., 1998), ROCK (RObust Clustering using linKs) (Guha et al., 2000), and Chameleon (Karypis et al., 1999), which will be discussed below.

3.4.1 BIRCH

The main advantages of BIRCH are two-fold, the ability to deal with large-scale data sets and the robustness to outliers (Zhang et al., 1996). The memory requirement and the frequent operations of a large volume of data cause a high computational burden, which could be reduced by operating on the summaries of the original data. The basic idea of BIRCH is to use a new data structure, a CF (Clustering Feature) tree, to store these summaries, which not only captures the important clustering information of the original data, but reduces the required storage.

In the context of the CF tree, for the ith cluster with N_i data points, the cluster is represented as a vector $\boldsymbol{CF}_i = (N_i, \boldsymbol{LS}_i, SS_i)$, where $\mathbf{LS}_i = \sum_{j=1}^{N_i} \mathbf{x}_j$ is the linear sum of the data points, and $SS_i = \sum_{j=1}^{N_i} \mathbf{x}^2$ is the square sum of the data points. This representation contains all necessary information for the clustering operations in BIRCH. For example, the **CF** vector of a new cluster, merged from the pair of clusters $\mathbf{CF}_i$ and $\mathbf{CF}_j$, is

$$\mathbf{CF}_{(ij)} = (N_i + N_j, \mathbf{LS}_i + \mathbf{LS}_j, SS_i + SS_j). \tag{3.19}$$

The CF tree is a height-balanced tree, with each internal vertex composed of entries defined as $[\mathbf{CF}_i, child_i]$, $i = 1, \ldots, B$, where $child_i$ is a pointer to the ith child node, $\mathbf{CF}_i$ is a representation of the cluster of this child node, and B is a threshold parameter that determines the maximum number of entries in the vertex, and each leaf composed of entries in the form of $[\mathbf{CF}_i]$, $i = 1, \ldots, L$, where $\mathbf{CF}_i$ is one of the sub-clusters that constitute the leaf node and L is the threshold parameter that controls the maximum number of entries in the leaf. Moreover, the leaves must follow the restriction that the diameter of each entry in the leaf, defined as

$$\gamma = \left[\sum_{l=1}^{N_i} \sum_{m=1}^{N_i} (\mathbf{x}_l - \mathbf{x}_m)^2 \Big/ N_i(N_i - 1) \right]^{1/2},$$

is less than a threshold T, which adjusts the size of the CF tree.

The general procedures of BIRCH are summarized in Fig. 3.4, which consists of four steps. Step 1 first constructs a **CF** tree with a prespecified value of T. If the tree size grows too big and causes the system to run out of memory, the value of T is increased to rebuild a new **CF** tree. The resulting tree may need to be condensed to a smaller one in step 2 in order to meet the input size requirement for the following clustering. Agglomerative hierarchical clustering can then work directly on the obtained sub-clusters, this is represented by the **CF** triplet. An additional step may be performed to refine the clusters by assigning a data point to its nearest cluster, so that the distance between the data point and the cluster centroid is minimized. During the tree's construction, outliers are eliminated from the summaries by identifying the objects sparsely distributed in the feature space. BIRCH can achieve a computational complexity of $O(N)$.

3.4.2 CURE and ROCK

As a midway between the single linkage (or complete linkage) method and the centroid linkage method (or k-means, introduced in Chapter 4), which use either all data points or one data point as the representatives for the generated cluster, CURE uses a set of well-scattered points to represent each cluster

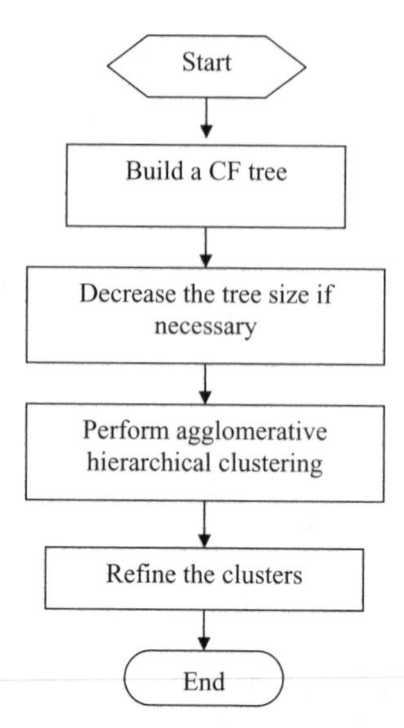

Fig. 3.4. Flowchart of the BIRCH algorithm. BIRCH consists of four steps. Steps 2 and 4 are optional depending on the size of the **CF** tree obtained in Step 1 and the requirement of the user, respectively.

(Guha et al., 1998). The representative points are further shrunk towards the cluster centroid according to an adjustable parameter α, which varies between 0 and 1, corresponding to all points or one point cases, in order to weaken the effects of outliers. This representation addresses the insufficiency of the above methods to handle outliers and is effective in exploring more sophisticated cluster shapes. Now, the distance between the pair of clusters C_i and C_j is determined as

$$D(C_i, C_j) = \min(d(\mathbf{x}_l, \mathbf{x}_m)), \mathbf{x}_l \in C_i^r, \mathbf{x}_m \in C_j^r, \tag{3.20}$$

where C_i^r and C_j^r are the representative points of the two clusters, respectively.

The basic procedure of CURE is depicted in Fig. 3.5. In order to deal with large-scale data sets, the first step of CURE is to generate a random sample from the original data. The sample size can be estimated based on the Chernoff bounds (Motwani and Raghavan, 1995). Given a cluster C with N_C data points, if the sample size s satisfies

$$s \geq \mu N + \frac{N}{N_C}\log\frac{1}{\varepsilon} + \frac{N}{N_C}\sqrt{\left(\log\frac{1}{\varepsilon}\right)^2 + 2\mu N_C \log\frac{1}{\varepsilon}}, \tag{3.21}$$

where μ is a constant in the range [0, 1], then the probability that the sample consists of fewer than μN_C data points in the cluster C is less than ε $(0 \leq \varepsilon \leq 1)$

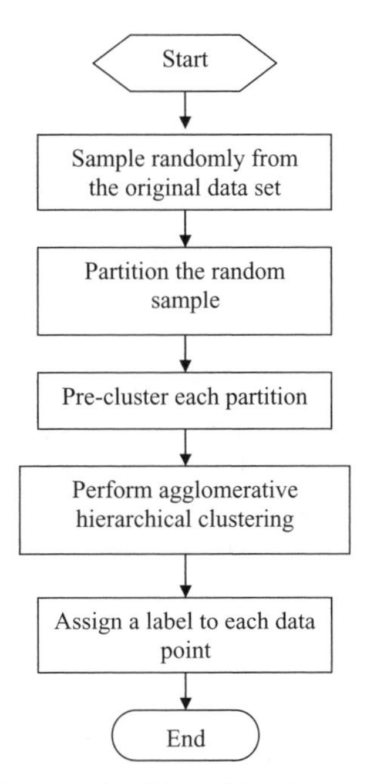

Fig. 3.5. Flowchart of the CURE algorithm. CURE uses both random sampling and pre-partitioning to deal with large-scale data sets.

(Guha et al., 1998). Another advantage of sampling is that a large portion of outliers will be screened out during the process. The sampled data are further divided into k_p partitions, with the equal size N/k_p. Each partition is pre-clustered in order to reduce computational time further. Agglomerative hierarchical clustering can then be applied to the pre-clustered data to obtain the final clusters. The clusters with few data points are treated as outliers. Finally, all data points are labeled based on their distance from the representative points of each cluster. The computational complexity for CURE is approximate to $O(N_{sample}^2)$ for low dimensionality, where N_{sample} is the number of sampling data points.

Another agglomerative hierarchical clustering algorithm proposed by Guha et al. (2000), called ROCK, follows a procedure similar to CURE. The major difference lies in the fact that the clustering objects are the data with qualitative attributes, instead of continuous variables. Correspondingly, at each step during the hierarchical clustering, the similarity between a pair of clusters C_i and C_j, which consist of n_i and n_j data points, respectively, is measured with a goodness function $g(C_i, C_j)$, defined as

$$g(C_i, C_j) = \frac{\sum_{\mathbf{x}_l \in C_i, \mathbf{x}_m \in C_j} link(\mathbf{x}_l, \mathbf{x}_m)}{(n_i + n_j)^{1+2f(\delta)} - n_i^{1+2f(\delta)} - n_j^{1+2f(\delta)}}. \quad (3.22)$$

where $link(\mathbf{x}_l, \mathbf{x}_m)$ is the link, introduced in Chapter 2, between the two data points, each of which comes from a different cluster, the denominator is the expected number of cross-links needed to normalize the interconnectivity measure, and $f(\cdot)$ is a function of a user-specified parameter δ, which controls the size of the neighborhood of a point, as defined in Chapter 2. Guha et al. (2000) used the form $f(\delta)=\frac{1-\delta}{1+\delta}$ in their experiments and claimed that performance was relatively robust to inaccuracies in the value of δ.

Now, the two clusters with the maximum value of the goodness function will be merged. Like CURE, the clustering algorithm is applied to a random sample in order to scale to large-scale data sets. Outliers are identified as the data points with very few neighbors. The worst-case time complexity for ROCK is $O(N_{sample}^2 \log N_{sample} + N_{sample}^2 + \kappa N_{sample})$, where κ is the product of the average number of neighbors and the maximum number of neighbors for a point.

3.4.3 Chameleon

Chameleon (Karypis et al., 1999) is a two-phase agglomerative hierarchical clustering algorithm (as depicted in Fig. 3.6), based on the k-nearest-neighbor graph, in which an edge is eliminated if both vertices are not within their k closest neighbors. During the first step, Chameleon divides the connectivity graph into a set of sub-clusters with the minimal edge cut. This is achieved with a graph-partitioning algorithm called hMetis (Karypis et al., 1999). Each sub-graph should contain enough nodes to achieve an effective similarity

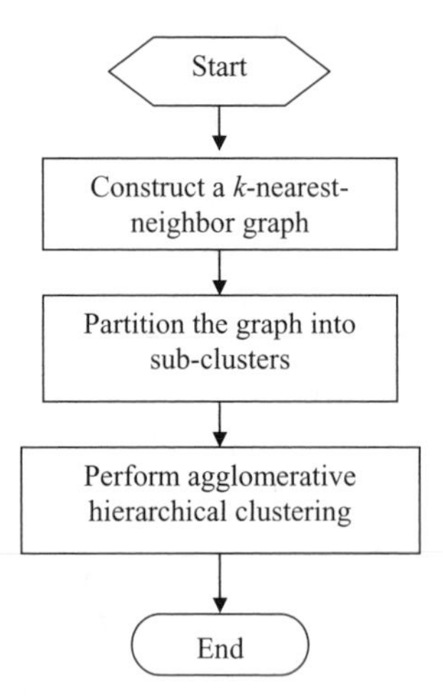

Fig. 3.6. Flowchart of the Chameleon algorithm. Chameleon performs a two-phase clustering. Data points are first divided into several sub-clusters using a graph-partitioning algorithm, and these sub-clusters are then merged by an agglomerative hierarchical clustering algorithm to generate the final clusters.

computation. Then, an agglomerative hierarchical clustering scheme is applied to these sub-clusters to form the ultimate clusters. The criterion or goodness function for instructing the merge of a pair of clusters C_i and C_j combines the effects of two factors, known as the relative interconnectivity and the relative closeness, which make Chameleon flexible enough to explore the characteristics of potential clusters. The relative interconnectivity is obtained by normalizing the sum of weights of the edges connecting the two clusters, denoted as $EC(C_i, C_j)$, over the internal connectivity of the clusters, defined as,

$$RI(C_i, C_j) = \frac{|EC(C_i, C_j)|}{(|EC(C_i)| + |EC(C_j)|)/2}, \tag{3.23}$$

where $EC(C_i)$ and $EC(C_j)$ are the minimum sum of the cut edges if we bisect the cluster C_i and C_j, respectively.

Similarly, the relative closeness is defined as

$$RC(C_i, C_j) = \frac{\overline{SEC}(C_i, C_j)}{\frac{N_i}{N_i + N_j}\overline{SEC}(C_i) + \frac{N_j}{N_i + N_j}\overline{SEC}(C_j)}, \tag{3.24}$$

where $\overline{SEC}(C_i, C_j)$ is the average weight of the edges connecting the two clusters, $\overline{SEC}(C_i)$ and $\overline{SEC}(C_j)$ are the average weights of the edges belonging in the min-cut bisector of the clusters; and N_i and N_j are the number of points in each cluster.

Now, the criterion function can be defined as

$$g(C_i, C_j) = RI(C_i, C_j) \times RC(C_i, C_j)^{\alpha}, \tag{3.25}$$

where α is a parameter trading the effect of interconnectivity and closeness.

For Chameleon, the hierarchical clustering operates on a k-nearest-neighbor graph representation. Another important graph representation for hierarchical clustering analysis is the Delaunay triangulation graph (DTG), which is the dual of the Voronoi diagram (Okabe et al., 2000). Cherng and Lo (2001) constructed a hypergraph (each edge is allowed to connect more than two vertices) from the DTG and used a two-phase algorithm that is similar to Chameleon to find clusters. Another DTG-based application, known as the AMOEBA algorithm, is presented by Estivill-Castro and Lee (1999).

3.4.4 Other Methods

Leung et al. (2000) proposed an interesting hierarchical clustering based on human visual system research. Under the framework of scale-space theory, clustering is interpreted as a blurring process in which each data point is regarded as a light point in an image, and a cluster is represented as a blob.

Therefore, the merge of smaller blobs, containing only one light point in the beginning, until the whole image becomes one light blob, corresponds to the process of agglomerative hierarchical clustering.

Li and Biswas (2002) extended agglomerative hierarchical clustering to deal with data having both continuous and nominal variables. The proposed algorithm, called similarity-based agglomerative clustering (SBAC), employs the Goodall similarity measure (Goodall, 1966), which pays extra attention to less common matches of feature values to calculate the proximity of mixed data types. In other words, for two pairs of data objects ($\mathbf{x}_i$, $\mathbf{x}_j$) and ($\mathbf{x}_m$, $\mathbf{x}_n$), each pair having an equal value $x_{(ij)l}$ and $x_{(mn)l}$ in feature d_l, but $x_{(ij)l} \neq x_{(mn)l}$, if $x_{(mn)l}$ is more frequent than $x_{(ij)l}$, the first pair of data objects is considered to be more similar to each other than the second pair. The algorithm further uses a UPGMA scheme to generate the clustering partition.

Basak and Krishnapuram (2005) introduced an unsupervised decision tree algorithm to perform top-down hierarchical clustering. The most prominent property of the decision tree is the interpretability of the results by virtue of a set of rules (Quinlan, 1993). Starting from the root node of the decision tree, data are split based on a specified feature, which is selected according to four different criteria, based on information theory (Gray, 1990). The process continues until the number of data points that a node contains is less than a prespecified threshold.

More exploration into hierarchical clustering includes the methods such as the agglomerative likelihood tree (ALT) and the Markov Chain Monte Carlo (MCMC)-based method (Castro et al., 2004), both of which are based on the maximum likelihood framework. For relative hierarchical clustering (RHC) (Mollineda and Vidal, 2000), the ratio of the internal distance (distance between a pair of clusters that are candidates to be merged) and the external distance (distance from the two clusters to the rest) is used to decide the proximities. Lee et al. (2005) proposed a divisive hierarchical clustering algorithm based on the optimization of the ordinal consistency between the hierarchical cluster structure and the data similarity. Parallel algorithms for hierarchical clustering are discussed by Dahlhaus (2000), Li (1990), Olson (1995), and Rajasekaran (2005), respectively.

3.5. APPLICATIONS

3.5.1 Gene Expression Data Analysis

Genome sequencing projects have achieved great advances in recent years. The first draft of the human genome sequence was completed in February 2001, several years earlier than expected (Consortium, I.H.G.S., 2001; Venter et al., 2001). However, these successes can only be regarded as the first step towards completely understanding the functions of genes and proteins and revealing the complicated gene regulatory interactions. Analyses based on traditional laboratory techniques are time-consuming and expensive, and they cannot meet the current requirements of understanding living systems. DNA microarray technologies provide an effective and efficient way to measure the expression levels of tens of thousands of genes simultaneously under different conditions, promising new insights into the investigation of gene functions and interactions (Eisen and Brown, 1999; Lipshutz et al., 1999; Lockhart et al., 1996; Moore, 2001; Moreau et al., 2002; Schena et al., 1995). As a matter of fact, DNA microarray technologies have already become an important platform benefiting genomic research, with a wide variety of applications, such as gene function identification, differential expression studies, disease diagnosis, therapy development and drug testing, toxic effect identification, and genetic regulatory network inference (Amaratunga and Cabrera, 2004; Lee, 2004; McLachlan et al., 2004; Stekel, 2003; Xu et al., 2007a, b). Gene expression data analysis is described as a three-level framework based on the complexity of the tasks (Baldi and Long, 2001). The low level focuses on the activities of single gene, while the intermediate level concentrates on the gene combinations. The goal of the top level is to infer the genetic regulatory networks and reveal the interactions between genes.

Different microarray technologies have been developed and generally fall into three categories based on the manufacturing technologies and the nature of DNA attached to the array: spotted cDNA microarray, spotted oligonucleotide microarray, and GeneChip (developed by Affymetrix Inc., Santa Clara, CA and also called in-situ oligonucleotide arrays). cDNA clones or expressed sequence tag (EST), which is a unique short subsequence for a certain gene, are usually derived from DNA collections or libraries, with a typical length ranging from 500 to 2,500 base pairs (Eisen and Brown, 1999; Schena et al., 1995). Oligonucleotides are shorter sequences taken from a gene, and the length of oligonucleotides used in GeneChip is 25 bases (Lipshutz et al., 1999; Lockhart et al., 1996). For both spotted cDNA and oligonucleotide microarrays, the DNA clones are either robotically spotted or ejected from a nozzle onto the support. The former is called contact printing and the latter is known as inkjet printing. Inkjet printing technology can also be used to eject the four different types of nucleotides and build the desired sequences using one nucleotide at a time. The GeneChip uses photolithography technology from the semiconductor industry to manufacture oligonucleotide microarrays. In photolithography, the substrate is covered with a photosensitive chemical, which will break down under illumination. The areas where some certain nucleotide will be attached are hit by light through a mask, which allows attachment of the nucleotides that are also coated with the covering chemical. The process repeats with different masks for the purpose of inserting different types of nucleotides until the desired sequence is built.

Typically, a microarray experiment consists of the following five steps (Amaratunga and Cabrera, 2004; Lee, 2004; McLachlan et al., 2004):

1. Microarray preparation. This step usually includes probe selection and microarray manufacturing. Here, probes refer to the DNA clones with

known sequences that are systematically immobilized to the microarray. They can be either cDNA or oligonucleotides. The arrays are usually made of a piece of glass, charged nylon, or silicon chip.

2. Sample preparation. mRNA specific to the study is extracted from the sample and is reverse-transcribed into cDNA or cRNA (synthesized by transcription from a single-stranded DNA template). The products are then labeled with fluorescent dyes for the purpose of detection, which are also called targets. Usually, a reference sample with a different fluorescent label is also used for comparison. The typical fluorophores used are Cyanine 5 (Cy5 in red) and Cyanine 3 (Cy3 in green).
3. Hybridization of the sample with probes on the microarray. Microarrays derive their power from hybridization, which refers to the binding of single-stranded DNA molecules from different sources (probes and targets) based on the complementary base pairing rule (Griffiths et al., 2000). A target will only bind to a probe whose strand of bases is complementary to its own strand.
4. Microarray scanning and image analysis. After the genes that do not hybridize with the probes are washed off, the microarray is scanned to produce an image of its surface, which is further processed with image analysis technologies to measure the fluorescence intensities of the labeled target at each spot.
5. Computational analysis of microarray data. After normalization and other statistical preprocessing of the fluorescence intensities, the gene expression profiles are represented as a matrix $E = (e_{ij})$, where e_{ij} is the expression level of the ith gene in the jth condition, tissue, or experimental stage. Computational methods rooted in statistics, computer science, machine learning, and other disciplines can be used to unveil the potential information in the data.

The overview for the basic procedures of cDNA microarray and Affymetrix oligonucleotide microarray are depicted in Fig. 3.7 and 3.8, respectively. Table 3.3 summarizes the major differences between these two technologies.

Along with the tremendous opportunities introduced by microarray technologies, they also bring many computational challenges as a result of the huge volume of data and the degree of complexity (Dopazo et al., 2001; McLachlan et al., 2004). For example, gene expression data usually have very high dimensions, causing the curse of dimensionality problem (Bellman, 1961). Also, the introduction of noise and variability during the different stages of microarray experiments demands extra attention during data analysis. Many computational approaches in machine learning, computer science, and statistics have already been applied in processing the generated expression data and accelerating the exploration of life science (Amaratunga and Cabrera, 2004; Dopazo et al., 2001; Lee, 2004; McLachlan et al., 2004). In particular, cluster analysis, with its applications in revealing the hidden structures of biological data, is

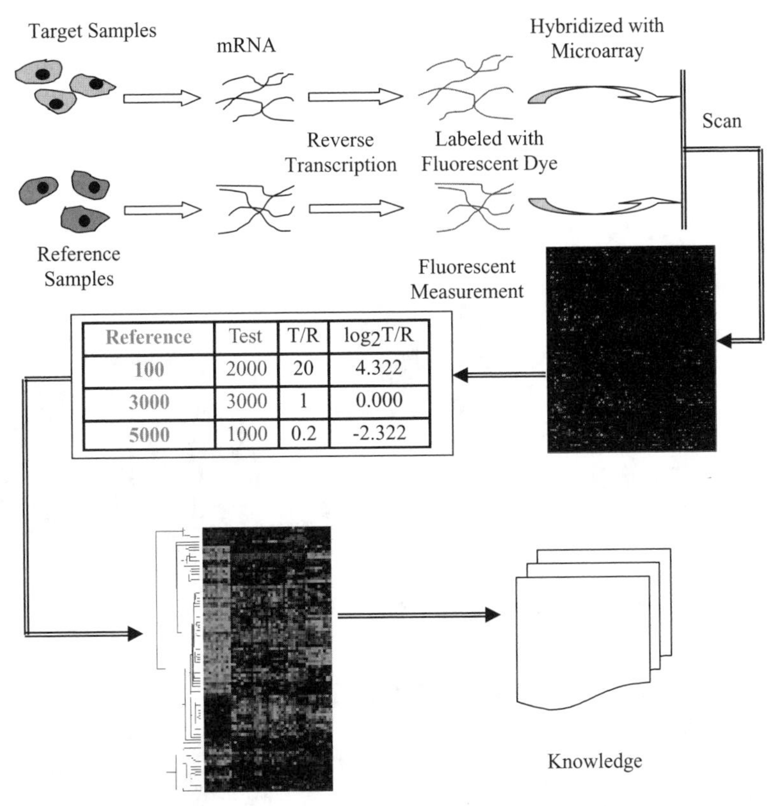

Reference	Test	T/R	$\log_2 T/R$
100	2000	20	4.322
3000	3000	1	0.000
5000	1000	0.2	-2.322

Fig. 3.7. Overview of cDNA microarray technology (Cummings and Relman, 2000). Fluorescently labeled cDNAs, obtained from target (in red) and reference (in green) samples through reverse transcription, are mixed and hybridized to the array, which is composed of a large amount of cDNA clones, placed by the robot spotters. The array is then scanned to detect both fluorophores. The red, green, and yellow spots in the hybridized array indicate increased expression levels in the test sample, increased expression levels in the reference sample, and similar expression levels in both samples. The final data are the ratios of the intensities of the test and reference samples and reflect relative levels of gene expression.

particularly important in helping biologists investigate and understand the activities of uncharacterized genes and proteins, and further, the systematic architecture of the entire genetic network (Jiang et al., 2004; Liew et al., 2005; Madeira and Oliveira, 2004; Tibshirani et al., 1999). As we have already seen, gene expression levels can be measured at a sequence of time points during a certain biological process or with samples under different conditions. Similarity in the expression behaviors of a cluster of genes often implies their similarity in biological functions (Eisen et al., 1998; Iyer et al., 1999). Thus, functionally known genes could provide meaningful insights in understanding the functions of the co-expressed, unannotated genes. Moreover, gene clustering could also suggest that genes in the same group share the same transcriptional regulation

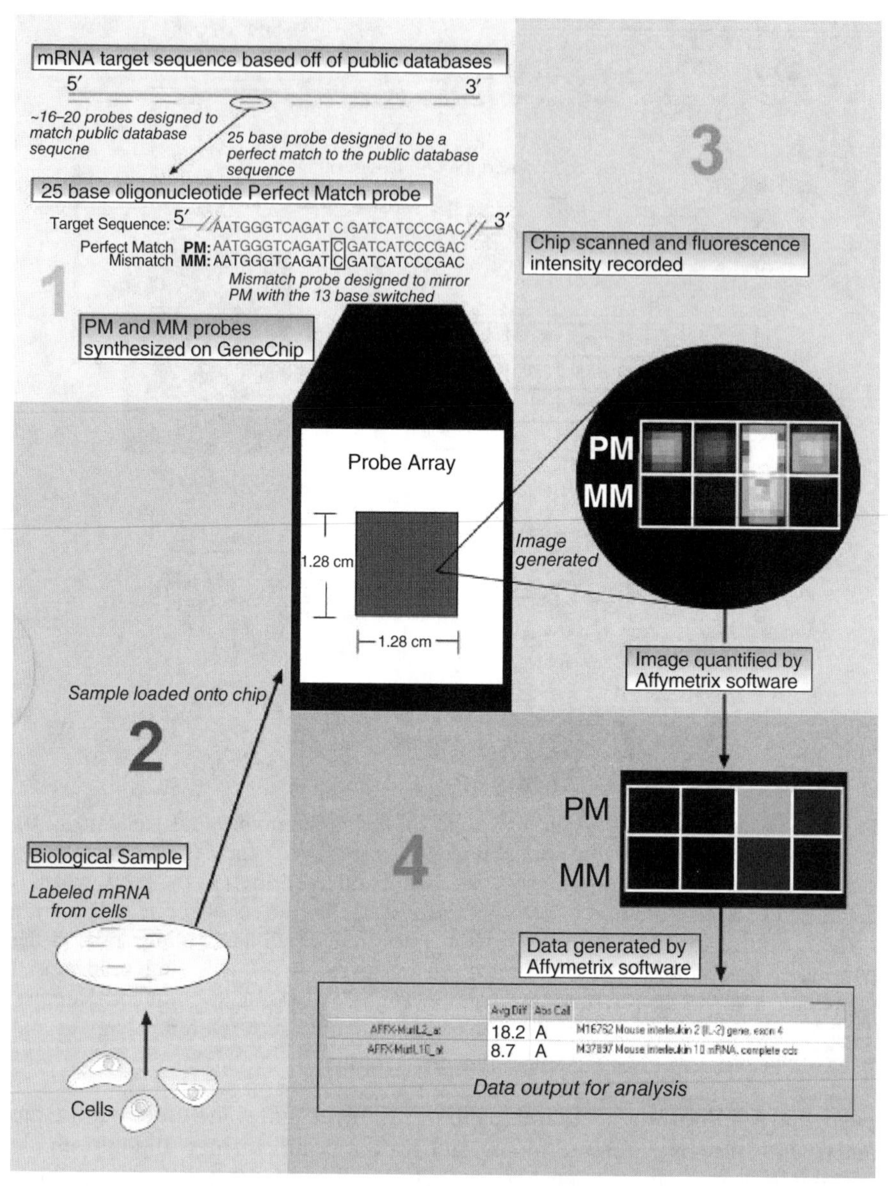

Fig. 3.8. Affymetrix GeneChip technology overview. Each gene in the array is designed to be represented by a set of oligonucleotide probes (usually 15–20), known as perfect-match (PM) probes. These probes are directly synthesized on the chip through photolithography technology. Each PM probe is also paired with a mismatch (MM) probe, which is similar to the PM probe except the central base, in order to prevent cross-hybridization and decrease false positives. The absolute expression levels are converted from the hybridization intensities with this technology. (From http://www.ohsu.edu/gmsr/amc/amc_technology.html. Reproduced with permission from the Gene Microarray Shared Resource at Oregon Health & Science University.)

TABLE 3.3. Comparison of cDNA microarray and Affymetrix oligonucleotide microarray technologies.

Microarray Technology	Length of Probes	Manufacturing Technology	Read-out	Pros	Cons
cDNA microarrays	Several hundred to a few thousand	Robotic spotting	Relative ratio	Rich cDNA library. Flexible in probe selections. Low cost.	Not uniform for spotting. Less scalable.
Affymetrix GeneChip	25 bases	Photolithography	Absolute expression levels	More accessible for hybridization. Offer highest density. Minimizing the potential for cross-hybridization.	High cost. No control of the probes for the researchers.

mechanism and participate in the same cellular process (Moreau et al., 2002; Spellman et al., 1998). On the other hand, the gene profiles could be used as fingerprints for different types of diseases, such as cancers (Alizadeh et al., 2000; Alon et al., 1999; Golub et al., 1999; Khan et al., 2001). The clusters of tissues could be valuable in discovering new disease types or subtypes and providing new approaches for disease diagnosis and drug development. We concentrate on the application of hierarchical clustering in gene expression data analysis in this section, and the practice of other clustering technologies in this field will be discussed in other chapters of the book.

Perhaps one of the earliest applications of clustering genes, based on the similarity in their expression patterns, was by performed by Eisen et al. (1998) on the study of budding yeast *Saccharomyces cerevisiae* and reaction of human fibroblasts to serum (Iyer et al., 1999). Cluster analysis has gradually become a very popular tool for investigating gene functions and genetic regulatory mechanisms (Cho et al., 1998; Spellman et al., 1998; Tavazoie et al., 1999). In this study, the average linkage hierarchical clustering algorithm was employed for clustering genes, with the Pearson correlation coefficient used to measure the similarity between a pair of genes. The clustering results are visualized as two parts (as shown in Fig. 3.9, which depicts the results of the data set consisting of the expression level measurements of 8,613 human genes at 12 time points using cDNA microarrays) using the software package TreeView. The left part of Fig. 3.9 represents the dendrogram, where the terminal branches are linked to each gene. The right part is called a heat map, which corresponds to the expression matrix, with rows representing genes and columns representing measurement conditions. The color scale ranging from green to red provides a good visualization of the patterns of gene expression. As can be seen from the large contiguous patches of color in Fig. 3.9, there exist some clusters of genes that display similar expression patterns over a set of conditions. More importantly, the similarity of a cluster of genes in expression patterns indicates a tendency for genes to share common functions. The five bars (A–E) in Fig. 3.9 correspond to five clusters that consist of genes with roles in (A) cholesterol biosynthesis, (B) the cell cycle, (C) the immediate-early response, (D) signaling and angiogenesis, and (E) wound healing and tissue remodeling. Such discovery becomes clearer in the analysis of the combined yeast *S. cerevisiae* data set because the functions of all genes used in the study are functionally annotated. The heat map of the entire clustering is shown in the left side of Fig. 3.10. Further examination of clusters B–F shows that genes (with full names shown in the figure) in the same cluster play similar roles in cellular processes of (B) spindle pole body assembly and function, (C) the proteasome, (D) mRNA splicing, (E) glycolysis, (F) the mitochondrial ribosome, (G) ATP synthesis, (H) chromatin structure, (I) the ribosome and translation, (J) DNA replication, and (K) the tricarboxylic acid cycle and respiration.

Hierarchical clustering algorithms have also been used to generate clusters of co-regulated genes, based on gene expression data, in order to elucidate the genetic regulation mechanism in a cell. By examining the corresponding

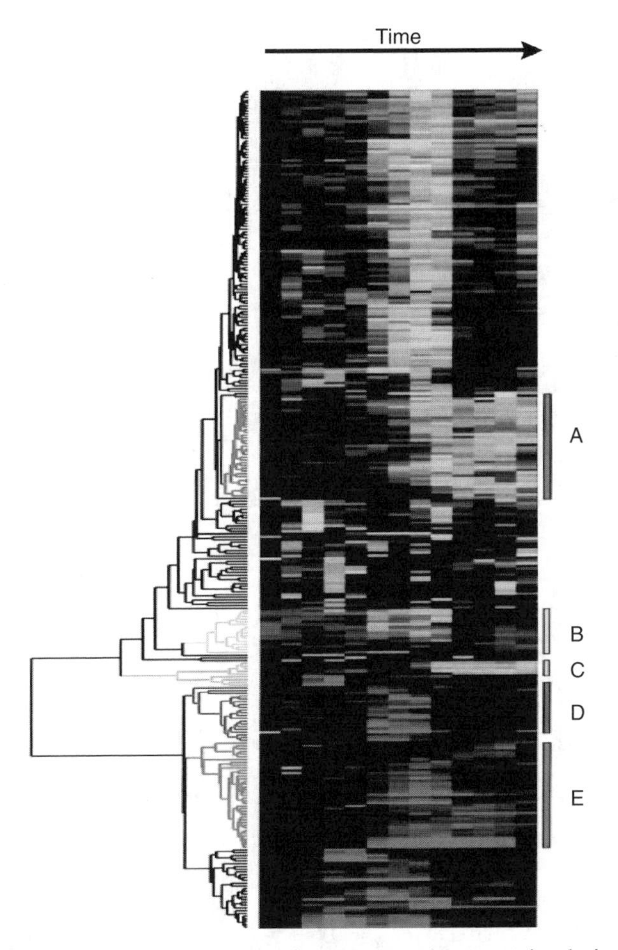

Fig. 3.9. Clustered display of data from time course of serum stimulation of primary human fibroblasts. The data set includes the expression level measurements of 8,613 human genes at 12 time points using cDNA microarrays. Each single row corresponds to a gene and each single column represents a time point. Five separate clusters are indicated by colored bars and by identical coloring of the corresponding region of the dendrogram. Genes that belong to these clusters are involved in (A) cholesterol biosynthesis, (B) the cell cycle, (C) the immediate-early response, (D) signaling and angiogenesis, and (E) wound healing and tissue remodeling. (From M. Eisen, P. Spellman, P. Brown, and D. Botstein. Cluster analysis and display of genome-wide expression patterns. Proceedings of National Academy of Sciences, vol. 95, pp. 14863–14868, 1998. Copyright © 1998 National Academy of Sciences, U.S.A. Reprinted with permission.)

DNA sequence in the control regions of each attained gene cluster, potential short and consensus sequence patterns, known as motifs, could be identified, and their interaction with transcriptional binding factors or regulators, leading to different gene activities, could also be investigated. Table 3.4 summarizes

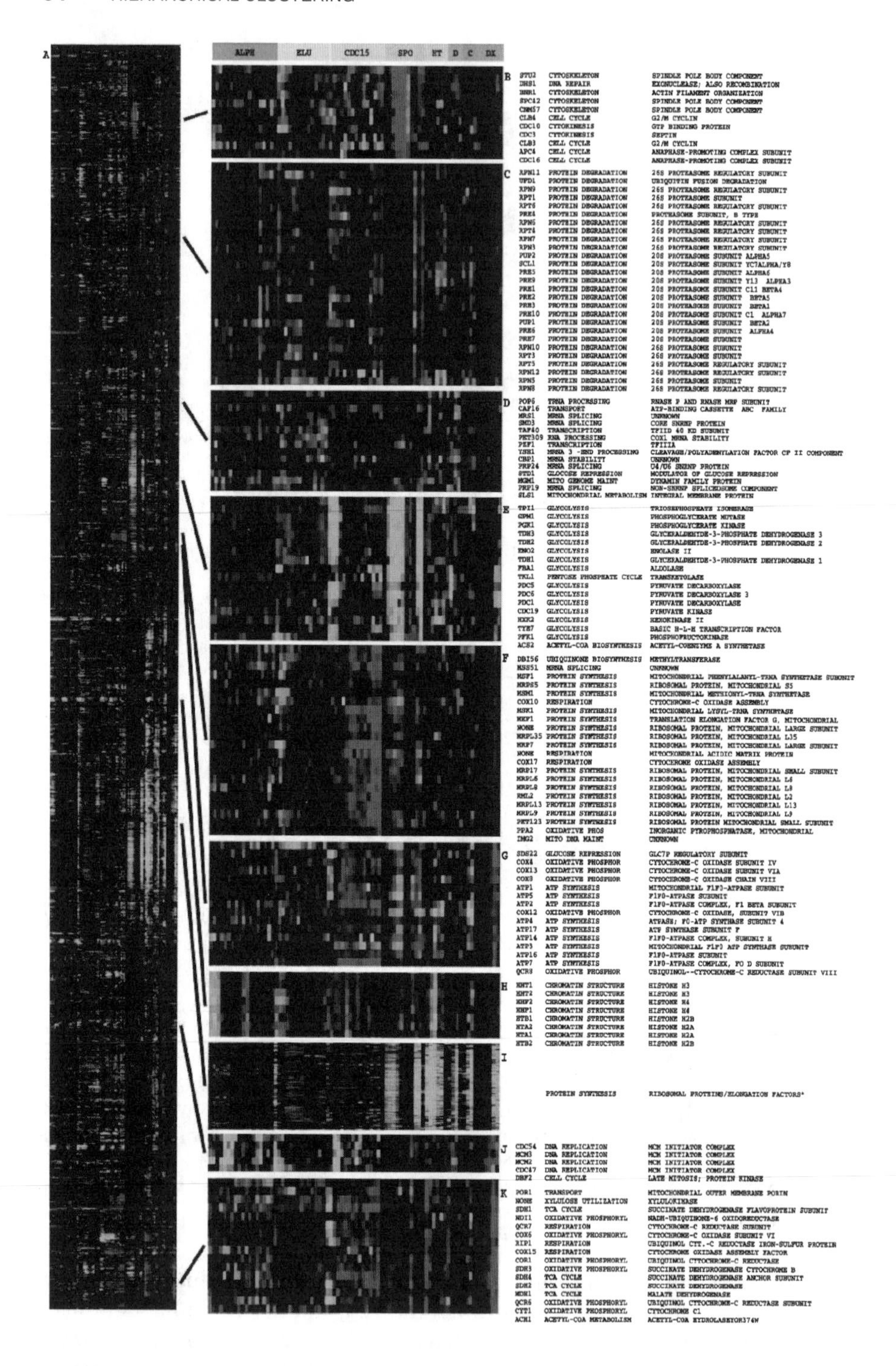
A
ALPH ELU CDC15 SPO HT D C DX
B
STU2 CYTOSKELETON SPINDLE POLE BODY COMPONENT
DHS1 DNA REPAIR EXONUCLEASE; ALSO RECOMBINATION
BNR1 CYTOSKELETON ACTIN FILAMENT ORGANIZATION
SPC42 CYTOSKELETON SPINDLE POLE BODY COMPONENT
CNM67 CYTOSKELETON SPINDLE POLE BODY COMPONENT
CLB4 CELL CYCLE G2/M CYCLIN
CDC10 CYTOKINESIS GTP BINDING PROTEIN
CDC3 CYTOKINESIS SEPTIN
CLB3 CELL CYCLE G2/M CYCLIN
APC4 CELL CYCLE ANAPHASE-PROMOTING COMPLEX SUBUNIT
CDC16 CELL CYCLE ANAPHASE-PROMOTING COMPLEX SUBUNIT
C
RPN11 PROTEIN DEGRADATION 26S PROTEASOME REGULATORY SUBUNIT
UFD1 PROTEIN DEGRADATION UBIQUITIN FUSION DEGRADATION
RPN9 PROTEIN DEGRADATION 26S PROTEASOME REGULATORY SUBUNIT
RPT1 PROTEIN DEGRADATION 26S PROTEASOME SUBUNIT
RPT6 PROTEIN DEGRADATION 26S PROTEASOME REGULATORY SUBUNIT
PRE4 PROTEIN DEGRADATION PROTEASOME SUBUNIT, B TYPE
RPN6 PROTEIN DEGRADATION 26S PROTEASOME REGULATORY SUBUNIT
RPT4 PROTEIN DEGRADATION 26S PROTEASOME REGULATORY SUBUNIT
RPN7 PROTEIN DEGRADATION 26S PROTEASOME REGULATORY SUBUNIT
RPN3 PROTEIN DEGRADATION 26S PROTEASOME REGULATORY SUBUNIT
PUP2 PROTEIN DEGRADATION 20S PROTEASOME SUBUNIT ALPHA5
SCL1 PROTEIN DEGRADATION 20S PROTEASOME SUBUNIT YC7ALPHA/Y8
PRE5 PROTEIN DEGRADATION 20S PROTEASOME SUBUNIT ALPHA6
PRE9 PROTEIN DEGRADATION 20S PROTEASOME SUBUNIT Y13 ALPHA3
PRE1 PROTEIN DEGRADATION 20S PROTEASOME SUBUNIT C11 BETA4
PRE2 PROTEIN DEGRADATION 20S PROTEASOME SUBUNIT BETA5
PRE3 PROTEIN DEGRADATION 20S PROTEASOME SUBUNIT BETA1
PRE10 PROTEIN DEGRADATION 20S PROTEASOME SUBUNIT C1 ALPHA7
PUP1 PROTEIN DEGRADATION 20S PROTEASOME SUBUNIT BETA2
PRE6 PROTEIN DEGRADATION 20S PROTEASOME SUBUNIT ALPHA4
PRE7 PROTEIN DEGRADATION 20S PROTEASOME SUBUNIT
RPN10 PROTEIN DEGRADATION 26S PROTEASOME SUBUNIT
RPT3 PROTEIN DEGRADATION 26S PROTEASOME SUBUNIT
RPT5 PROTEIN DEGRADATION 26S PROTEASOME REGULATORY SUBUNIT
RPN12 PROTEIN DEGRADATION 26S PROTEASOME REGULATORY SUBUNIT
RPN5 PROTEIN DEGRADATION 26S PROTEASOME SUBUNIT
RPN8 PROTEIN DEGRADATION 26S PROTEASOME REGULATORY SUBUNIT
D
POP6 TRNA PROCESSING RNASE P AND RNASE MRP SUBUNIT
CAF16 TRANSPORT ATP-BINDING CASSETTE ABC FAMILY
MRS1 MRNA SPLICING UNKNOWN
SMD3 MRNA SPLICING CORE SNRNP PROTEIN
TAF40 TRANSCRIPTION TFIID 40 KD SUBUNIT
PET309 RNA PROCESSING COX1 MRNA STABILITY
PZF1 TRANSCRIPTION TFIIIA
YSH1 MRNA 3 -END PROCESSING CLEAVAGE/POLYADENYLATION FACTOR CF II COMPONENT
CBP1 MRNA STABILITY UNKNOWN
PRP24 MRNA SPLICING U4/U6 SNRNP PROTEIN
STD1 GLUCOSE REPRESSION MODULATOR OF GLUCOSE REPRESSION
MGM1 MITO GENOME MAINT DYNAMIN FAMILY PROTEIN
PRP19 MRNA SPLICING NON-SNRNP SPLICEOSOME COMPONENT
SLS1 MITOCHONDRIAL METABOLISM INTEGRAL MEMBRANE PROTEIN
E
TPI1 GLYCOLYSIS TRIOSEPHOSPHATE ISOMERASE
GPM1 GLYCOLYSIS PHOSPHOGLYCERATE MUTASE
PGK1 GLYCOLYSIS PHOSPHOGLYCERATE KINASE
TDH3 GLYCOLYSIS GLYCERALDEHYDE-3-PHOSPHATE DEHYDROGENASE 3
TDH2 GLYCOLYSIS GLYCERALDEHYDE-3-PHOSPHATE DEHYDROGENASE 2
ENO2 GLYCOLYSIS ENOLASE II
TDH1 GLYCOLYSIS GLYCERALDEHYDE-3-PHOSPHATE DEHYDROGENASE 1
FBA1 GLYCOLYSIS ALDOLASE
TKL1 PENTOSE PHOSPHATE CYCLE TRANSKETOLASE
PDC5 GLYCOLYSIS PYRUVATE DECARBOXYLASE
PDC6 GLYCOLYSIS PYRUVATE DECARBOXYLASE 3
PDC1 GLYCOLYSIS PYRUVATE DECARBOXYLASE
CDC19 GLYCOLYSIS PYRUVATE KINASE
HXK2 GLYCOLYSIS HEXOKINASE II
TYE7 GLYCOLYSIS BASIC H-L-H TRANSCRIPTION FACTOR
PFK1 GLYCOLYSIS PHOSPHOFRUCTOKINASE
ACS2 ACETYL-COA BIOSYNTHESIS ACETYL-COENZYME A SYNTHETASE
F
DBI56 UBIQUINONE BIOSYNTHESIS METHYLTRANSFERASE
MSS51 MRNA SPLICING UNKNOWN
MSF1 PROTEIN SYNTHESIS MITOCHONDRIAL PHENYLALANYL-TRNA SYNTHETASE SUBUNIT
MRPS5 PROTEIN SYNTHESIS RIBOSOMAL PROTEIN, MITOCHONDRIAL S5
MSM1 PROTEIN SYNTHESIS MITOCHONDRIAL METHIONYL-TRNA SYNTHETASE
COX10 RESPIRATION CYTOCHROME-C OXIDASE ASSEMBLY
MSK1 PROTEIN SYNTHESIS MITOCHONDRIAL LYSYL-TRNA SYNTHETASE
MEF1 PROTEIN SYNTHESIS TRANSLATION ELONGATION FACTOR G, MITOCHONDRIAL
NONE PROTEIN SYNTHESIS RIBOSOMAL PROTEIN, MITOCHONDRIAL LARGE SUBUNIT
MRPL35 PROTEIN SYNTHESIS RIBOSOMAL PROTEIN, MITOCHONDRIAL L35
MRP7 PROTEIN SYNTHESIS RIBOSOMAL PROTEIN, MITOCHONDRIAL LARGE SUBUNIT
NONE RESPIRATION MITOCHONDRIAL ACIDIC MATRIX PROTEIN
COX17 RESPIRATION CYTOCHROME OXIDASE ASSEMBLY
MRP17 PROTEIN SYNTHESIS RIBOSOMAL PROTEIN, MITOCHONDRIAL SMALL SUBUNIT
MRPL6 PROTEIN SYNTHESIS RIBOSOMAL PROTEIN, MITOCHONDRIAL L6
MRPL8 PROTEIN SYNTHESIS RIBOSOMAL PROTEIN, MITOCHONDRIAL L8
RML2 PROTEIN SYNTHESIS RIBOSOMAL PROTEIN, MITOCHONDRIAL L2
MRPL13 PROTEIN SYNTHESIS RIBOSOMAL PROTEIN, MITOCHONDRIAL L13
MRPL9 PROTEIN SYNTHESIS RIBOSOMAL PROTEIN, MITOCHONDRIAL L9
PET123 PROTEIN SYNTHESIS RIBOSOMAL PROTEIN MITOCHONDRIAL SMALL SUBUNIT
PPA2 OXIDATIVE PHOS INORGANIC PYROPHOSPHATASE, MITOCHONDRIAL
IMG2 MITO DNA MAINT UNKNOWN
G
SDS22 GLUCOSE REPRESSION GLC7P REGULATORY SUBUNIT
COX4 OXIDATIVE PHOSPHOR CYTOCHROME-C OXIDASE SUBUNIT IV
COX13 OXIDATIVE PHOSPHOR CYTOCHROME-C OXIDASE SUBUNIT VIA
COX8 OXIDATIVE PHOSPHOR CYTOCHROME-C OXIDASE CHAIN VIII
ATP1 ATP SYNTHESIS MITOCHONDRIAL F1F0-ATPASE SUBUNIT
ATP5 ATP SYNTHESIS F1F0-ATPASE SUBUNIT
ATP2 ATP SYNTHESIS F1F0-ATPASE COMPLEX, F1 BETA SUBUNIT
COX12 OXIDATIVE PHOSPHOR CYTOCHROME-C OXIDASE, SUBUNIT VIB
ATP4 ATP SYNTHESIS ATPASE; F0-ATP SYNTHASE SUBUNIT 4
ATP17 ATP SYNTHESIS ATP SYNTHASE SUBUNIT F
ATP14 ATP SYNTHESIS F1F0-ATPASE COMPLEX, SUBUNIT H
ATP3 ATP SYNTHESIS MITOCHONDRIAL F1F0 ATP SYNTHASE SUBUNIT
ATP16 ATP SYNTHESIS F1F0-ATPASE SUBUNIT
ATP7 ATP SYNTHESIS F1F0-ATPASE COMPLEX, F0 D SUBUNIT
QCR8 OXIDATIVE PHOSPHOR UBIQUINOL--CYTOCHROME-C REDUCTASE SUBUNIT VIII
H
HHT1 CHROMATIN STRUCTURE HISTONE H3
HHT2 CHROMATIN STRUCTURE HISTONE H3
HHF2 CHROMATIN STRUCTURE HISTONE H4
HHF1 CHROMATIN STRUCTURE HISTONE H4
HTB1 CHROMATIN STRUCTURE HISTONE H2B
HTA2 CHROMATIN STRUCTURE HISTONE H2A
HTA1 CHROMATIN STRUCTURE HISTONE H2A
HTB2 CHROMATIN STRUCTURE HISTONE H2B
I
PROTEIN SYNTHESIS RIBOSOMAL PROTEINS/ELONGATION FACTORS*
J
CDC54 DNA REPLICATION MCM INITIATOR COMPLEX
MCM3 DNA REPLICATION MCM INITIATOR COMPLEX
MCM2 DNA REPLICATION MCM INITIATOR COMPLEX
CDC47 DNA REPLICATION MCM INITIATOR COMPLEX
DBF2 CELL CYCLE LATE MITOSIS; PROTEIN KINASE
K
POR1 TRANSPORT MITOCHONDRIAL OUTER MEMBRANE PORIN
NONE XYLULOSE UTILIZATION XYLULOKINASE
SDH1 TCA CYCLE SUCCINATE DEHYDROGENASE FLAVOPROTEIN SUBUNIT
NDI1 OXIDATIVE PHOSPHORYL NADH-UBIQUINONE-6 OXIDOREDUCTASE
QCR7 RESPIRATION CYTOCHROME-C REDUCTASE SUBUNIT
COX6 OXIDATIVE PHOSPHORYL CYTOCHROME-C OXIDASE SUBUNIT VI
RIP1 RESPIRATION UBIQUINOL CYT.-C REDUCTASE IRON-SULFUR PROTEIN
COX15 RESPIRATION CYTOCHROME OXIDASE ASSEMBLY FACTOR
COR1 OXIDATIVE PHOSPHORYL UBIQUINOL CYTOCHROME-C REDUCTASE
SDH3 OXIDATIVE PHOSPHORYL SUCCINATE DEHYDROGENASE CYTOCHROME B
SDH4 TCA CYCLE SUCCINATE DEHYDROGENASE ANCHOR SUBUNIT
SDH2 TCA CYCLE SUCCINATE DEHYDROGENASE
MDH1 TCA CYCLE MALATE DEHYDROGENASE
QCR6 OXIDATIVE PHOSPHORYL UBIQUINOL CYTOCHROME-C REDUCTASE SUBUNIT
CYT1 OXIDATIVE PHOSPHORYL CYTOCHROME C1
ACH1 ACETYL-COA METABOLISM ACETYL-COA HYDROLASEYOR374W

the seven major clusters using the average linkage hierarchical clustering algorithm to the expression data of 800 yeast genes (Spellman et al., 1998). Genes that belong to the same cluster are found to share common upstream sequence patterns. For example, most genes in cluster MCM contain binding sites for the regulator Mcm1p. The motif ACGCGT can be searched in more than 50% of genes grouped in cluster CLN2.

TABLE 3.4. Cluster summary. G1, S, M, and M/G1 in Column 5 represent phase groups of genes, which are created by identifying the time of peak expression for each gene and ordering all genes by this time.

Cluster	No. of genes	Binding site	Regulator	Peak expression
CLN2	119	ACGCGT	MBF, SBF	G1
Y′	26	Unknown	Unknown	G1
FKS1	92	ACRMSAAA	SBF, (MBF?)	G1
Histone	10	ATGCGAAR	Unknown	S
MET	20	AAACTGTGG	Met31p, Met32p	S
CLB2	35	MCM1+SFF	Mcm1p+SFF	M
MCM	34	MCM1	Mcm1p	M/G1
SIC1	27	RRCCAGCR	Swi5p/Ace2p	M/G1
Total	363			

The FKS1 set of genes is not a cluster, and the CLN2 cluster includes genes outside the core cluster that is shown. Consensus binding sites defined in the literature for various factors are as follows: MCM1, TTACCNAATTNGGTAA; SFF, GTMAACAA.

(Reproduced with permission of The American Society for Cell Biology, from Comprehensive identification of cell cycle-regulated genes of the yeast *saccharomyces cerevisiae* by microarray hybridization, P. Spellman, G. Sherlock, M. Zhang, V. Iyer, K. Anders, M. Eisen, P. Brown, D. Botstein, and B. Futcher, vol. 9, pp. 3273–3291, Copyright © 1998; permission conveyed through Copyright Clearance Center, Inc.)

Fig. 3.10. Cluster analysis of combined yeast data sets. The data sets include the expression level measurements of 2,467 genes from time courses during the diauxic shift (DX 7 time points); the cell division cycle after synchronization by alpha factor arrest (ALPH; 18 time points); centrifugal elutriation (ELU; 14 time points), and with a temperature-sensitive cdc15 mutant (CDC15; 15 time points); sporulation (SPO; 7 time points plus 4 additional samples); and shock by high temperature (HT; 6 time points); reducing agents (D; 4 time points) and low temperature (C; 4 time points). Full gene names are shown for representative clusters containing functionally related genes involved in (B) spindle pole body assembly and function, (C) the proteasome, (D) mRNA splicing, (E) glycolysis, (F) the mitochondrial ribosome, (G) ATP synthesis, (H) chromatin structure, (I) the ribosome and translation, (J) DNA replication, and (K) the tricarboxylic acid cycle and respiration. Gene name, functional category, and specific function are from the *Saccharomyces* Genome Database. (From M. Eisen, P. Spellman, P. Brown, and D. Botstein. Cluster analysis and display of genome-wide expression patterns. Proceedings of National Academy of Sciences, vol. 95, pp. 14863–14868, 1998. Copyright © 1998 National Academy of Sciences, U.S.A. Reprinted with permission.)

Another important application of hierarchical clustering algorithms in gene expression data analysis involves molecular clustering of cancer samples, where genes are regarded as features. Figure 3.11 illustrates one such example, based on the study of human lung cancer (Garber et al., 2001). In Fig 3.11(a), a dendrogram is shown for 67 lung tumors belonging to 4 different types: squamous, large cell, small cell, and adenocarcinoma, plus 5 normal tissues and a fetal lung tissue. Apparently, all normal tissues (including the fetal lung tissue) and small cell tumors, and most squamous lung tumors and large cell tumors, are clustered within their respective groups. Adenocarcinomas display more heterogeneity in terms of their expression patterns, and most of them fall into three clusters. Further examination of several clusters of genes (Fig. 3.11(b)) that are either strongly expressed (in red) or weakly expressed (in green), relevant to large cell tumors, small cell tumors, and squamous lung tumors, suggests the possible roles of these genes in tumor discrimination. The gene expression profiles can also be used to perform patient survival analysis in the early stage of cancers (Rosenwald et al., 2002). The Kaplan-Meier curves shown in Fig. 3.11(c) indicate a significant difference in survival between the patients in adenocarcinoma subgroup 1 and 3 (Fig. 3.11(a)). Since all patients are alive for subgroup 2, there is no such comparison available. This difference, due to the heterogeneity of tumors, raises the requirement of extra caution in clinical practice.

By using the centroid average hierarchical clustering algorithm, Alizadeh et al. (2000) successfully distinguished two molecularly distinct subtypes of diffuse large B-cell lymphoma, which cannot be discriminated by traditional morphological appearance-based methods, thus causing a high percentage of failure in clinical treatment. As stated in Alizadeh et al. (2000), "the new methods of gene expression profiling call for a revised definition of what is deemed a disease," and, "gene expression profiling presents a new way of approaching cancer therapeutics." This point is supported by much other research, such as the study of the relationship between gene expression patterns and drug sensitivity, where gene expression and molecular pharmacology databases are integrated to provide molecularly relevant criteria for therapy selection and drug discovery (Scherf et al., 2000).

It is important to point out here that in most of the current gene profiling-based cancer research, the microarrays include measurements of a large volume of genes, but the number of cancer samples is very limited (Alizadeh et al., 2000; Alon et al., 1999; Golub et al., 1999). Collecting samples is expensive—and those few that are collected are of high dimensionality. This is another form of the aforementioned curse of dimensionality. Often, only a small subset of genes is related to some specific cancer types. The introduction of many irrelevant genes (features) to tumor discrimination not only increases the computational complexity and demands for more samples, but impairs the analysis of the relevant ones. Questions about how many genes are really

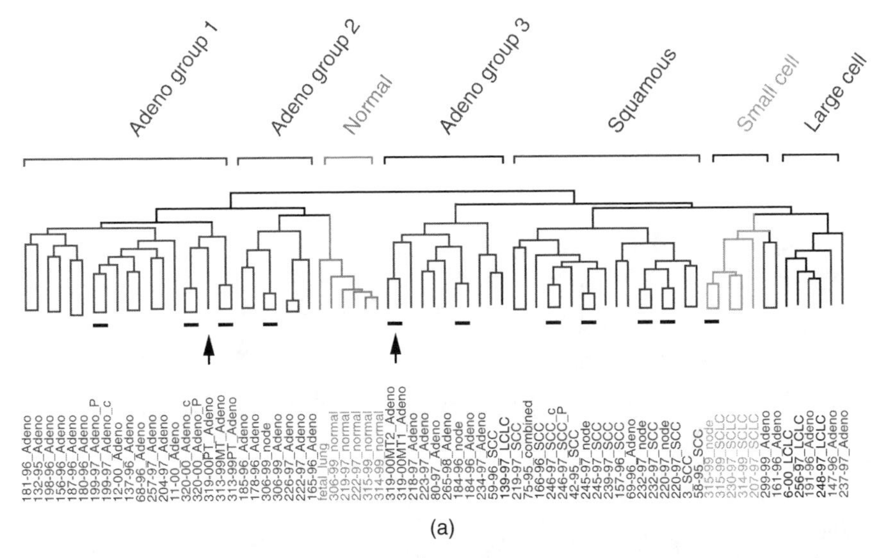

Fig. 3.11. Cluster analysis of human lung tumors. (a) Patterns of gene expression correspond to the major morphological classes of lung tumors. The dendrogram depicts the hierarchical clustering result of 67 lung tumors plus 5 normal tissues and a fetal lung tissue based on 918 cDNA clones representing 835 unique genes.

needed for disease diagnosis and whether or not the selected gene subsets are biologically meaningful still remain open. Another problem that needs extra caution involves cluster validation. As we discussed in Chapter 1, different clustering algorithms usually produce different partitions for the same data. How to evaluate the quality of the generated clusters of genes, and how to choose appropriate algorithms for a specified application, are particularly crucial questions for gene expression data research because sometimes even biologists cannot identify the real patterns from the artifacts of the clustering algorithms due to the limitations of biological knowledge. Some relevant discussions can be found in Ben-Dor et al. (1999) and Yeung et al. (2001).

3.5.2 Extensible Markup Language (XML) Document Clustering

Document clustering continues to play an important role in information retrieval and text mining due to its applications in extracting useful information from a huge amount of documents, organizing query results returned from a search engine, and so on (Cai et al., 2005; Hammouda and Kamel, 2004; Osi ski and Weiss, 2005). A comparison of agglomerative and divisive (based on bisecting *K*-means) hierarchical clustering in clustering document data sets

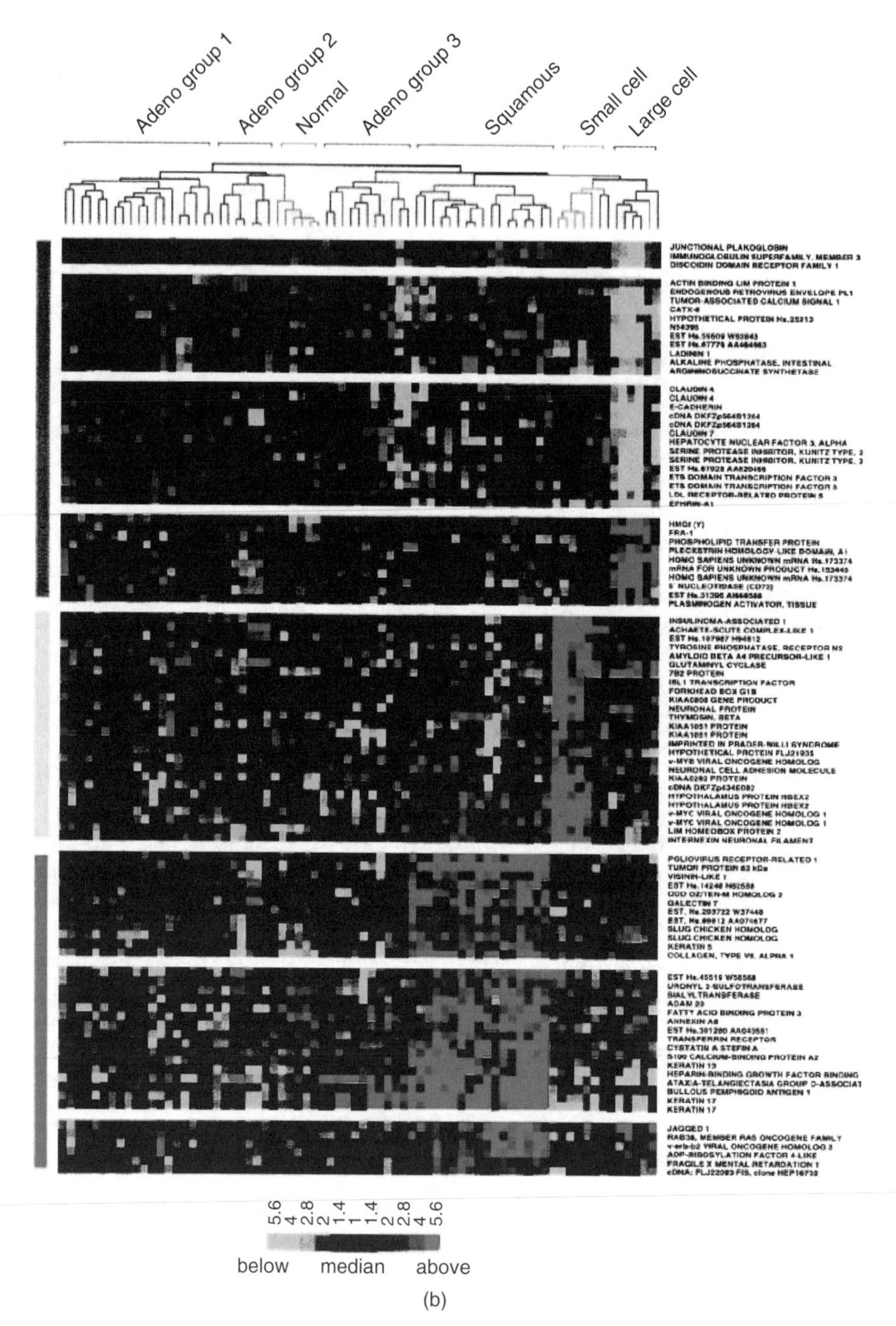

Fig. 3.11. Continued. (b) Squamous, small cell, and large cell lung tumors express a unique set of genes. Gene clusters are relevant to large cell tumors (blue bar), small cell tumors (yellow bar), and squamous lung tumors (red bar). Each row represents a specific gene, with its name listed on the right. The scale bar reflects the fold increase (red) or decrease (green) for any given gene relative to the median level of expression across all samples.

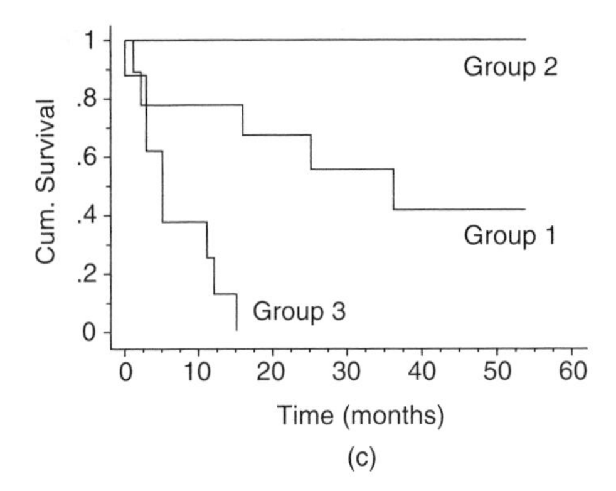

Fig. 3.11. Continued. (c) Kaplan-Meier curves show differences in survival for adenocarcinoma subgroups. The cumulative survival of the Kaplan-Meier curves represents percentage of patients living for the indicated times. (From M. Garber, O. Troyanskaya, K. Schluens, S. Petersen, Z. Thaesler, M. Pacyna-Gengelbach, M. van de Rijn, G. Rosen, C. Perou, R. Whyte, R. Altman, P. Brown, D. Botstein, and I. Petersen. Diversity of gene expression in adenocarcinoma of the lung. Proceedings of National Academy of Sciences, vol. 98, pp. 13784–13789, 2001. Copyright © 2001 National Academy of Sciences, U.S.A. Reprinted with permission.)

is given in Zhao and Karypis (2005). This study investigated six partitional criterion functions and three clustering linkages.

Among the available documents, those in an XML format have increased rapidly as the convenience of XML in facilitating data sharing and exchange via the Internet has increased. Lian et al. (2004) proposed the S-GRACE algorithm to cluster XML documents based on their structural information. In order to measure the distance between XML documents, a directed graph $sg(X) = (V, E)$, called an s-graph (structure-graph), is defined on a set of XML documents X. Here, V represents the set of all the elements and features in the documents in X, and an edge between nodes a and b is added into E if, and only if, a is a parent element of b or b is a feature of element a in some document in X. Specifically, the distance between a pair of documents X_1 and X_2 is given as,

$$D(X_1, X_2) = 1 - \frac{|sg(X_1) \cap sg(X_2)|}{\max(|sg(X_1)|, |sg(X_2)|)}, \tag{3.26}$$

where $|sg(X_i)|$, $i = 1, 2$ is the number of edges in the corresponding graph, and the union ($\cap$) of two documents creates a set of common edges of them. In the implementation of the clustering framework, each s-graph is encoded as a binary string, and the algorithm ROCK (Guha et al., 2000) is adopted to cluster the s-graphs.

TABLE 3.5. Average query response time for Q1–Q5 in milliseconds. Scheme-mapping and structure-mapping methods, with partitioned and unpartitioned schema, are used.

Methods	Q1	Q2	Q3	Q4	Q5
Unpartitioned schema with schema mapping	12,713	17,682	23,720	42,248	92,3691
Partitioned schema with schema mapping	2,460	4,787	6,952	13,462	177,697
Unpartitioned schema with structure mapping	137,336	2,693	29,934	54,984	4,560,031
Partitioned schema with structure mapping	12,344	533	4,500	13,527	761,327

Q1: $/A_1/A_2/\ldots/A_k$.
Q2: $/A_1/A_2/\ldots/A_k$[*text*() = "*value*"].
Q3: $/A_1/A_2/\ldots/A_k$[*contains*(.,"*substring*")].
Q4: find the titles of articles published in the VLDB journal in 1999.
Q5: find the names of authors that have at least one journal article and one conference paper.
(From W. Lian, D. Cheung, N. Mamoulis, and S. Yiu. An efficient and scalable algorithm for clustering XML documents by structure. IEEE Transactions on Knowledge and Data Engineering, vol. 16, pp. 82–96, 2004.)

This algorithm was applied to the data set, with 200,000 SML documents composed of 36 elements, from the DBLP database (www.acm.org/sigmod/dblp/db/index.html), which consists of bibliography entries in computer science. These documents are described as inproceedings, articles, postgraduate students' theses, white papers, and so on. Four clusters are identified by the S-GRACE algorithm. For example, 80,000 articles are grouped into a cluster, while 73,000 and 39,000 inproceedings form other two clusters. The remaining cluster is regarded as an outlier set, with about 7,000 documents. Particularly, the two clusters with inproceedings are hard to discriminate manually. Furthermore, the partitioned schema from the S-GRACE algorithm, with both schema mapping and structure mapping, is compared with the original unpartitioned schema in terms of query performance. The results summarized in Table 3.5 show that the partitioned schema derived from the clustering has achieved significant speed-up in all query cases for both mapping techniques.

3.5.3 Concurrent Engineering Team Structuring

Hierarchical clustering algorithms, including single linkage, complete linkage, average linkage, centroid linkage, and Ward's method, have been used in concurrent engineering team structuring based on projected communication flows between design tasks and the disciplines required for the tasks (Componation and Byrd, 2000). This approach was applied to a project on the development of a medical waste disposal unit, in which the product develop-

TABLE 3.6. Clusters of design tasks.

Team Structure	Design Tasks Assigned
Sub-Team #1	Chamber (A1), Sprayer (A2), Mixer (A3), Blade (B1), Rotor Shaft (B2)
Sub-Team #2	Disinfectant (A4), Auger Material (B3)
Sub-Team #3	Screen (B4), Auger Discharge (B5), Hopper (E1), Feed (E2), Metal Detectors (E3), Waster Discharge (F1), Waste Discharge (F2)
Sub-Team #4	Build Up (C1), Release (C2), Controls (C3), Heat System Source (D1), Heat System Controls (D2)

(From P. Componation and J. Byrd, Jr. Utilizing cluster analysis to structure concurrent engineering teams. IEEE Transactions on Engineering Management, vol. 47, pp. 269–280, 2000. Copyright © 2000 IEEE.)

ment team consisted of two chemists, three material scientists, three controls engineers, four manufacturing engineers, five quality assurances, and six mechanical engineers. The following results are based on the analysis using average linkage, which produces the team structure with shorter duration and lower risk than the other hierarchical clustering algorithms aforementioned. At the beginning, the requirements for completing the design tasks (19 in total) are determined using the process of quality function deployment (QFD). The design tasks, represented by 25 obtained variables, are then clustered into 4 groups, as summarized in Table 3.6. Here, the cutting point of the dendrogram is determined based on the estimation of two statistics and the data properties. Team members are assigned to the four subteams based on the specific requirements of the tasks in these groups. The resulting project work schedule is given in Table 3.7, together with the risk levels. Compared with the traditional sequential scheduling approaches that lead to a project duration of 17 weeks with a risk level of 2,532, the project schedule in Table 3.7 is preferable, with a project duration of 14 weeks and a risk level of 2040.

3.6. SUMMARY

Hierarchical clustering algorithms, including both bottom-up agglomerative hierarchical clustering and top-down divisive hierarchical clustering, produce a sequence of nested partitions. In practice, agglomerative hierarchical clustering is more commonly used while divisive hierarchical clustering is ignored due to its prohibitive computational burden. The clustering output of hierarchical clustering is usually represented as a dendrogram, which clearly describes the proximity among data objects and their clusters and provides good visualization. Although the classical hierarchical clustering methods are conceptually easy to understand, they suffer the disadvantages of quadratic computational complexity and the sensitivity to outliers, among others, which motivate the

TABLE 3.7. Project work schedule and resultant level of risk.

Week	Sub-Team #1	Sub-Team #2	Sub-Team #3	Sub-Team #4	Tasks Completed	Level
0						
1	B2	A4	E2	D1	B2/E2	312
2	B1	A4	B5	D1	B1/B5/D1	300
3		A4	B4/E3	C1	B4	244
4		A4	E3	C1		232
5		A4	E3	C1	C1	232
6		A4	E3/E1	D2	A4/E1/E3	228
7	A3	B3	F2	D2/C2	D2	168
8	A3/A2	B3	F2	C2	A3/F2	168
9	A2	B3		C2	A2/B3	108
10	A1			C2/C3	C2	24
11	A1			C3	A1/C3	24
12			F1			0
13			F1			0
14			F1		F1	0
15						0
16						0
17						0
18						0

(From P. Componation and J. Byrd, Jr. Utilizing cluster analysis to structure concurrent engineering teams. IEEE Transactions on Engineering Management, vol. 47, pp. 269–280, 2000.

development of more efficient algorithms such as BIRCH and CURE. These methods use preprocessing technologies, e.g., sampling a subset of the original data or pre-cluster, in order to process large-scale data. They are also equipped with specialized mechanisms in order to deal with outliers and increase system robustness.

CHAPTER 4

PARTITIONAL CLUSTERING

4.1. INTRODUCTION

In contrast to hierarchical clustering, which yields a successive level of clusters by iterative fusions or divisions, partitional clustering assigns a set of data points into K clusters without any hierarchical structure. This process usually accompanies the optimization of a criterion function. More specifically, given a set of data points $\mathbf{x}_i \in \Re^d$, $i = 1, \ldots, N$, partitional clustering algorithms aim to organize them into K clusters $\{C_1, \ldots, C_K\}$ while maximizing or minimizing a prespecified criterion function J. In principle, the optimal partition, based on the criterion function J, can be found by enumerating all possibilities. However, this brute force method is infeasible in practice due to the extremely expensive computation, as given by the formula (Liu, 1968):

$$P(N,K) = \frac{1}{K!} \sum_{m=1}^{K} (-1)^{K-m} C_K^m m^N. \tag{4.1}$$

Obviously, even for a small-scale clustering problem, simple enumeration is impossible. For example, in order to group 30 objects into 3 clusters, the number of possible partitions is approximatly 2×10^{14}. Therefore, heuristic algorithms seek approximate solutions.

This chapter starts with an introduction to the criterion functions for data partitioning. One of the widely used iterative optimization methods, the

Clustering, by Rui Xu and Donald C. Wunsch, II

K-means algorithm, based on the sum-of-squared-error criterion, is discussed in Section 4.3. The section also discusses its disadvantages and the algorithm's variants used to address them. Section 4.4 considers partitional clustering under a probabilistic framework, with each cluster represented by a probability density. Section 4.5 describes the application of graph theory in partitional clustering. Fuzzy clustering, which is useful when the clusters are not well separated and the boundaries are ambiguous, is introduced in Section 4.6. Under this framework, a data object is allowed to belong to more than one cluster with a degree of membership in each. Finally, Section 4.7 focuses on the search techniques that explore the optimal solutions of the clustering problem.

4.2. CLUSTERING CRITERIA

Cluster analysis aims to seek a partition of the data in which data objects in the same clusters are homogenous while data objects in different groups are well separated. This homogeneity and separation are evaluated through the criterion functions. One such criterion function is the sum-of-squared-error criterion, which is the most widely used criterion function in partitional clustering practice (Duda et al., 2001; Jain and Dubes, 1988).

Suppose we have a set of objects $\mathbf{x}_j \in \Re^d$, $j = 1, \ldots, N$, and we want to organize them into K clusters $C = \{C_1, \ldots, C_K\}$. The sum-of-squared-error criterion then is defined as

$$J_s(\mathbf{\Gamma}, \mathbf{M}) = \sum_{i=1}^{K}\sum_{j=1}^{N} \gamma_{ij}\|\mathbf{x}_j - \mathbf{m}_i\|^2 = \sum_{i=1}^{K}\sum_{j=1}^{N} \gamma_{ij}(\mathbf{x}_j - \mathbf{m}_i)^T(\mathbf{x}_j - \mathbf{m}_i), \quad (4.2)$$

where

$\mathbf{\Gamma} = \{\gamma_{ij}\}$ is a partition matrix, $\gamma_{ij} = \begin{cases} 1 & \text{if } \mathbf{x}_j \in \text{cluster } i \\ 0 & \text{otherwise} \end{cases}$ with $\sum_{i=1}^{K} \gamma_{ij} = 1 \; \forall j$;

$\mathbf{M} = [\mathbf{m}_1, \ldots, \mathbf{m}_K]$ is the cluster prototype or centroid (means) matrix; and

$\mathbf{m}_i = \frac{1}{N_i}\sum_{j=1}^{N} \gamma_{ij}\mathbf{x}_j$ is the sample mean for the ith cluster with N_i objects.

The partition that minimizes the sum-of-squared-error criterion is regarded as optimal and is called the minimum variance partition. The sum-of-squared-error criterion is appropriate for the clusters that are compact and well separated. However, this criterion may be sensitive to the existence of outliers and therefore may incorrectly divide a large cluster into small pieces, as indicated by Duda et al. (2001).

The sum-of-squared-error criterion can also be derived from the scatter matrices defined in multi-class discriminant analysis (Duda et al., 2001),

$$\mathbf{S}_T = \mathbf{S}_W + \mathbf{S}_B, \tag{4.3}$$

where

$\mathbf{S}_T = \sum_{j=1}^{N} (\mathbf{x}_j - \mathbf{m})(\mathbf{x}_j - \mathbf{m})^T$ is the total scatter matrix;

$\mathbf{S}_W = \sum_{i=1}^{K} \sum_{j=1}^{N} \gamma_{ij}(\mathbf{x}_j - \mathbf{m}_i)(\mathbf{x}_j - \mathbf{m}_i)^T$ is the within-cluster scatter matrix;

$\mathbf{S}_B = \sum_{i=1}^{K} N_i(\mathbf{m}_i - \mathbf{m})(\mathbf{m}_i - \mathbf{m})^T$ is the between-cluster scatter matrix; and

$\mathbf{m} = \frac{1}{N} \sum_{i=1}^{K} N_i \mathbf{m}_i$ is the total mean vector for the entire data set.

It can be seen that the total scatter matrix is independent of the partition while the other two matrices are related to each other. In the case of univariate analysis, the decrease of $\mathbf{S}_W$ is equivalent to the increase of $\mathbf{S}_B$. For data that are more than one dimension, both matrices can be scalarized by means of the measures such as trace and determinant. Specifically, the minimization of the trace of $\mathbf{S}_W$ is equivalent to the minimization of the sum-of-squared-error criterion, since,

$$\mathrm{tr}(\mathbf{S}_W) = \sum_{i=1}^{K} \sum_{j=1}^{N} \gamma_{ij}(\mathbf{x}_j - \mathbf{m}_i)^T(\mathbf{x}_j - \mathbf{m}_i) = J_s(\mathbf{\Gamma}, \mathbf{M}). \tag{4.4}$$

At the same time, the minimization of the sum-of-squared-error is also equivalent to the maximization of the trace of $\mathbf{S}_B$, since $\mathrm{tr}(\mathbf{S}_T) = \mathrm{tr}(\mathbf{S}_W) + \mathrm{tr}(\mathbf{S}_B)$.

On the other hand, when the determinants of the scatter matrices are considered, the selection of $|\mathbf{S}_B|$ is usually avoided, since $\mathbf{S}_B$ will become singular when the number of clusters is less than or equal to the data dimensionality. As a result, the criterion based on determinants aims to minimize the determinant of $\mathbf{S}_W$, as defined below,

$$J_d(\mathbf{\Gamma}, \mathbf{M}) = |\mathbf{S}_W| = \left| \sum_{i=1}^{K} \sum_{j=1}^{N} \gamma_{ij}(\mathbf{x}_j - \mathbf{m}_i)(\mathbf{x}_j - \mathbf{m}_i)^T \right|. \tag{4.5}$$

Unlike the trace criterion, this criterion is independent of the scaling, which may cause quite different solutions for data before and after the linear transformations when $\mathrm{tr}(\mathbf{S}_W)$ is minimized.

Another criterion that is immune from this disadvantage is based on the maximization of the trace of the matrix from the product of between-cluster matrix and within-cluster matrix,

$$\mathrm{tr}\left(\mathbf{S}_W^{-1}\mathbf{S}_B\right)=\sum_{i=1}^{d}\lambda_i, \tag{4.6}$$

where $\lambda_1, \ldots, \lambda_d$ are the eigenvalues of $\mathbf{S}_W^{-1}\mathbf{S}_B$. Similarly, by considering $\mathbf{S}_B$, we can obtain a rich class of scale-independent criterion functions (Duda et al., 2001; Everitt et al., 2001):

$$\mathrm{tr}\left(\mathbf{S}_T^{-1}\mathbf{S}_W\right)=\sum_{i=1}^{d}\frac{1}{1+\lambda_i}, \tag{4.7}$$

and

$$\frac{|\mathbf{S}_T|}{|\mathbf{S}_W|}=\prod_{i=1}^{d}(1+\lambda_i). \tag{4.8}$$

Now, we reconsider the sum-of-squared-error criterion by removing the mean vectors from Eq. 4.2, which becomes

$$J_s(\mathbf{\Gamma})=\sum_{i=1}^{K}\frac{1}{N_i}\sum_{\mathbf{x}_m\in C_i}\sum_{\mathbf{x}_j\in C_i}\|\mathbf{x}_m-\mathbf{x}_j\|^2. \tag{4.9}$$

Eq. 4.9 can be generalized further as

$$J_s(\mathbf{\Gamma})=\sum_{i=1}^{K}\psi_1(i), \tag{4.10}$$

where

$$\psi_1(i)=\frac{1}{N_i}\sum_{\mathbf{x}_m\in C_i}\sum_{\mathbf{x}_j\in C_i}D_{mj} \tag{4.11}$$

measures the homogeneity of the cluster C_i and D_{mj} defines the distance between a pair of objects $\mathbf{x}_m$ and $\mathbf{x}_j$ in this cluster. It is easy to see that this criterion measures the dissimilarity of data objects within a certain cluster.

Other homogeneity-based indices include (Everitt et al., 2001; Hansen and Jaumard, 1997):

1. the diameter of C_i, which measures the maximum distance between objects in C_i:

$$\psi_2(i) = \max_{\mathbf{x}_m \in C_i, \mathbf{x}_j \in C_i, m \neq j} (D_{mj}); \tag{4.12}$$

2. the star index, which measures the minimum sum of the distances between a data object $\mathbf{x}_j$ of C_i and all other members:

$$\psi_3(i) = \min_{\mathbf{x}_m \in C_i} \left(\sum_{\mathbf{x}_j \in C_i} D_{mj} \right); \tag{4.13}$$

3. the radius of C_i, which measures the minimum for all objects $\mathbf{x}_j$ of C_i of the maximum distance between $\mathbf{x}_j$ and the others:

$$\psi_4(i) = \min_{\mathbf{x}_m \in C_i} \left(\max_{\mathbf{x}_j \in C_i} D_{mj} \right). \tag{4.14}$$

Since cluster analysis emphasizes both internal homogeneity and external separation, similar indices can also be constructed by considering the maximum separation of different clusters:

4. the cut index of C_i, which measures the sum of the distances between an object of C_i and all other objects falling out of C_i:

$$\psi_5(i) = \sum_{\mathbf{x}_m \in C_i} \sum_{\mathbf{x}_j \notin C_i} D_{mj}; \tag{4.15}$$

5. the split index of C_i, which measures the minimum distance between an object of C_i and all other objects falling out of C_i:

$$\psi_6(i) = \min_{\mathbf{x}_m \in C_i, \mathbf{x}_j \notin C_i} (D_{mj}). \tag{4.16}$$

Ultimately, each measure can be aggregated over all clusters to form the partition criterion function, as defined in Eq. 4.10. Similar constructions can be obtained by only examining the clusters having extreme index values, as described below:

$$J_w(\mathbf{\Gamma}) = \max_{i=1,\mathrm{K},K} (\psi(i)), \tag{4.17}$$

$$J_b(\mathbf{\Gamma}) = \min_{i=1,\mathrm{K},K} (\psi(i)). \tag{4.18}$$

4.3. *K*-MEANS ALGORITHM

4.3.1 *K*-Means Algorithm

The *K*-means algorithm (Forgy, 1965; MacQueen, 1967) is one of the best-known and most popular clustering algorithms (Duda et al., 2001; Theodoridis and Koutroumbas, 2006). K-means seeks an optimal partition of the data by minimizing the sum-of-squared-error criterion in Eq. 4.2 with an iterative optimization procedure, which belongs to the category of hill-climbing algorithms. The basic clustering procedure of K-means is summarized as follows:

1. Initialize a K-partition randomly or based on some prior knowledge. Calculate the cluster prototype matrix $\mathbf{M} = [\mathbf{m}_1, \ldots, \mathbf{m}_K]$;
2. Assign each object in the data set to the nearest cluster C_l, i.e.,

$$\mathbf{x}_j \in C_l, \text{ if } \|\mathbf{x}_j - \mathbf{m}_l\| < \|\mathbf{x}_j - \mathbf{m}_i\| \quad \text{for } j = 1, \ldots, N, i \neq l, \text{ and } i = 1, \ldots, K; \tag{4.19}$$

3. Recalculate the cluster prototype matrix based on the current partition,

$$\mathbf{m}_i = \frac{1}{N_i} \sum_{\mathbf{x}_j \in C_i} \mathbf{x}_j; \tag{4.20}$$

4. Repeat steps 2 and 3 until there is no change for each cluster.

Note that the partition in step 2 follows the nearest-neighbor rule and therefore is a Voronoi partition (Gersho and Gray, 1992). The input space is divided into Voronoi regions corresponding to a set of prototype vectors or Voronoi vectors. Each point in a certain Voronoi region is closer to its Voronoi vector than any other ones.

The algorithm described above performs batch mode learning, since the update of the prototypes occurs after all the data points have been processed. Correspondingly, the on-line or incremental mode K-means adjusts the cluster centroids each time a data point is presented,

$$\mathbf{m}^{new} = \mathbf{m}^{old} + \eta(\mathbf{x} - \mathbf{m}^{old}), \tag{4.21}$$

where η is the learning rate. We will discuss more on the online version of K-means algorithm in Chapter 5, under the framework of hard competitive learning clustering.

It is worthwhile to mention that, in the community of communications, similar algorithms were suggested such as the Lloyd algorithm (Lloyd, 1982), which was further generalized for vector quantization (VQ), known as generalized Lloyd algorithm or Linde-Buzo-Gray (LBG) algorithm (Gersho and Gray, 1992; Linde et al., 1980), in order to achieve data and signal compression. In the context of VQ, the prototype vectors are called code words, which constitute a code book. VQ aims to obtain a representation of the data distribution with a reduced number of elements while minimizing the information loss, measured with a criterion function called a distortion function. This makes the objective of VQ different from cluster analysis. However, this difference becomes negligible in real practice since "the two approaches roughly execute the same operations: grouping data into a certain number of groups so that a loss (or error) function is minimized" (Patanè and Russo, 2002). An enhanced LBG (ELBG) algorithm was proposed by Patanè and Russo (2001) with an introduction of an additional step for the purpose of shifting the badly positioned code words, evaluated in terms of the concept of utility. They further extended the concept to an algorithm, called a fully automatic clustering system (FACS), which can dynamically determine the number of code words with regard to a certain distortion value (Patanè and Russo, 2002). More discussion on the advancement and applications of vector quantization are given in Bracco et al. (2003), Cosman et al. (1993), Fowler et al. (1995), Han and Kim (2006), Nguyen (2005), and Pan (1997).

K-means is regarded as a staple of clustering methods (Duda et al., 2001), due to its ease of implementation. It works well for many practical problems, particularly when the resulting clusters are compact and hyperspherical in shape. The time complexity of K-means is $O(NKdT)$, where T is the number of iterations. Since K, d, and T are usually much less than N, the time complexity of K-means is approximately linear. Therefore, K-means is a good selection for clustering large-scale data sets. Moreover, several methods have been proposed to speed up K-means (Pelleg and Moore, 1999; Stoffel and Belkoniene, 1999). A spherical K-means (*spkmeans*) was introduced to address high-dimensional and sparse data objects, particularly for document clustering, which uses a concept vector containing semantic information to represent each cluster (Dhillon and Modha, 2001). The spkmeans algorithm is further cast into the framework of a maximum likelihood estimation of a mixture of K von Mises-Fisher distributions (Banerjee and Ghosh, 2004).

While K-means has these desirable properties, it also suffers several major drawbacks, particularly the inherent limitations when hill-climbing methods are used for optimization. These disadvantages of K-means attract a great deal of effort from different communities, and as a result, many variants of K-means have appeared to address these obstacles. In the following discussion, we summarize the major disadvantages of K-means with an emphasis on the corresponding approaches and strategies for improvement.

4.3.2 Advancement of *K*-Means

4.3.2.1 Convergence and Initial Partition The first problem that plagues K-means is that the iteratively optimal procedure cannot guarantee the convergence to a global optimum, although the convergence of K-means was proved (Selim and Ismail, 1984). Since K-means can converge to a local optimum, different initial points generally lead to different convergence centroids, which makes it important to start with a reasonable initial partition in order to achieve high quality clustering solutions. However, in theory, there exist no efficient and universal methods for determining such initial partitions.

Stochastic optimal search techniques, such as simulated annealing and genetic algorithms, which are further discussed in Section 4.7 in this chapter, provide a possible way to search the complicated problem space more effectively and find the global or approximately global optimum. Krishna and Murty (1999) proposed the genetic K-means (GKA) algorithm, integrating genetic algorithm (also introduced in Section 4.7) with K-means, in order to achieve a global search and fast convergence. They replaced the computationally expensive crossover operator with the K-means operator, which is realized as a one-step K-means algorithm. The aim of the mutation operator is to avoid getting the solutions stuck in local minima, and therefore, it is designed based on the Euclidean distance between an object and the centroids. The proposed ELBG algorithm adopts a roulette mechanism typical of genetic algorithms to become near-optimal; therefore, it is not as sensitive to initialization (Patanè and Russo, 2001).

In order to examine the sensitivity of generated partitions to the starting points, a general strategy is to run the algorithm several times with random initial partitions (Jain and Dubes, 1988). The clustering results of these different runs provide some insights into the quality of the ultimate clusters. Peña et al. (1999) compared the random method with three other initial partition methods by Forgy (1965), Kaufman and Rousseeuw (1990), and MacQueen (1967), based on the effectiveness, robustness, and convergence speed criteria. Forgy's method generates the initial partition by first randomly selecting K points as prototypes and then separating the remaining points based on their distance from these seeds. MacQueen's method is similar to Forgy's method except that it works in an online mode and therefore is dependent on the presentation order of the data points. Kaufman and Rousseeuw's method builds the centroids successively. The first centroid is the most centrally located and has the smallest sum of distance to other points. The point that achieves the most decrease of the cost function is selected as the second centroid. This process iterates until all K centroids are determined. The four methods are compared on three data sets from the Machine Learning Repository (Asuncion and Newman, 2007). According to their experimental results, the random method and Kaufman and Rousseeuw's method work much better than the other two methods under the effectiveness and sensitivity criteria, while MacQueen's method stands out when only the convergence speed is considered. Overall, they recommend Kaufman and Rousseeuw's method. It is certainly possible that different data sets would yield different relative performance results.

Likas et al. (2003) proposed a global K-means algorithm consisting of a series of K-means clustering procedures with the number of clusters varying from 1 to K. After the identification of the centroid when only one cluster is considered, at each iteration k, $k = 2, \ldots, K$, the previous k-1 centroids are fixed and the new centroid is selected by examining all data points, which are used as the initial centroids. The authors claimed that the algorithm is independent of the initial partitions and provided accelerating strategies. The intermediate cluster results can also be used to estimate the actual number of clusters. One disadvantage of the algorithm lies in the requirement for executing K-means N times for each value of k, which causes high computational burden for large data sets.

Bradley and Fayyad (1998) argued that, "the solutions obtained by clustering over a small subsample may provide good refined initial estimates of the true means, or centroids, in the data". They presented a refinement algorithm that first utilizes K-means M times to M random subsets sampled from the original data. The set formed from the union of the obtained $M \times K$ clustering centroids is then clustered M times to generate multiple candidate solutions, setting each subset solution as the initial guess for each clustering performance. The starting points for the entire data set are determined by choosing the ones with the minimal sum of squared distances. Zhong et al. (2005) introduced a successive method for K-means initialization and applied it to sequence motifs clustering. The original K-means is run first to ignite the algorithm, and only those candidate points that lead to satisfying structural similarity and are also not close to the prototypes already selected are regarded as the initial centroids.

4.3.2.2 The Number of K

K-means assumes that the number of clusters K is already known by the users, which, unfortunately, usually is not true in practice. Like the situation for cluster initialization, there are also no efficient and universal methods for the selection of K. Therefore, identifying K in advance becomes a very important topic in cluster validity (Dubes, 1993). Here, we describe several heuristics that are directly related to K-means, and further discussion is elaborated in Chapter 10.

The ISODATA (Iterative Self-Organizing Data Analysis Technique) algorithm, developed by Ball and Hall (1967), deals with the dynamic estimation of K. ISODATA adaptively adjusts the number of clusters by merging and splitting clusters according to some predefined thresholds. In this sense, the problem of identifying the initial number of clusters becomes that of parameter (threshold) tweaking. The splitting criterion is based on the within-class variability while the merge of two clusters is dependent on the distance of their centroids. The new K is used as the expected number of clusters for the next iteration.

For incremental cases, a new cluster is created only when no existing ones can represent the new data object. This is achieved through the introduction of a threshold, which determines whether the similarity between the winning prototype and the data object meets some expectation. Specifically, the cluster C_l not only is the nearest cluster to the new data object $\mathbf{x}_j$, in order to represent $\mathbf{x}_j$, but also follows an extra constraint, i.e.,

$$\|\mathbf{x}_j - \mathbf{m}_l\| < \rho, \tag{4.22}$$

where ρ is the threshold. Obviously, the smaller ρ is, the more the number of clusters and the less data points in each cluster, and vice versa. If the condition in Eq. 4.22 does not hold, a new cluster is generated to represent the data point. This algorithm is known as the leader algorithm (Duda et al., 2001; Hartigan, 1975), and a similar strategy is used in Adaptive Resonance Theory (Carpenter and Grossberg, 1987a), discussed in Chapter 5, which also generates new clusters when needed (see also Moore (1989)).

4.3.2.3 Robustness K-means is sensitive to outliers and noise. The calculation of the means considers all the data objects in the cluster, including the outliers. Even if an object is quite far away from the cluster centroids, it is still forced into a cluster and used to calculate the prototype representation, which therefore distorts the cluster shapes.

ISODATA (Ball and Hall, 1967) and PAM (Partitioning Around Medoids) (Kaufman and Rousseeuw, 1990) both consider the effect of outliers in clustering procedures. ISODATA discards the clusters in which the number of data points is below some threshold. The splitting operation of ISODATA eliminates the possibility of elongated clusters typical of K-means. Rather than utilizing the calculated means, PAM utilizes real data points, called medoids, as the cluster prototypes, and it avoids the effect of outliers to the resulting prototypes. A medoid is a point that has the minimal average distance to all other objects in the same cluster. Starting from the same strategy, a K-medoids based algorithm is presented by Estivill-Castro and Yang (2000), which uses the discrete medians as the cluster centroids. This algorithm is slower than K-means and has a time complexity of $O(N\log N)$.

4.3.2.4 Extension of the Definition of Means The definition of means limits the application of K-means only to numerical variables, while leaving the categorical variables unhandled. Moreover, even for the numerical variables, the obtained means may not have the physical meaning or may be difficult to interpret. The aforementioned K-medoids algorithm is a natural choice when the computation of means is unavailable because the medoids do not need any calculation and always exist (Kaufman and Rousseeuw, 1990).

Ordonez (2003) discussed K-means in binary data clustering and suggested three variants. It is indicated that binary data can also be used to represent categorical data. Both Huang (1998) and Gupta et al. (1999) defined different dissimilarity measures to extend K-means to categorical variables. Particularly, for Huang's method, the distance between a pair of categorical data points is measured in terms of the total number of mismatches of the features,

$$D(\mathbf{x}_i, \mathbf{x}_j) = \sum_{l=1}^{d} \delta(x_{il}, x_{jl}), \tag{4.23}$$

where

$$\delta(x_{il}, x_{jl}) = \begin{cases} 1 & x_{il} \neq x_{jl} \\ 0 & x_{il} = x_{jl} \end{cases}, \tag{4.24}$$

or the weighted total number of mismatches,

$$D(\mathbf{x}_i, \mathbf{x}_j) = \sum_{l=1}^{d} \frac{n_{x_{il}} + n_{x_{jl}}}{n_{x_{il}} n_{x_{jl}}} \delta(x_{il}, x_{jl}), \tag{4.25}$$

where $n_{x_{il}}$ and $n_{x_{jl}}$ are the frequencies of the occurrence of x_{il} and x_{jl}.

Accordingly, the clustering goal is formulated so as to minimize the cost function

$$J(\mathbf{\Gamma}, \mathbf{Q}) = \sum_{i=1}^{K} \sum_{j=1}^{N} \gamma_{ij} D(\mathbf{x}_j, \mathbf{Q}_i), \tag{4.26}$$

with a set of d-dimensional vectors $\mathbf{Q} = \{\mathbf{Q}_1, \ldots, \mathbf{Q}_K\}$, $\mathbf{Q}_j = (\mathbf{Q}_{j1}, \ldots, \mathbf{Q}_{jd})$, which are used to represent the clusters instead of the means in the standard version. Each vector $\mathbf{Q}_j$ is known as a mode and is defined to minimize the sum of distances

$$J(\mathbf{Q}) = \sum_{i=1}^{N} D(\mathbf{x}_i, \mathbf{Q}_j). \tag{4.27}$$

The modes of the clusters are determined using a frequency-based approach, and the proposed K-modes algorithm operates in an iterative way as K-means.

More recent discussions on K-means, its variants, and other squared-error based clustering algorithms with their applications can be found in (Bradley et al., 2000a, Charalampidis, 2005; Hansen and Mladenović, 2001; Huang et al., 2005; Lingras and West, 2004; Peters, 2006; Su and Chou, 2001; Wagstaff et al., 2001; Zha et al., 2001).

4.4. MIXTURE DENSITY-BASED CLUSTERING

4.4.1 Introduction

In the probabilistic view, each data object is assumed to be generated from one of K underlying probability distributions. Data points in different clusters are drawn from different probability distributions. These probability sources can take different functional forms, such as multivariate Gaussian or t-distributions, or come from the same families but with different parameters. Usually, the forms of the mixture densities are assumed to be known, which makes the process of finding the clusters of a given data set equivalent to estimating the parameters of the K underlying models. It is worth mentioning that the mixture densities are useful in supervised classification (Fraley and Raftery, 2002) because, "mixtures can also be seen as a class of models that are able to represent arbitrarily complex probability density functions" (Jain et al., 2000). An application of mixtures to feature selection is illustrated by Law et al. (2004).

Let us consider the following procedure to generate a set of data objects. Assume the number of clusters or mixture densities K is already known, which can also be estimated as discussed in Chapter 10. Each time, a data object $\mathbf{x}$ is generated according to a class-conditional probability density $p(\mathbf{x}|C_i, \boldsymbol{\theta}_i)$ for the cluster C_i, where $\boldsymbol{\theta}_i$ is the unknown parameter vector. The class-conditional probability density is also known as the component density. The chance of $\mathbf{x}$ coming from C_i is dependent on the prior probability $P(C_i)$, also called the mixing parameter. Both the class-conditional probability density and the prior probability are known. Then, the mixture probability density for the entire data set can be expressed as

$$p(\mathbf{x}|\boldsymbol{\theta}) = \sum_{i=1}^{K} p(\mathbf{x}|C_i, \boldsymbol{\theta}_i) P(C_i), \tag{4.28}$$

where $\boldsymbol{\theta} = (\boldsymbol{\theta}_1, \ldots, \boldsymbol{\theta}_K)$ and $\sum_{i=1}^{K} P(C_i) = 1$. Here, the mixtures can be constructed with any types of components theoretically, but more commonly, all components are assumed to take the same forms, particularly the multivariate Gaussian densities because of its complete theory, analytical tractability, and natural occurrence in many situations (Duda et al., 2001; Everitt et al., 2001; Frey and Jojic, 2003; Roberts et al., 1998; Titterington et al., 1985; Zhuang et al., 1996).

Moreover, as long as the parameter vector $\boldsymbol{\theta}$ is decided, it is not difficult to see that the posterior probability for assigning a data object to a cluster can be calculated using Bayes formula,

$$P\left(C_i|\mathbf{x}, \hat{\boldsymbol{\theta}}\right) = \frac{P(C_i) p\left(\mathbf{x}|C_i, \hat{\boldsymbol{\theta}}_i\right)}{p\left(\mathbf{x}|\hat{\boldsymbol{\theta}}\right)}, \tag{4.29}$$

where $\hat{\boldsymbol{\theta}}$ is the vector of estimated parameters.

Before further estimating the parameter vector $\boldsymbol{\theta}$, it is natural to investigate whether the mixtures are identifiable or whether there exists a unique solution $\boldsymbol{\theta}$ that generates the obtained observances (Duda et al., 2001; Teicher, 1961). If there are multiple solutions $\boldsymbol{\theta}$, the component distribution cannot be inferred, and the clustering becomes impossible theoretically. Duda et al. (2001) illustrate an example where the mixture is completely unidentifiable, which is reproduced here. The data point x takes the value of either 1 or 0, and the mixture density is

$$P(x|\boldsymbol{\theta}) = \frac{1}{2}\theta_1^x(1-\theta_1)^{1-x} + \frac{1}{2}\theta_2^x(1-\theta_2)^{1-x}$$
$$= \begin{cases} \frac{1}{2}(\theta_1+\theta_2) & \text{if } x=1 \\ 1-\frac{1}{2}(\theta_1+\theta_2) & \text{if } x=0 \end{cases}. \quad (4.30)$$

Assume that we know the density $P(x|\boldsymbol{\theta})$, for example, we have $P(x = 1|\boldsymbol{\theta}) = 0.6$ and $P(x = 0|\boldsymbol{\theta}) = 0.4$. The most information we can obtain is $\theta_1 + \theta_2 = 1.2$, and the individual component cannot be identified. Duda et al. (2001) further pointed out that, "most mixtures of commonly encountered density functions" and, "most complex or high-dimensional density functions encountered in real-world problems" are identifiable. However, we should always keep in mind that identifiability may cause a problem during the mixture density based clustering practice.

4.4.2 Maximum Likelihood Estimation and Expectation Maximization

Maximum likelihood (ML) estimation is one of the most important statistical approaches for parameter estimation (Duda et al., 2001), and ML considers the best estimate $\hat{\boldsymbol{\theta}}$ as the one that maximizes the probability of producing all the observations $\mathbf{X} = \{\mathbf{x}_1, \ldots, \mathbf{x}_N\}$, which is given by the joint density function,

$$p(\mathbf{X}|\boldsymbol{\theta}) = \prod_{j=1}^{N} p(\mathbf{x}_j|\boldsymbol{\theta}), \quad (4.31)$$

or, in a logarithm form,

$$l(\boldsymbol{\theta}) = \sum_{j=1}^{N} \ln p(\mathbf{x}_j|\boldsymbol{\theta}). \quad (4.32)$$

Thus, the best estimate $\hat{\boldsymbol{\theta}}$ can be achieved by solving the log-likelihood equations,

$$\frac{\partial l(\hat{\boldsymbol{\theta}})}{\partial \hat{\boldsymbol{\theta}}_i} = 0, \quad i = 1, \cdots, K. \quad (4.33)$$

By using Eq. 4.28, 4.29, and 4.32, we further have

$$\sum_{j=1}^{N} P\left(C_i|\mathbf{x}_j, \hat{\boldsymbol{\theta}}\left(\partial \ln p\left(\mathbf{x}_j\middle|C_i, \hat{\boldsymbol{\theta}}_i\right)\middle/\partial \hat{\boldsymbol{\theta}}_i\right) = 0, \quad i = 1, \cdots, K. \quad (4.34)$$

The prior probabilities $P(C_i)$ are considered to be known during the previous discussion. When $P(C_i)$ are unknown, they can be estimated in terms of the estimated posterior probabilities,

$$\hat{P}(C_i) = \frac{1}{N}\sum_{j=1}^{N} \hat{P}\left(C_i|\mathbf{x}_j, \hat{\boldsymbol{\theta}}\right), \tag{4.35}$$

where

$$\hat{P}\left(C_i|\mathbf{x}_j, \hat{\boldsymbol{\theta}}\right) = \frac{\hat{P}(C_i)\, p\left(\mathbf{x}_j|C_i, \hat{\boldsymbol{\theta}}_i\right)}{\sum_{i=1}^{K} \hat{P}(C_i)\, p\left(\mathbf{x}_j|C_i, \hat{\boldsymbol{\theta}}_i\right)}. \tag{4.36}$$

Accordingly, the log-likelihood equations are

$$\sum_{j=1}^{N} \hat{P}\left(C_i|\mathbf{x}_j, \hat{\boldsymbol{\theta}}\right)\left(\partial \ln p\left(\mathbf{x}_j|C_i, \hat{\boldsymbol{\theta}}_i\right)/\partial\hat{\boldsymbol{\theta}}_i\right) = 0, \quad i = 1, \cdots, K. \tag{4.37}$$

As a specific example, when the component densities $p(\mathbf{x}|\boldsymbol{\theta}_i)$ take the form of multivariate Gaussian densities,

$$p(\mathbf{x}|\boldsymbol{\theta}_i) = \frac{1}{(2\pi)^{D/2}|\boldsymbol{\Sigma}_i|^{1/2}} \exp\left[-\frac{1}{2}(\mathbf{x}-\boldsymbol{\mu}_i)^T \boldsymbol{\Sigma}_i^{-1}(\mathbf{x}-\boldsymbol{\mu}_i)\right], \tag{4.38}$$

where $\boldsymbol{\mu}$ is the mean vector and $\boldsymbol{\Sigma}$ is covariance matrix, and all these parameters are unknown, the following equations can be obtained (Duda et al., 2001; Roberts, 1997):

$$\hat{P}(C_i) = \frac{1}{N}\sum_{j=1}^{N} \hat{P}\left(C_i|\mathbf{x}_j, \hat{\boldsymbol{\theta}}\right), \tag{4.39}$$

$$\hat{\boldsymbol{\mu}}_i = \frac{\sum_{j=1}^{N} \hat{P}\left(C_i|\mathbf{x}_j, \hat{\boldsymbol{\theta}}\right)\mathbf{x}_j}{\sum_{j=1}^{N} \hat{P}\left(C_i|\mathbf{x}_j, \hat{\boldsymbol{\theta}}\right)}, \tag{4.40}$$

$$\hat{\boldsymbol{\Sigma}}_i = \frac{\sum_{j=1}^{N} \hat{P}\left(C_i|\mathbf{x}_j, \hat{\boldsymbol{\theta}}\right)(\mathbf{x}_j - \hat{\boldsymbol{\mu}}_i)(\mathbf{x}_j - \hat{\boldsymbol{\mu}}_i)^T}{\sum_{j=1}^{N} \hat{P}\left(C_i|\mathbf{x}_j, \hat{\boldsymbol{\theta}}\right)}, \tag{4.41}$$

where

$$\hat{P}\left(C_i \middle| \mathbf{x}_j, \hat{\boldsymbol{\theta}}\right) = \frac{\left|\hat{\boldsymbol{\Sigma}}_i\right|^{-\frac{1}{2}} \exp\left[-\frac{1}{2}(\mathbf{x}_j - \hat{\boldsymbol{\mu}}_i)^T \hat{\boldsymbol{\Sigma}}_i^{-1}(\mathbf{x}_j - \hat{\boldsymbol{\mu}}_i)\right] \hat{P}(C_i)}{\sum_{l=1}^{K} \left|\hat{\boldsymbol{\Sigma}}_l\right|^{-\frac{1}{2}} \exp\left[-\frac{1}{2}(\mathbf{x}_j - \hat{\boldsymbol{\mu}}_l)^T \hat{\boldsymbol{\Sigma}}_l^{-1}(\mathbf{x}_j - \hat{\boldsymbol{\mu}}_l)\right] \hat{P}(C_l)}. \tag{4.42}$$

In most circumstances, the solutions to the likelihood equations cannot be obtained analytically. A similar situation also holds for the Bayesian methods (Duda et al., 2001; Richardson and Green, 1997), which usually involve the more complicated multidimensional integration operations. The common strategy to deal with this problem is based on iteratively suboptimal approaches, which can be used to approximate the ML estimates (Figueiredo and Jain, 2002; McLachlan and Peel, 2000). Among these methods, the Expectation-Maximization (EM) algorithm is the most popular (Dempster et al., 1977; McLachlan and Krishnan, 1997).

EM regards the data set as incomplete and divides each data point $\mathbf{x}_j$ into two parts, $\mathbf{x}_j = \{\mathbf{x}_j^g, \mathbf{x}_j^m\}$, where $\mathbf{x}_j^g$ represents the observable features, $\mathbf{x}_j^m = (x_{j1}^m, \cdots, x_{jK}^m)$ is the missing data, and x_{ji}^m takes the value of 1 or 0 according to whether $\mathbf{x}_j$ belongs to the component i or not. Thus, the complete data log-likelihood is

$$l(\boldsymbol{\theta}) = \sum_{j=1}^{N} \sum_{i=1}^{K} x_{ji}^m \log\left[P(C_i)\, p\left(\mathbf{x}_j^g | \boldsymbol{\theta}_i\right)\right]. \tag{4.43}$$

The standard EM algorithm generates a series of parameter estimates $\{\hat{\boldsymbol{\theta}}^0, \hat{\boldsymbol{\theta}}^1, \ldots, \hat{\boldsymbol{\theta}}^T\}$ until the convergence criterion is reached at the final step T. The basic steps of the EM algorithm are summarized as follows, which proceed by alternating the conditional expectation computation and the parameter estimates:

1. Initialize $\hat{\boldsymbol{\theta}}^0$ and set $t = 0$;
2. E-step: Compute Q, which is the conditional expectation of the complete data log-likelihood, given the current estimate,

$$Q\left(\boldsymbol{\theta}, \hat{\boldsymbol{\theta}}^t\right) = E\left[\log p(\mathbf{x}^g, \mathbf{x}^m | \boldsymbol{\theta}) | \mathbf{x}^g, \hat{\boldsymbol{\theta}}^t\right], \tag{4.44}$$

and

$$E\left[x_{ji}^m | \mathbf{x}^g, \hat{\boldsymbol{\theta}}^t\right] = \frac{\hat{P}(C_i)\, p\left(\mathbf{x}_j^g \middle| \hat{\boldsymbol{\theta}}_i^t\right)}{\sum_{l=1}^{K} P(C_l)\, p\left(\mathbf{x}_j^g \middle| \hat{\boldsymbol{\theta}}_l^t\right)}; \tag{4.45}$$

3. M-step: Select a new parameter estimate that maximizes the Q-function,

$$\hat{\boldsymbol{\theta}}^{t+1} = \arg\max_{\boldsymbol{\theta}} Q\left(\boldsymbol{\theta}, \hat{\boldsymbol{\theta}}^{t}\right); \quad (4.46)$$

4. Increase $t = t + 1$. Repeat steps 2 and 3 until the convergence condition is satisfied.

It is interesting to see that *K*-means can be regarded as a special case of EM under a spherical Gaussian mixture, where the dimensions have the same variance (Celeux and Govaert, 1992; Roweis and Ghahramani, 1999). *K*-means proceeds by alternately applying the E- and M-step until the clusters become stable. During the E-step, the data points are assigned to the closest cluster, while during the M-step, the prototypes of the clusters are updated according to the current partition.

Similar to other iterative optimization technologies, the EM algorithm suffers from sensitivity to parameter initialization, convergence to the local optima, the effect of a singular covariance matrix, and a slow convergence rate (Fraley and Raftery, 2002; McLachlan and Krishnan, 1997). Ueda et al. (2000) introduced an SMEM algorithm that integrates split and merge operations in order to avoid the local optima. The split criterion for a cluster is based on the distance between its local data density and the density obtained from the current parameter estimate. The merge criterion considers combining a pair of clusters for which each of many data points has almost equal posterior probability. Figueiredo and Jain (2002) proposed an EM variant working directly on the minimum message length (MML) criterion (Oliver et al., 1996) in order to reduce the reliance of EM on the initialization and to prevent EM from converging to the boundary of the parameter space. More discussions on the variants of EM are provided in McLachlan and Krishnan (1997) and McLachlan and Peel (2000).

4.4.3 Finite Mixture Density Based Clustering Algorithms

Fraley and Raftery (2002) described a comprehensive mixture-model based clustering scheme, which was implemented as a software package, known as MCLUST (Fraley and Raftery, 1999). Under this framework, the component density is assumed to be multivariate Gaussian, with a mean vector μ and a covariance matrix $\boldsymbol{\Sigma}$ as the parameters to be estimated, though other mixtures could also be used. The covariance matrix $\boldsymbol{\Sigma}_i$ for each component i can further be parameterized in terms of eigenvalue decomposition, represented as,

$$\boldsymbol{\Sigma}_i = \lambda_i \boldsymbol{\Pi}_i \mathbf{A}_i \boldsymbol{\Pi}_i^T, \quad (4.47)$$

where λ_i is a scalar, $\boldsymbol{\Pi}_i$ is the orthogonal matrix of eigenvectors, and $\mathbf{A}_i$ is the diagonal matrix with its elements proportional to the eigenvalues of $\boldsymbol{\Sigma}_i$ (Banfield and Raftery, 1993; Fraley and Raftery, 2002). These three parameters

TABLE 4.1. Parameterizations of the covariance matrix Σ_i in the Gaussian model and their geometric interpretation.

Σ_i	Distribution	Volume	Shape	Orientation
$\lambda\mathbf{I}$	Spherical	Fixed	Fixed	NA
$\lambda_i\mathbf{I}$	Spherical	Variable	Fixed	NA
$\lambda\mathbf{\Pi}\mathbf{A}\mathbf{\Pi}$	Elliptical	Fixed	Fixed	Fixed
$\lambda_i\mathbf{\Pi}_i\mathbf{A}_i\mathbf{\Pi}_i$	Elliptical	Variable	Variable	Variable
$\lambda\mathbf{\Pi}_i\mathbf{A}\mathbf{\Pi}_i$	Elliptical	Fixed	Fixed	Variable
$\lambda_i\mathbf{\Pi}_i\mathbf{A}\mathbf{\Pi}_i$	Elliptical	Variable	Fixed	Variable

(Modified from Fraley and Raftery, How many clusters? Which clustering method?—Answers via model-based cluster analysis, The Computer Journal, 1998, vol. 41, pp. 578–588, by permission of Oxford University Press.)

determine the geometric properties of each component, including the orientation, the shape, and the volume, as summarized in Table 4.1.

The clustering process starts with a model-based agglomerative hierarchical clustering (Fraley and Raftery, 2002), which, at each step, merges a pair of clusters leading to the largest increase of the classification likelihood,

$$l(\boldsymbol{\theta}, \boldsymbol{\omega}) = \sum_{j=1}^{N} p(\mathbf{x}_j | \boldsymbol{\theta}_{\omega_j}), \tag{4.48}$$

where $\boldsymbol{\omega} = (\omega_1, \ldots, \omega_N)$ are the labels for each data point for the corresponding cluster. The obtained hierarchical partitions, ranging from 2 to the maximum number of clusters, were then used to ignite the EM algorithm for a set of candidate models. The optimal clustering result is achieved by evaluating the Bayesian Information Criterion (BIC) (Schwarz, 1978).

Gaussian Mixture Density Decomposition (GMDD) is also based on multivariate Gaussian densities (Zhuang et al., 1996). In contrast with the other techniques, GMDD considers the Gaussian density contaminated, represented as,

$$p(\mathbf{x}|\boldsymbol{\theta}_i) = \frac{1-\varepsilon}{(2\pi)^{D/2}|\boldsymbol{\Sigma}_i|^{1/2}} \exp\left[-\frac{1}{2}(\mathbf{x}-\boldsymbol{\mu}_i)^T \boldsymbol{\Sigma}_i^{-1}(\mathbf{x}-\boldsymbol{\mu}_i)\right] + \varepsilon h(\mathbf{x}), \tag{4.49}$$

where ε is the probability that the data point $\mathbf{x}$ is generated with a noise distribution $h(\cdot)$. Accordingly, the probability for generating $\mathbf{x}$ from the multivariate normal distribution is $1 - \varepsilon$. Based on this idea, GMDD works as a recursive algorithm that sequentially estimates each component. At each step, data points that are not generated from the contaminated Gaussian distribution are regarded as noise, and the contaminated density is constructed with an enhanced model-fitting estimator. The data points associated with the currently estimated density are then removed from the data set, and the algorithm continues to proceed until the data set becomes too small. In this way, it is not necessary to identify the number of components in advance.

Multivariate Gaussian mixtures consider data with continuous variables and have been widely used in modeling real data, owing to their computational convenience and their natural occurrence in many problems. However, Gaussian distributions may not always be capable enough to model the data encountered. For example, if the data distribution from a cluster has longer tails than the Gaussian density, or there exist atypical data points or outliers in the cluster, normal mixtures may not be as suitable as other distributions, such as *t*-distributions (McLachlan and Peel, 1998; Peel and McLachlan, 2000), which are realized as the EMMIX algorithm (McLachlan et al., 1999). The multivariate *t*-distributions provide a more robust approach to the fitting of the mixture models, since outliers to a certain cluster are given less weight during the estimation of the corresponding parameters, which are still achieved by maximum likelihood via the EM algorithm.

Another situation that is beyond the capability of Gaussian densities occurs when the data contain categorical variables. In this case, discrete component densities, such as multivariate Bernoulli distributions and Poisson distributions, can be used to model the categorical data (Celeux and Govaert, 1991; Cheeseman and Stutz, 1996). Consider that each object in the data is described by D categorical variables, with the number of categories $\sigma_1, \ldots, \sigma_D$. Supposing we have the probability δ_{jlc} that variable j belongs to category l, given the object falling into the cluster c, then the component density can be represented as,

$$p(\mathbf{x}|\boldsymbol{\theta}_i) = \prod_{j=1}^{D} \prod_{l} \{\delta_{jlc} : 1 \le l \le \sigma_j, l = x_j\}. \tag{4.50}$$

AutoClass (Cheeseman and Stutz, 1996) considers a wide variety of probability distributions, such as Bernoulli, Poisson, Gaussian, and log-Gaussian densities, for different data types. AutoClass uses a Bayesian approach to find the optimal partition of the given data based on the prior probabilities. Specifically, AutoClass consists of a two-level search: the parameter level search and the model level search. For any given functional form of the probability density, AutoClass seeks the parameter estimation through the maximum *a posteriori* (MAP) (Duda et al., 2001). On the other hand, AutoClass searches for the best cluster models and the number of clusters from a possible candidate set, regardless of any parameters. Cheeseman and Stutz (1996) also stated that Bayesian parameter priors can prevent the overfitting that plagues maximum likelihood. A parallel realization of AutoClass, known as P-AutoClass, is described in Pizzuti and Talia (2003). Wallace and Dowe (1994) presented the SNOB program, which uses the minimum message length principle for parameter estimation, to model data from both continuous Gaussian or von Mises circular distributions and discrete multi-state or Poisson distributions. Another important finite mixtures program for continuous and binary variables is called MULTIMIX, proposed by Hunt and Jorgensen (1999).

4.5. GRAPH THEORY-BASED CLUSTERING

Graph theory can also be used for nonhierarchical clusters. As introduced in Chapter 3, for a weighted graph G, each node $v_i \in V$ corresponds to a data point in the feature space, while each edge $e_{ij} \in E$ between a pair of nodes v_i and v_j represents their proximity. In order to seek a partition of the data, the constructed graph structures must be capable of detecting the inconsistent edges that reflect the division of the different clusters (Jain and Dubes, 1988). Zahn's clustering algorithm (Zahn, 1971) uses the concept of a minimum spanning tree (West, 2001), which is defined as a tree with the minimal sum of the edge weights among all the trees that contain all vertices. The edges with weights larger than the average of the edges in the neighborhood in an important level are treated as inconsistent, and the removal of these edges separates the graph into a set of connected components, regarded as clusters. Hartuv and Shamir (2000) treated clusters as highly connected sub-graphs (HCS), where "highly connected" means that the connectivity, defined as the minimum number of edges for disconnecting a graph, of the sub-graph is at least half as great as the number of vertices. A minimum weight cut procedure (a set of edges whose removal disconnects a graph is called a cut), which aims to separate a graph with a minimum number of edges, is used to identify these HCSs recursively.

Likely, the algorithm CLICK (Clustering Identification via Connectivity Kernels) is also based on the calculation of the minimum weight cut in order to generate clusters (Sharan and Shamir, 2000). Here, the graph is weighted with the similarity value between a pair of data points, and the edge weights are assigned a new interpretation by combining probability and graph theory. Specifically, the edge weight between nodes v_i and v_j is defined as

$$e_{ij} = \log \frac{P_0 p(S_{ij}|v_i, v_j \in C_l)}{(1-P_0)\, p(S_{ij}|v_i \in C_m, v_j \in C_l, m \neq l)}, \tag{4.51}$$

where S_{ij} represents the similarity between the two nodes and P_0 is the probability that two randomly selected data points are from the same cluster. CLICK further assumes that the similarity values within clusters and between clusters follow Gaussian distributions with different means and variances, respectively, represented as

$$p(S_{ij}|v_i, v_j \in C_l) = \frac{1}{\sqrt{2\pi}\sigma_W} \exp\left(-\frac{(S_{ij} - \mu_W)^2}{2\sigma_W^2}\right), \tag{4.52}$$

$$p(S_{ij}|v_i \in C_m, v_j \in C_l, m \neq l) = \frac{1}{\sqrt{2\pi}\sigma_B} \exp\left(-\frac{(S_{ij}-\mu_B)^2}{2\sigma_B^2}\right), \tag{4.53}$$

where μ_B, σ_B^2, μ_W, σ_W^2 are the means and variances for between-cluster similarities and within-cluster similarities, respectively. These parameters can be estimated either from prior knowledge or by using parameter estimation methods, such as the EM algorithm introduced in Section 4.4 (Duda et al., 2001). For similarity values that are not normally distributed, the distribution can be approximated.

Therefore, with simple mathematical manipulation, Eq. 4.51 can be rewritten as:

$$e_{ij} = \log \frac{P_0 \sigma_B}{(1-P_0)\sigma_W} + \frac{(S_{ij}-\mu_B)^2}{2\sigma_B^2} - \frac{(S_{ij}-\mu_W)^2}{2\sigma_W^2}. \tag{4.54}$$

CLICK recursively checks the current sub-graph and generates a kernel list, which consists of the components that satisfy the following criterion function,

$$W(E_c) + W^*(E_c) > 0, \tag{4.55}$$

where E_c is a cut of the graph, and $W(E_c)$ and $W^*(E_c)$ are the weights of E_c in a complete and incomplete graph, respectively. Supposing H_0 is the null hypothesis that the cut E_c only has edges between the nodes from different clusters and H_1 is the alternate hypothesis that the cut only has edges between the nodes from the same clusters, $W(E_c)$ can be calculated as

$$W(E_c) = \log \frac{P(H_1|E_c)}{P(H_0|E_c)}. \tag{4.56}$$

Similarly, $W^*(E_c)$ can be defined as

$$W^*(E_c) = (r - |E_c|) \log \frac{P_0 \Phi((t_\theta - \mu_W)/\sigma_W)}{(1-P_0)\Phi((t_\theta - \mu_B)/\sigma_B)}, \tag{4.57}$$

where $r = |G_1||G_2|$, $G \rightarrow (G_1, G_2) \backslash E_c$, Φ is the cumulative standard normal distribution function, and t_θ is a threshold value. This criterion corresponds to the fact that the probability of the cut containing only the edges between nodes from the same clusters is larger than the probability of the cut containing only edges between nodes from different clusters.

On the other hand, sub-graphs that include only one node are regarded as singletons and are separated into the singleton list for further operation. Using kernels as the basic clusters, CLICK carries out a series of singleton adoptions and cluster merges to generate the resulting clusters. Additional heuristics, such as screening low-weight nodes and computing a minimum s-t cut, are provided to accelerate the algorithm performance.

The algorithm CAST (Cluster Affinity Search Technique) also considers a probabilistic model in achieving a partition of the data (Ben-Dor et al., 1999). Clusters are modeled as corrupted clique graphs, which, in ideal conditions, are regarded as a set of disjointed cliques. The effect of noise is incorporated by adding or removing edges from the ideal model, with a probability P_α. Proofs were given for recovering the uncorrupted graph with a high probability. CAST is the heuristic implementation of the original theoretical version. CAST creates clusters sequentially, and each cluster begins with a random and unassigned data point. The relation between a data point $\mathbf{x}_i$ and a cluster C_j being built is determined by the affinity, defined as

$$a(\mathbf{x}_i) = \sum_{l \in C_j} S_{il}, \tag{4.58}$$

and the affinity threshold parameter t_v. If $a(\mathbf{x}_i) \geq t_v|C_j|$, then the data point is highly related to the cluster; therefore, the data point is absorbed into the cluster. On the other hand, a data point with affinity $a(\mathbf{x}_i) < t_v|C_j|$ is said to have low affinity and is removed from the cluster. CAST alternately adds high-affinity data points and deletes low-affinity data points from the cluster until no more changes occur. The flowchart of CAST is summarized in Fig. 4.1.

More discussions of graph theory-based clustering are offered by Delattre and Hansen (1980), Hansen and Jaumard (1997), Jain and Dubes (1988), Jenssen et al. (2003), Jenssen et al. (2007), Johnson et al. (1993), Kannan et al. (2000), Shi and Malik (2000), Urquhart (1982), and Wu and Leahy (1993).

4.6. FUZZY CLUSTERING

4.6.1 Fuzzy *c*-means

Fuzzy logic (Zadeh, 1965) creates intermediate classifications, rather than binary, or "crisp" ones. So far, the clustering techniques we have discussed are referred to as hard or crisp clustering, which means that each data object is assigned to only one cluster. For fuzzy clustering, this restriction is relaxed, and the object can belong to all of the clusters with a certain degree of membership (Bezdek, 1981). This is particularly useful when the boundaries between clusters are ambiguous and not well separated. Moreover, the memberships may help the users discover more sophisticated relationshps between a given object and the disclosed clusters.

Fuzzy *c*-Means (FCM) (Bezdek, 1981; Höppner et al., 1999) can be regarded as a generalization of ISODATA (Dunn, 1974a) and was realized by Bezdek (1981). FCM attempts to find a partition, represented as *c* fuzzy

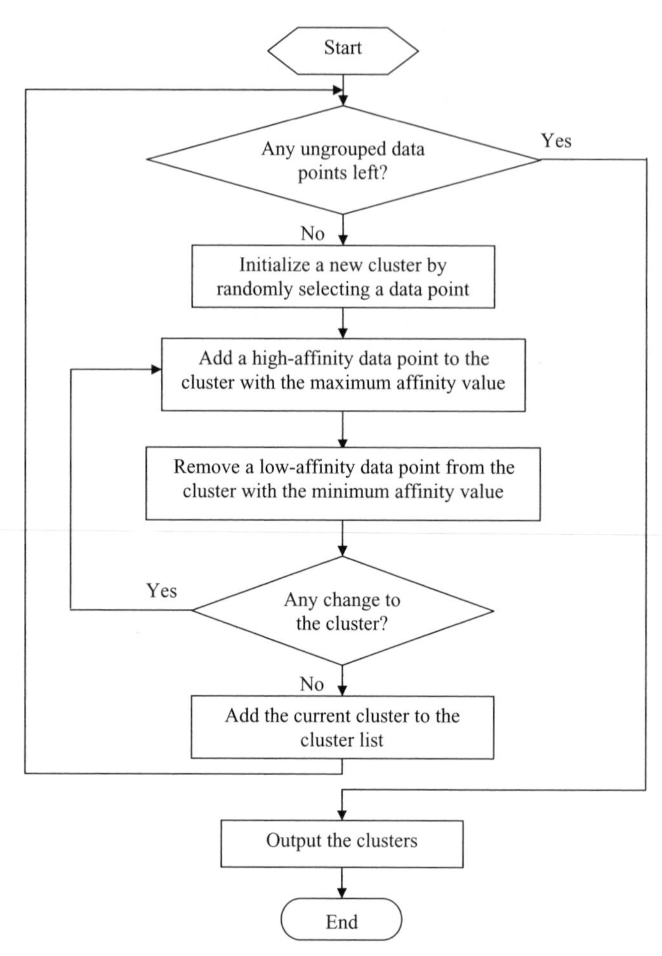

Fig. 4.1. Flowchart of CAST. Clusters are constructed incrementally, and CAST alternately adds high-affinity data points to the current cluster and removes low-affinity data points from it.

clusters, for a set of data objects $\mathbf{x}_j \in \Re^d, j = 1, \ldots, N$, while minimizing the cost function

$$J(\mathbf{U}, \mathbf{M}) = \sum_{i=1}^{c} \sum_{j=1}^{N} (u_{ij})^m D_{ij}^2, \tag{4.59}$$

where

- $\mathbf{U} = [u_{ij}]_{c \times N}$ is the fuzzy partition matrix and $u_{ij} \in [0, 1]$ is the membership coefficient of the jth object in the ith cluster that satisfies the following two constraints:

$$\sum_{i=1}^{c} u_{ij} = 1, \forall j, \tag{4.60}$$

which assures the same overall weight for every data point (when fuzzy clustering meets this constraint, it is also called probabilistic clustering), and

$$0 < \sum_{j=1}^{N} u_{ij} < N, \forall i, \tag{4.61}$$

which assures no empty clusters;

- $\mathbf{M} = [\mathbf{m}_1, \ldots, \mathbf{m}_c]$ is the cluster prototype (mean or center) matrix;
- $m \in [1, \infty)$ is the fuzzification parameter and a larger m favors fuzzier clusters. m is usually set to 2 (Hathaway and Bezdek, 2001);
- $Dij = D(\mathbf{x}_j, \mathbf{m}_i)$ is the distance between $\mathbf{x}_j$ and $\mathbf{m}_i$, and with FCM, $D_{ij}^2 = \|\mathbf{x}_j - \mathbf{m}_i\|_2^2$ i.e., the Euclidean or L_2 norm distance function is used.

The criterion function in Eq. 4.59 can be optimized with an iterative procedure that leads to the standard FCM algorithm. The membership and prototype matrix update equations are obtained through alternating optimization. A more detailed discussion of the convergence properties of FCM is given in Bezdek (1981), Höppner and Klawonn (2003), and Gröll and Jäkel (2005). Many FCM variants have also been proposed as a result of the intensive investigation of the distance measure functions, the effect of weighting exponents on fuzziness control, the optimization approaches for fuzzy partitioning, the speeding up of FCM, and the improvements of the drawbacks of FCM (Bobrowski and Bezdek, 1991; Cannon et al., 1986; Cheng et al., 1998; Eschrich et al., 2003; Hathaway and Bezdek, 2001; Honda and Ichihashi, 2005; Höppner et al., 1999; Leski, 2003; Wong et al., 2001). The basic steps of the FCM algorithm are summarized as follows:

1. Select appropriate values for m, c, and a small positive number ε. Initialize the prototype matrix $\mathbf{M}$ randomly. Set step variable $t = 0$;
2. Update the membership matrix $\mathbf{U}$ by

$$u_{ij}^{(t+1)} = \begin{cases} 1\Big/\left(\sum_{l=1}^{c} (D_{lj}/D_{ij})^{2/(1-m)}\right), & \text{if } I_j = \varnothing \\ 1/|I_j|, & \text{if } I_j \neq \varnothing, i \in I_j \\ 0, & \text{if } I_j \neq \varnothing, i \notin I_j \end{cases} \quad \text{for } i = 1, \ldots, c, \text{and } j = 1, \ldots, N, \tag{4.62}$$

where $I_j = \{i | i \in [1, c], \mathbf{x}_j = \mathbf{m}_i\}$;

3. Update the prototype matrix **M** by

$$\mathbf{m}_i^{(t+1)} = \left(\sum_{j=1}^{N} \left(u_{ij}^{(t+1)}\right)^m \mathbf{x}_j\right) \Bigg/ \left(\sum_{j=1}^{N} \left(u_{ij}^{(t+1)}\right)^m\right), \quad \text{for } i = 1, \ldots, c; \tag{4.63}$$

4. Repeat steps 2 and 3 until $\|\mathbf{M}^{(t+1)} - \mathbf{M}^{(t)}\| < \varepsilon$.

Like its hard counterpart, FCM also suffers from two major problems: the lack of guidance to identify an appropriate initial partition and the presence of noise and outliers. Yager and Filev (1994) proposed a mountain method (MM) to estimate the cluster centers and generate an initial partition. Candidate centers consist of a set of vertices that are formed by building a grid on the feature space. The density of the grid is not necessarily equivalent across the entire space and can vary based on the properties of the data. Therefore, the finer the grid, the more potential cluster centers, and the higher the likelihood of obtaining a good partition. However, the computational burden also becomes more intensive. Following the discretization, a mountain function, defined as

$$M(\mathbf{v}_i) = \sum_{j=1}^{N} e^{-\alpha D(\mathbf{x}_j, \mathbf{v}_i)}, \tag{4.64}$$

where $D(\mathbf{x}_j, \mathbf{v}_i)$ is the distance between the jth data point and the ith node, and α is a positive constant, is constructed to capture the data distribution. This can be regarded as a total potential of all data points (Davé and Krishnapuram, 1997). A data object that is closer to a vertex will contribute more to the mountain function value of that vertex, which reflects the data density in the neighborhood of the vertex. Finally, the cluster centers are selected sequentially based on the value of the mountain function. A mountain destruction procedure is performed in order to eliminate the effects of the selected center. In other words, at iteration t, the vertex $\mathbf{v}_t^*$ with the maximum value of the mountain function is selected as the cluster center. The mountain function value for each of the remaining vertices is updated by subtracting the original one with an amount dependent on the current maximal and the distance between the vertex and the selected center,

$$M^{t+1}(\mathbf{v}_j) = M^t(\mathbf{v}_j) - M(\mathbf{v}_t^*)\sum_{t=1}^{N} \exp(-\beta D(\mathbf{v}_t^*, \mathbf{v}_j)), \tag{4.65}$$

where β is a constant. An additional constraint can be added to bound the new mountain function by zero. The process iterates until the ratio between the current maximum $M(\mathbf{v}_t^*)$ and the maximum at the first iteration $M(\mathbf{v}_1^*)$ is below some threshold. The relation of MM with some other clustering techniques was further discussed by Davé and Krishnapuram (1997).

Gath and Geva (1989) addressed the initialization problem by dynamically adding cluster prototypes, which are located in the space that is not represented well by the previously generated centers. A two-layer clustering scheme was proposed, in which FCM and fuzzy maximum likelihood estimation are effectively combined. Hung and Yang (2001) suggested the use of a k-d tree to partition the data into sub-regions and simplified the original data with the centroids of the sub-blocks. Pre-clustering is performed on the reduced data to provide an initial partition. Since the number of sub-regions is much less than the number of original data points, the computational burden is also reduced.

Kersten (1997) suggested that the use of fuzzy medians with a city block distance (or L_1 norm), rather than means with L_2 norm, for cluster representation could improve the robustness of FCM to outliers. Two strategies were also proposed to approximate the fuzzy medians in order to reduce the computational burden for large data sets. An earlier discussion on the L_1 norm in fuzzy clustering can be found in Trauwaert (1987) and Bobrowski and Bezdek (1991). Bobrowski and Bezdek (1991) also investigated the properties of the L_∞ norm in fuzzy clustering. Furthermore, Hathaway et al. (2000) extended FCM to a more universal situation by using the Minkowski distance (or L_p norm, $p \geq 1$) and semi-norm ($0 < p < 1$) for the models that operate either directly on the data points or indirectly on the dissimilarity measures, called relational data. According to their empirical results, both L_1 and L_2 norms are recommended for direct data point clustering. Particularly, the L_1 norm is useful in dealing with data mixed with noise. This robustness of the L_1 norm holds for relational data, too. They also pointed out the possible improvement of models for other L_p norms with the price of more complicated optimization operations with regard to the data points.

The possibilistic *c*-means clustering algorithm (PCM) is another approach for abating the effect of noise and outliers, based on the absolute typicality (Krishnapuram and Keller, 1993). Under this model, the memberships are interpreted with a possibilistic view, i.e., "the compatibilities of the points with the class prototypes" (Krishnapuram and Keller, 1993), rather than as the degree of membership of a certain object belonging to a certain cluster, following the probabilistic constraint in Eq. 4.60. The problems related to this constraint are illustrated in Fig. 4.2, which is reproduced from the examples in Krishnapuram and Keller (1993). To address these problems, the probabilistic constraint of the membership coefficient is relaxed to

$$\max_i u_{ij} > 0, \forall j. \tag{4.66}$$

This condition now simply ensures that each data point belongs to at least one cluster. Also, the possibility that a data point falls into a cluster is independent of its relations with other clusters.

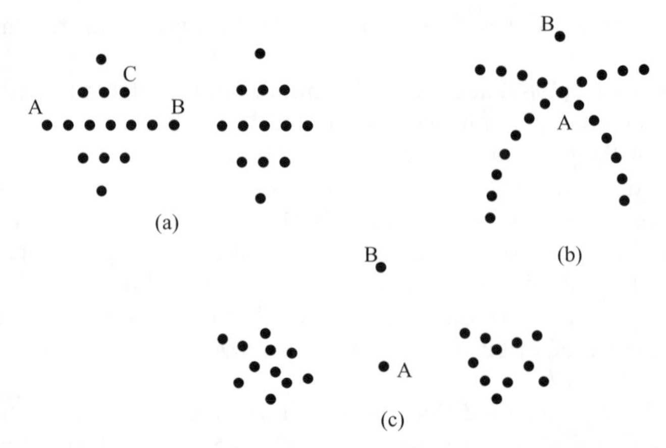

Fig. 4.2. Examples of problems with the probabilistic constraint. (a) FCM would generate different memberships for points A and B in the left cluster, even though they are equally typical of this cluster. However, points A and C would be given equal memberships, even though point C is more typical than A. (b) FCM would give memberships of 0.5 to both points A and B, even though point A is a real member of both clusters while point B is an outlier of the clusters. (c) FCM would give memberships of 0.5 to both points A and B, even though point B is a more extremely atypical member than point A. (From R. Krishnapuram and J. Keller. A possibilistic approach to clustering. IEEE Transactions on Fuzzy Systems, vol. 1, pp. 98–110, 1993. Copyright © 1993 IEEE.)

Accordingly, the cost function is reformulated as

$$J(\mathbf{U},\mathbf{M})=\sum_{i=1}^{c}\sum_{j=1}^{N}(u_{ij})^m D_{ij}^2+\sum_{i=1}^{c}\eta_i\sum_{j=1}^{N}(1-u_{ij})^m, \tag{4.67}$$

where η_i are some positive constants, in order to avoid the trivial zero solution. The additional term tends to assign credits to memberships with large values, and its effect to the cost function is balanced by η_i, whose value determines the distance where the membership of a data point in a cluster becomes 0.5. η_i can be estimated with the following formulas:

$$\eta_i=\varsigma\frac{\sum_{j=1}^{N}(u_{ij})^m D_{ij}^2}{\sum_{j=1}^{N}(u_{ij})^m}, \tag{4.68}$$

where ς is a constant and usually is set to 1 (Krishnapuram and Keller, 1993), or

$$\eta_i = \varsigma \frac{\sum_{\mathbf{x}_j \in (\Pi_i)\alpha} D_{ij}^2}{|(\Pi_i)_\alpha|}, \tag{4.69}$$

where $(\Pi_i)_\alpha$ is a suitable α-cut of Π_i, which is the possibility distribution for the cluster c_i over the domain of discourse including all data points $\mathbf{x}_j$.

The minimization of the $J(\mathbf{U}, \mathbf{M})$ with respect to $\mathbf{U}$ can lead to the following update equation for membership:

$$u_{ij} = \frac{1}{1 + \left(\frac{D_{ij}^2}{\eta_i} \right)^{\frac{1}{m-1}}}. \tag{4.70}$$

Obviously, the membership u_{ij} is inversely proportional to the distance between the data point $\mathbf{x}_j$ and the prototype of the cluster c_i. Furthermore, the prototypes are updated in a similar way with FCM, as described in Eq. 4.63. The final PCM algorithm is depicted in Fig. 4.3. Other possibilistic clustering algorithms can also be derived by considering different distance measures (Krishnapuram and Keller, 1993).

Krishnapuram and Keller (1996) further investigated the effect of parameters m and η_i and proposed a modified version of the cost function, without the involvement of m:

$$J(\mathbf{U}, \mathbf{M}) = \sum_{i=1}^{c} \sum_{j=1}^{N} (u_{ij}) D_{ij}^2 + \sum_{i=1}^{c} \eta_i \sum_{j=1}^{N} (u_{ij} \log(u_{ij}) - u_{ij}). \tag{4.71}$$

The corresponding updated equation for membership now becomes

$$u_{ij} = \exp\left(-\frac{D_{ij}^2}{\eta_i} \right). \tag{4.72}$$

As pointed out by Barni et al. (1996), PCM has the tendency to generate coincident clusters, and the resulting clusters are sensitive to initialization. The answer to this problem lies in the integration of both fuzzy membership and possibilistic values, as in the possibilistic fuzzy c-means (PFCM) model proposed by Pal et al. (2005). Another combined methods approach was proposed by Zhang and Leung (2004), which improves the performance of PCM and its modified version. See also Yang and Wu (2006) for a new possibilistic clustering algorithm, which generalizes the existing validity indexes to possibilistic clustering and integrates them into the cost function.

Davé and Krishnapuram (1997) further elaborated the discussion of fuzzy clustering robustness and indicated its connection with robust statistics. They investigated prototyped-based robust clustering algorithms (Ohashi, 1984),

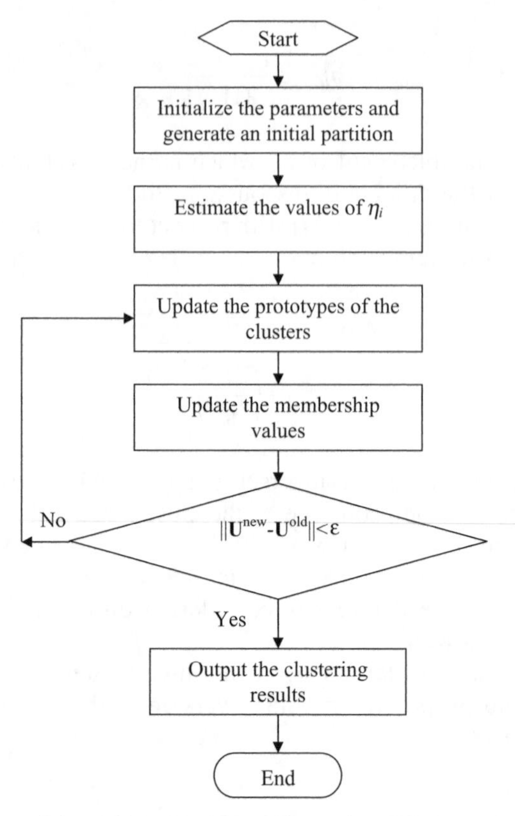

Fig. 4.3. Flowchart of the PCM algorithm. The major difference between PCM and FCM is that PCM relaxes the constraint of the membership coefficient and only requires that each data point is associated with at least one cluster.

the noise clustering algorithm (Davé, 1991), and PCM (Krishnapuram and Keller, 1993) and exposed their similarities by terms of a special case, called the noise/possibilistic clustering (N/PC1) approach, which seeks only one cluster at a time. Under this framework, their relations with other fuzzy clustering methods, such as the mountain method (Yager and Filev, 1994), the least-biased fuzzy clustering algorithm based on deterministic annealing (Rose et al., 1993), and the iteratively reweighted least-squares approach (Whaite and Ferrie, 1991), and some robust statistical methods, such as the generalized minimum volume ellipsoid algorithm (Jolion et al., 1991) and cooperative robust estimation (Darrell and Pentland, 1995), were further discussed and reviewed. Davé and Krishnapuram (1997) reached a unified framework as the conclusion for the above discussion and proposed generic algorithms for robust clustering.

The standard FCM alternates the calculation of the membership and prototype matrix, which causes a computational burden for large-scale data sets. Kolen and Hutcheson (2002) accelerated the computation by combining updates of the two matrices within one step. Cheng et al. (1998) introduced a multistage random sampling FCM algorithm, which can normally achieve a 2–3 speedup factor. Eschrich et al. (2003) focused on data reduction while keeping the cluster quality, and they suggested a two-step preprocessing method consisting of quantization and aggregation. FCM variants were also developed to deal with data types besides continuous ones, such as symbolic data (El-Sonbaty and Ismail, 1998), and other special occasions, such as data with missing values (Hathaway and Bezdek, 2001) or data mixed with both labeled and unlabeled objects, which is also called semi-supervised clustering (Grira et al., 2004) or partial supervised learning (Pedrycz and Waletzky, 1997).

4.6.2 Fuzzy *c*-Shells Clustering Algorithms

A family of fuzzy *c*-shells clustering algorithms has also appeared to detect different types of cluster shapes, especially contours (lines, circles, ellipses, rings, rectangles, hyperbolas) with hollow interiors in a two-dimensional data space, as shown in Fig. 4.4. They use the "shells" (curved surfaces) (Davé and Bhaswan, 1992) as the cluster prototypes instead of the points or hyperplanes used in traditional fuzzy clustering algorithms. According to Davé and Bhaswan (1992), "in principle, any curved surface may be utilized as a prototype as long as a meaningful definition of distance measurement can be given."

In the case of adaptive fuzzy *c*-shells (AFCS) (Davé and Bhaswan, 1992), the cluster prototype is represented as a d-dimensional hyperellipsoidal shell,

$$\mathbf{m}(\mathbf{v}, r, \mathbf{A}) = \left\{\mathbf{x} \in \Re^d \middle| (\mathbf{x}-\mathbf{v})^T \mathbf{A}(\mathbf{x}-\mathbf{v}) = r^2\right\}, \tag{4.73}$$

where $\mathbf{v} \in \Re^d$ is the center, $r \in \Re$ is the radius, and $\mathbf{A}$ is a symmetric positive definite matrix. Note that when $\mathbf{A}$ becomes an identity matrix, the definition generates a hyperspherical shell prototype.

A distance measure between a data object $\mathbf{x}_i$ and the prototype $\mathbf{m}_j$ is given as

$$D_{ij} = \left((\mathbf{x}_i - \mathbf{m}_j)^T \mathbf{A}_j (\mathbf{x}_i - \mathbf{m}_j)\right)^{1/2} - r_j. \tag{4.74}$$

Fig. 4.4. Examples of curvilinear clusters.

The prototype for each cluster is obtained by solving nonlinear equations using Newton's method. Davé and Bhaswan (1992) also introduced two other variants that performed better on the digital image experiment analysis. Krishnapuram et al. (1992) suggested the fuzzy c-spherical shells (FCSS) algorithm, which eliminates the requirement for dealing with nonlinear equations. Bezdek and Hathaway (1992) further investigated the convergence properties of the original version of AFCS (Davé, 1990) by means of general convergence theory for grouped coordinate minimization.

Similarly, other cluster shapes can be achieved by defining appropriate prototypes and corresponding distance functions, examples of which include fuzzy c-rings (FCR) (Man and Gath, 1994), fuzzy c-quadratic shells (FCQS) (Krishnapuram et al., 1995), and fuzzy c-rectangular shells (FCRS) (Hoeppner, 1997). See Höppner et al. (1999) for further details.

It is worth mentioning that fuzzy set theories can also be used to generate hierarchical cluster structures. Geva (1999) proposed a hierarchical unsupervised fuzzy clustering (HUFC) algorithm, which, like hierarchical clustering, can effectively explore data structures at different levels, while also establishing connections between each object and cluster in the hierarchy with the memberships. This design allows HUFC to overcome one of the major disadvantages of hierarchical clustering, i.e., hierarchical clustering cannot reassign an object once it is designated into a cluster. Another fuzzy hierarchical clustering algorithm with its application in document clustering is given in Horng et al. (2005). Some recent advances in fuzzy clustering can be found in Bargiela et al. (2004), Chiang et al. (2004), Kumar et al. (2006), Liu et al. (2005), Nascimento et al. (2003), and Suh et al. (1999).

4.7. SEARCH TECHNIQUES-BASED CLUSTERING ALGORITHMS

The basic goal of search techniques is to seek the global or approximate global optimum for combinatorial optimization problems, which usually have NP-hard complexity and require a search of an exponentially large solution space. As introduced early in this chapter, clustering can be regarded as a category of optimization problems with maximizing or minimizing a specified criterion function. According to Eq. 4.1, even for small clustering problems, the computational complexity is already very expensive, not to mention the large-scale clustering problems frequently encountered in recent decades. Simple local search techniques, like hill-climbing algorithms introduced in Section 4.3, are used to find the partitions, but they are easily stuck in local minima and therefore cannot guarantee optimality. In this section, we focus on the more complex search methods, such as evolutionary algorithms (EAs) (Fogel, 1994, 2005), simulated annealing (SA) (Aarts and Korst, 1989; Kirkpatrick et al., 1983), and tabu search (TS) (Glover, 1989; Glover and Laguna, 1997), known as stochastic optimization methods, and deterministic annealing (DA) (Hofmann and Buhmann, 1997; Rose, 1998), one of the most typical deterministic search techniques, in exploring the solution space.

4.7.1 Genetic Algorithms

Inspired by the natural evolutionary process, evolutionary computation, including genetic algorithms (GAs) (Holland, 1975), evolution strategies (ESs) (Rechenberg, 1973), evolutionary programming (EP) (Fogel et al., 1966), and genetic programming (GP) (Koza, 1992, 1994), optimizes a population of structure by using a set of evolutionary operators (Fogel, 1994, 2005). Evolutionary algorithms maintain a population of structures consisting of a set of individuals. Each individual is evaluated by an optimization function, called the fitness function, and evolves via the processes of selection, recombination, and mutation. The selection operator ensures the continuity of the population by favoring the best individuals in the next generation. The recombination and mutation operators support the diversity of the population by exerting perturbations on the individuals. Among these evolutionary algorithms, GAs are the most popular approaches applied in cluster analysis (Babu and Murty, 1993; Hall et al., 1999; Sheng et al., 2005). We will focus on GA-based clustering algorithms in the following discussion, and clustering algorithms based on ESs and EP are described by Babu and Murty (1994) and Ghozeil and Fogel (1996), respectively.

In GAs, each individual, representing a candidate solution, is encoded as a binary bit string, also known as a chromosome. After an initial population is generated either randomly or according to some heuristic rules, a series of operations, including selection, crossover and mutation, as depicted in Fig. 4.5, are iteratively applied to the population until the stop condition is satisfied.

Generally, in the context of GA-based clustering, each chromosome in the population encodes the information for a valid partition of the data. These partitions are altered with evolutionary operators, and the best ones with high scores of the fitness function are retained. The process iterates until a satisfactory partition is achieved. In the following discussion, we use the genetically guided algorithm (GGA), proposed by Hall et al. (1999), as an example to show the basic procedures of GA-based clustering. GGA can be regarded as a general scheme for center-based (hard or fuzzy) clustering problems.

GGA only works with the prototype matrix without considering the partition matrix. In order to adapt the change of the construction of the optimization problem, the fitness functions are reformulated from the standard sum-of-squared-error criterion functions defined in Eq. 4.75 for hard clustering and Eq. 4.76 for fuzzy clustering:

$$J(\mathbf{M}) = \sum_{j=1}^{N} \min(D_{1j}, D_{2j}, \cdots, D_{Kj}), \quad \text{for hard clustering,} \tag{4.75}$$

$$J(\mathbf{M}) = \sum_{j=1}^{N} \left(\sum_{i=1}^{K} D_{ij}^{1/(1-m)} \right)^{1-m}, \quad \text{for fuzzy clustering,} \tag{4.76}$$

where D_{ij}, $i = 1, \ldots, K$, $j = 1, \ldots, N$ is the distance between the ith cluster prototype vector and the jth data object, and m is the fuzzification parameter.

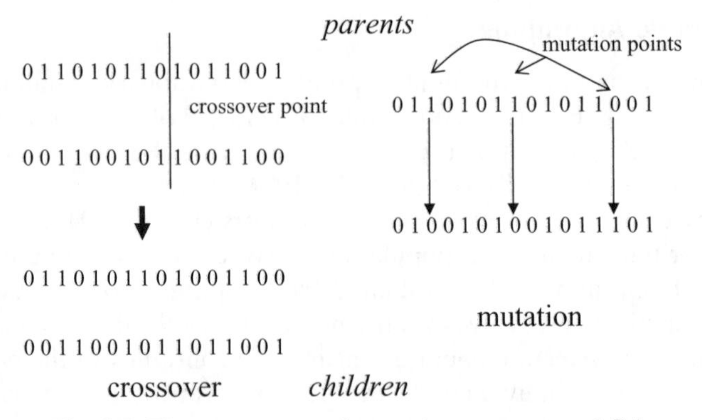

Fig. 4.5. The crossover and mutation operators of GAs.

GGA proceeds with the following steps:

1. Choose appropriate parameters for the algorithm; initialize the population randomly with P individuals, each of which represents a $K \times d$ prototype matrix and is encoded as gray codes; calculate the fitness value for each individual;
2. Use a selection (tournament selection) operator to choose parental members for reproduction;
3. Use crossover (two-point crossover) and mutation (bitwise mutation) operators to generate offspring from the individuals chosen in step 2;
4. Determine the next generation by keeping the individuals with the highest fitness.
5. Repeat steps 2–4 until the termination condition is satisfied.

In order to avoid the occurrence of an empty row in the prototype vector, the criterion function is modified as

$$J'(\mathbf{M}) = J(\mathbf{M}) + \beta \times J(\mathbf{M}), \tag{4.77}$$

where $\beta \in [0, K]$ is the number of empty clusters.

Other GA-based clustering algorithms have appeared based on a similar framework. They mostly differ in the meaning of an individual in the population, encoding methods, fitness function definition, and evolutionary operators (Cowgill et al., 1999; Maulik and Bandyopadhyay, 2000; Tseng and Yang, 2001; Sheng et al., 2005). Sheng et al. (2005) constructed the criterion function as a weighted sum of six cluster validity functions, which is further optimized by the hybrid niching genetic algorithm (HNGA). A niching method is developed in order to avoid premature convergence. The clustering algorithm introduced by Tseng and Yang (2001) uses a single-linkage algorithm to divide data into small subsets, before GA-based clustering, in order to reduce the computational complexity. The fitness function was designed to adjust the different effects of the within-class and between-class distances. Furthermore, Lozano and Larrañaga (1999) suggested an application of GAs in hierarchical clustering. They discussed the properties of ultrametric distance (Hartigan, 1975) and reformulated the hierarchical clustering as an optimization problem that tries to find the closest ultrametic distance for a given dissimilarity with Euclidean norm. They proposed an order-based GA to solve the problem.

GAs are very useful for improving the performance of K-means algorithms. Babu and Murty (1993) used GAs to find good initial partitions. Krishna and Murty (1999) combined GAs with K-means and developed a GKA algorithm to explore the global optimum. Laszlo and Mukherjee (2006) constructed a hyper-quadtree data structure to represent cluster prototypes (centroids), which aims to improve the efficiency of the crossover operator and assures the maintenance of good centroids. As indicated in Section 4.3, the algorithm ELBG (Patanè and Russo, 2001) uses the roulette mechanism to address problems due to bad initialization. It is worthwhile to note that ELBG is equivalent to another algorithm, FACS (fully automatic clustering system) (Patanè and Russo, 2002), in terms of quantization level detection. The difference lies in the input parameters employed, i.e., ELBG adapts the number of quantization levels, while FACS uses the desired distortion error.

4.7.2 Simulated Annealing

Simulated annealing (SA) is a sequential and global search technique that is motivated by the annealing process in physics and chemistry (Aarts and Korst, 1989; Kirkpatrick et al., 1983). At the first stage of annealing, the solid consisting of many molecules or components is heated to a high temperature so that each component is distributed in a random fashion. The system gradually lowers the temperature towards zero, which means no randomness, while maintaining the thermal equilibrium at each new temperature. At any step, the thermal equilibrium state follows the Boltzmann distribution, which governs the probability P of a specific configuration with energy E at temperature T,

$$P = \gamma \exp(-E/T), \tag{4.78}$$

where γ is a constant.

SA is based on the Metropolis algorithm (Metropolis et al., 1953), which investigates the equilibrium properties of the interacting components. A candidate configuration is obtained by a small random perturbation to the current configuration, which causes the change of the energy corresponding to the two configurations, represented as

$$\Delta E = E_n - E_c, \tag{4.79}$$

where E_n is the energy for the new configuration and E_c is for the current configuration. If $\Delta E < 0$, which means the new configuration has a lower energy, the candidate configuration is accepted. However, if the candidate configuration has a higher energy, it is accepted with a probability

$$P_c = \exp(-\Delta E/T). \tag{4.80}$$

In other words, SA allows the search process to accept a worse solution with a certain probability that is controlled by the temperature. Since the temperature goes through an annealing schedule from initial high to ultimate low values, SA attempts to explore solution space more completely at high temperatures while favoring the solutions that lead to lower energy at low temperatures.

The general procedure of SA-based clustering is described as follows (Brown and Huntley, 1992; Klein and Dubes, 1989; Selim and Alsultan, 1991):

1. Choose appropriate parameters for the algorithm; generate an initial partition randomly and calculate the corresponding energy (criterion function);
2. Generate a candidate partition by exerting some perturbation on the current partition, and calculate the change of energy ΔE in Eq. 4.79;
3. a). If $\Delta E < 0$, accept the candidate partition. Go to step 4;
   b). Otherwise, accept the candidate partition if $P_c < \lambda$, where λ is a random number generated uniformly in the interval [0, 1]. Go to step 4;
4. a). If the maximum number of iterations is not met, go to step 1;
   b). Otherwise, lower the temperature as $T = \tau T$, where $0 < \tau < 1$. If T reaches the final temperature, stop; otherwise, go to step 2.

Bandyopadhyay (2005) proposed an SA-RJMCMC (reversible jump Markov chain Monte Carlo) algorithm to achieve fuzzy partitions. Unlike the classical fuzzy c-means algorithm, the number of clusters is independent of the users and determined through the RJMCMC algorithm. Correspondingly, the criterion function used in this method is the Xie-Beni index (Xie and Beni, 1991) for accompanying such a change. The cluster centers are dynamically adjusted with five different operators, chosen with equal probability. Among these, the birth and death operators and the split and merge operators form two pairs of reversible moves, while the other provides a symmetrical perturbation to the cluster centers.

4.7.3 Deterministic Annealing

Similar to SA, deterministic annealing (DA) also performs annealing with a gradually decreasing temperature. The difference is that "it (DA) minimizes the free energy directly rather than via stochastic simulation of the system dynamics" (Rose, 1998). The energy function is obtained through this expectation.

Given a data point $\mathbf{x}_i$, $i = 1, \ldots, N$ and a prototype vector $\mathbf{m}_j$, $j = 1, \ldots, K$, the relation between them is described by the conditional probability $p(\mathbf{m}_j|\mathbf{x}_i)$, also known as the association probability (Rose et al., 1992). Therefore, the data point is assigned to a cluster with a certain probability, which is similar to the fuzzy clustering introduced in Section 4.6. By defining the criterion function or expected distortion J as

$$J = \sum_i \sum_j p(\mathbf{x}_i, \mathbf{m}_j) D(\mathbf{x}_i, \mathbf{m}_j), \tag{4.81}$$

where $p(\mathbf{x}_i, \mathbf{m}_j)$ is the joint probability distribution and $D(\cdot, \cdot)$ is the distance function, and the Shannon entropy H as

$$H = -\sum_i \sum_j p(\mathbf{x}_i, \mathbf{m}_j) \log p(\mathbf{x}_i, \mathbf{m}_j), \tag{4.82}$$

the free energy of the system is achieved as

$$E = J - TH, \tag{4.83}$$

where T is the temperature.

The association probability is calculated by minimizing the energy, which is achieved at an equilibrium following the Gibbs distribution,

$$p(\mathbf{m}_j|\mathbf{x}_i) = \frac{\exp\left(-\frac{D(\mathbf{x}_i, \mathbf{m}_j)}{T}\right)}{Z_{\mathbf{x}_i}}, \tag{4.84}$$

where $Z_{\mathbf{x}_i} = \sum_l \exp\left(-\frac{D(\mathbf{x}_i, \mathbf{m}_l)}{T}\right)$ is the normalization factor. As T approaches infinity, the probability of assigning a point to any cluster is equal. However, with T equal to 0, each point is associated with only one cluster with probability 1. Therefore, the cooling process beginning from a high temperature corresponds to the tradeoff between the global and local influence of a data point.

The new prototype vectors satisfy the equation

$$\sum_i p(\mathbf{m}_j, \mathbf{x}_i) \frac{d}{d\mathbf{m}_j} D(\mathbf{x}_i, \mathbf{m}_j) = 0. \tag{4.85}$$

4.7.4 Particle Swarm Optimization

PSO is based on the simulation of complex social behavior, such as bird flocking or fish schooling, and it was originally intended to explore optimal or near-optimal solutions in difficult continuous spaces (Clerc and Kennedy, 2002; Kennedy et al., 2001; Shi and Eberhart, 1998). PSO consists of a swarm of particles, each of which consists of a set of prototype vectors or cluster centers that represent the clusters and are denoted as $\mathbf{z}_i = (\mathbf{m}_1, \dots, \mathbf{m}_C)$ (Chen and Ye, 2004; Merwe and Engelbrecht, 2003). Each particle moves in the multidimensional problem space with a corresponding velocity $\mathbf{v}$. The basic idea of PSO is to accelerate each particle i towards its two best locations at each time step, which is also described in Fig. 4.6. One is the particle's own previous best position, recorded as a vector $\mathbf{p}_i$, based on the fitness value, calculated using the criterion functions defined in Section 4.2; the other is the best position in the whole swarm, represented as $\mathbf{p}_g$. Also, $\mathbf{p}_g$ can be replaced with a local best solution obtained within a certain local topological neighborhood (Kennedy and Mendes, 2002). Therefore, for each iteration, the velocity and the position vector are updated as

$$\mathbf{v}_i(t+1) = W_I \times \mathbf{v}_i(t) + c_1 \times \varphi_1 \times (\mathbf{p}_i - \mathbf{z}_i(t)) + c_2 \times \varphi_2 \times (\mathbf{p}_g - \mathbf{z}_i(t)), \quad (4.86)$$

$$\mathbf{z}_i(t+1) = \mathbf{z}_i(t) + \mathbf{v}_i(t), \quad (4.87)$$

where W_I is the inertia weight, c_1 and c_2 are the acceleration constants, and φ_1 and φ_2 are the uniform random functions.

PSO has the properties of ease of implementation, fast convergence to high-quality solutions, and flexibility in balancing global and local exploration. Moreover, the memory mechanism of PSO can keep track of previous best solutions and therefore can avoid the possible loss of previously learned knowledge.

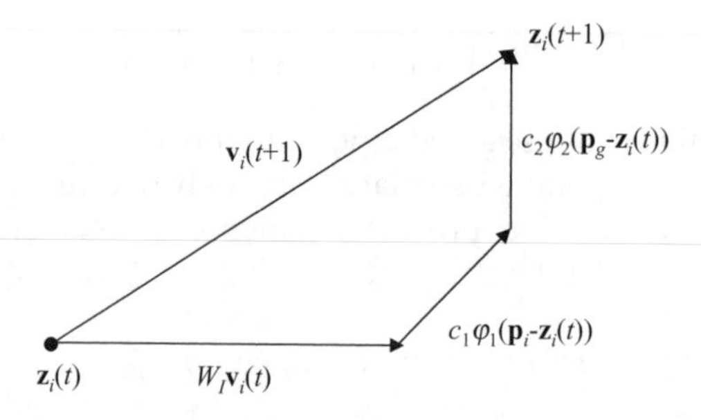

Fig. 4.6. Concept of a swarm particle's position. The new velocity $\mathbf{v}_i(t+1)$ is determined by the momentum part $W_I\mathbf{v}_i(t)$, cognitive part $c_1\varphi_1(\mathbf{p}_i - \mathbf{z}_i(t))$, and social part $c_2\varphi_2(\mathbf{p}_g - \mathbf{z}_i(t))$.

4.7.5 Tabu Search

Tabu search (TS) is a combinatory search technique that uses the tabu list to guide the search process consisting of a sequence of moves, each of which corresponds to a procedure of obtaining a new solution from the current one (Glover, 1989; Glover and Laguna, 1997). The tabu list stores part or all of the previously selected moves depending on its size. These moves are temporarily forbidden in the current search and are called tabu. Therefore, only candidate moves that are not tabu are eligible for current consideration. This condition is overruled for moves that satisfy the aspiration criteria, which make the attractive moves allowable for selection. When the tabu list size is reached, the first move on the list is released from the tabu so that a new tabu can be added into the list.

In the TS clustering algorithm developed by Al-Sultan (1995), a set of candidate solutions are generated from the current solution. Each candidate solution represents the allocations of N data objects in K clusters. The candidate with the optimal fitness function is selected as the current solution and appended to the tabu list, if it is not already in the tabu list, or if it is in, it meets the aspiration criterion. Otherwise, the remaining candidates are evaluated in order of their fitness function values until all these conditions are satisfied. In the case that all the candidates are tabu, a new set of candidate solutions are created, followed by the same search process. The search process proceeds until the maximum number of iterations is reached. The flowchart of the algorithm is summarized in Fig. 4.7. Sung and Jin (2000) suggested a more elaborate search process with packing and releasing procedures. Data objects packed or grouped together make the same move in the search. This strategy improves search efficiency and reduces the computational requirement. The opposite process of releasing separates the grouped objects in the posterior stage in order to achieve a complete search. The algorithm also implements a secondary tabu list in order to keep the search from getting trapped in the potential cycles. Delgado et al. (1997) discussed a TS application in fuzzy clustering, and Xu et al. (2002) combined fuzzy logic in TS for parameter adaptation.

Hybrid approaches that combine these search techniques are also proposed. Scott et al. (2001) combined TS in a GA clustering algorithm for population variety and computational efficiency. The chromosomes that have appeared in the previous steps are added to the tabu list. An application of SA for improving TS was reported by Chu and Roddick (2000). The function of SA is to determine the current best solution.

4.8. APPLICATIONS

4.8.1 Gene Expression Data Analysis

In a study of transcriptional regulatory sub-networks in yeast *Saccharomyces cerevisiae*, the K-means algorithm was used to partition 3,000 genes into 30

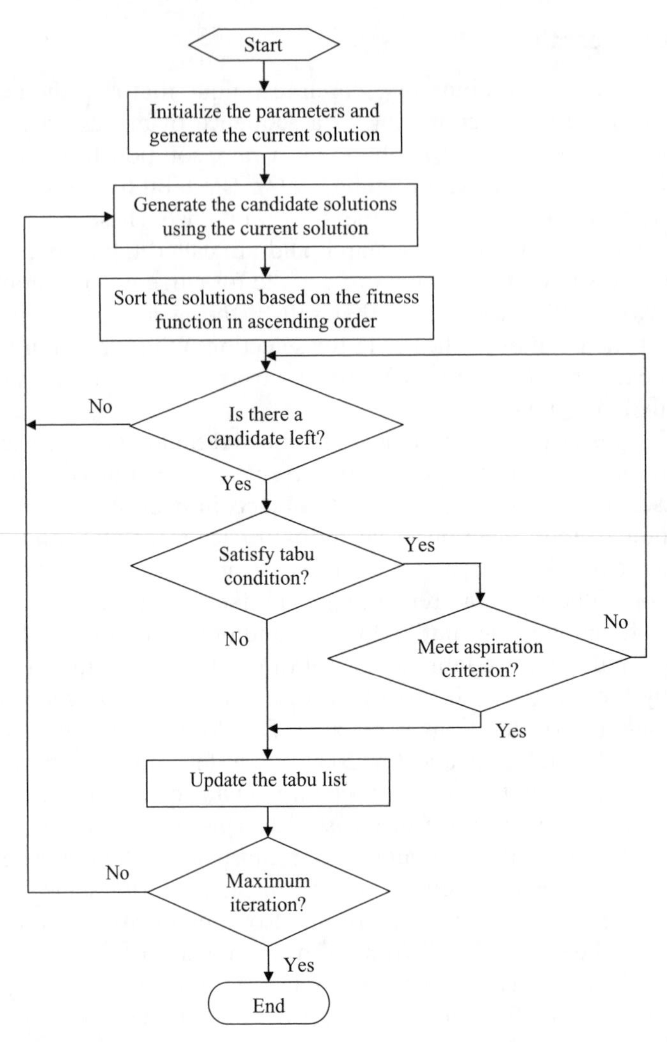

Figure 4.7 Flowchart of the TS-based clustering algorithm. Candidate solutions that represent clustering of data objects are generated from the current solution, which has the highest fitness and is added to the tabu list. The algorithm iteratively searches the optimal candidate solution that is not in the tabu list or meets the aspiration criterion as the current solution until the maximum number of iterations is reached.

clusters based on their expression level measurements, as shown in Figs. 4.8 and 4.9 (Tavazoie et al., 1999). Each cluster is considered to consist of a set of co-regulated genes, and 600 base pairs of upstream sequences of the genes within the same cluster were searched for specific motifs using the program AlignACE (Roth et al., 1998). According to the results reported, 18 motifs were identified from 12 clusters, among which, 7 motifs, such as STRE in Fig. 4.9, can be verified via the previous empirical results in the literature.

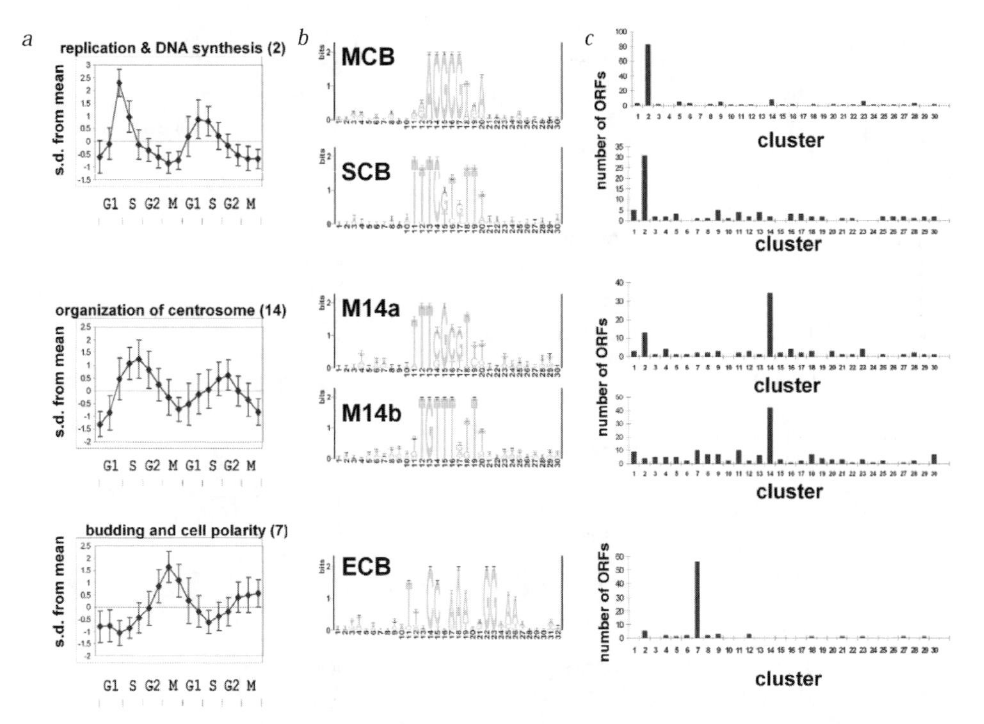

Fig. 4.8. Top periodic clusters, their motifs, and overall distribution in all clusters. (a) Mean temporal profile of a cluster, named according to the biological functions for which it is most highly enriched (with the numerical designation of the cluster in parentheses). Error bars represent the standard deviation of the members of each cluster about the mean of the particular time point. (b) Sequence logo representation of the motif(s) discovered within the clusters. The height of each letter is proportional to its frequency. The letters are sorted with the most frequent one on top. The overall height of the stack signifies information content of the sequence at that position (0–2 bits). Motifs M14a and M14b were identified in this study. (c) The occurrence of the motifs across all 30 clusters. (Reprinted by permission from Macmillan Publishers Ltd: Nature Genetics, S. Tavazoie, J. Hughes, M. Campbell, R. Cho, and G. Church, Systematic determination of genetic network architecture, vol. 22, pp. 281–285, 1999, Copyright © 1999.)

New motifs, e.g., M14a and M14b in Fig. 4.8, were also discovered in this study. In Figs 4.8 and 4.9, it also can be seen that the motifs are highly selective for the specific cluster where they are identified. For instance, more than 40 members of cluster 8 contain the STRE motif, while the other clusters include less than 5 members having this motif. Another application of an incremental K-means algorithm in clustering a set of 2,029 human cDNA clones was presented in Ralf-Herwig et al. (1999), where mutual information was used as the similarity measure and the number of clustering centroids can also be dynamically adjusted.

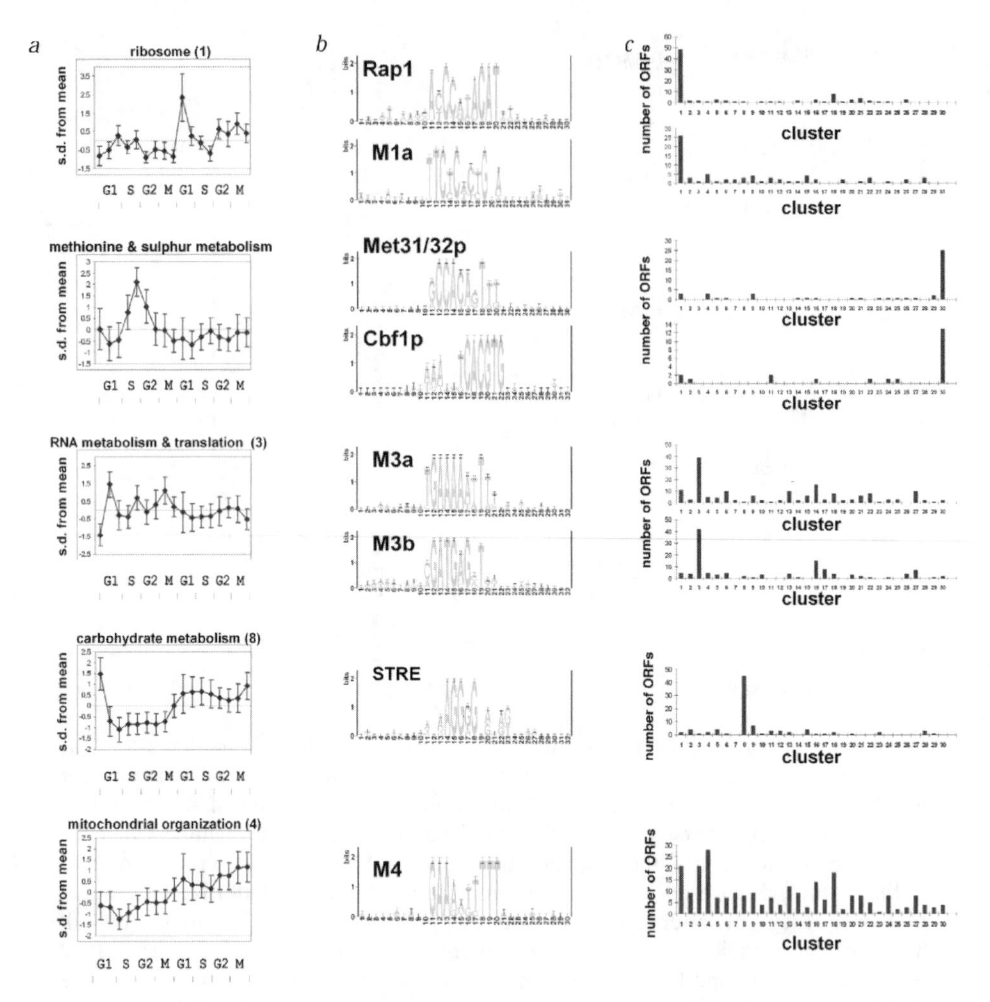

Fig. 4.9. "Nonperiodic" clusters, their motifs and overall distribution in all clusters. (a) Mean temporal profile of a cluster. (b) Sequence logo representation of the motifs. Motifs M1a, M3a, M3b, and M4 were identified in this study. (c) The occurrence of the motifs across all 30 clusters. (Reprinted by permission from Macmillan Publishers Ltd: Nature Genetics, S. Tavazoie, J. Hughes, M. Campbell, R. Cho, and G. Church, Systematic determination of genetic network architecture, vol. 22, pp. 281–285, 1999, Copyright © 1999.)

Since many genes do not display just one biological function in the cellular process or are not co-regulated with only one cluster of other genes, fuzzy clustering could provide a clearer picture in unveiling the different features of these genes' functions and regulation (Dembélé and Kastner, 2003; Gasch and Eisen, 2002). In this framework, each gene is associated with all clusters with a fuzzy membership function, and genes that have expression patterns similar to more than one gene cluster will lead to significant membership values for these clusters. The resulting partition of the genes can be formed by just setting a cutoff threshold and assigning genes with membership values above the threshold to the corresponding clusters. Dembélé and Kastner (2003) investigated the effect of fuzzification parameter m of fuzzy c-means algorithm in expression data clustering. It is observed that the commonly recommended value $m = 2$ was ineffective in several case studies. Alternately, an empirical was presented to estimate an appropriate value of m using the information on the distance distribution of genes in the given data set. Results of FCM on a yeast data set with 2,945 genes (Tavazoie et al., 1999) indicate increased biological relevance of the genes assigned to the same clusters (Dembélé and Kastner, 2003). Another application of FCM to a yeast data set with 6,153 genes shows that 17% of assigned genes are related to more than one cluster with threshold set at 0.1, and the number increases to 64% when the threshold is set at 0.04 (Gasch and Eisen, 2002).

Graph theories based clustering algorithms, such as CAST (Ben-Dor et al., 1999) minimum spanning trees (Xu et al., 2002) and CLICK (Sharan and Shamir, 2000), have also shown very promising performances in tackling different types of gene expression data. Fig. 4.10 depicts the clustering results of a set of 698 yeast genes across 72 conditions (Spellman et al., 1998) using the CLICK algorithm (Shamir and Sharan, 2002). The comparison of the clustering solution quality between CLICK and K-means or self-organizing feature maps in terms of homogeneity and separation shows a more preferred performance of CLICK (Shamir and Sharan, 2002).

There are many other partitional clustering algorithms applied in gene expression data analysis. Typical examples include mixtures of multivariate Gaussian distributions (Ghosh and Chinnaiyan, 2002), evolutionary clustering algorithms (Ma et al., 2006), and simulated annealing-based clustering algorithms (Lukashin and Fuchs, 2001).

4.8.2 Quality Assessment of Thematic Maps

It is important to compare sites on the thematic maps generated from remote sensing images to reference samples in order to evaluate the map's accuracy quantitatively. However, obtaining such ground truth knowledge is not always a simple and low-cost task, which causes problems, for example, when a set of competing thematic maps generated from the same image needs to be compared, or the generalization capability of several different classifiers has to be evaluated. In order to address these issues, an original data-driven thematic map quality assessment (DAMA) strategy was developed, with its block diagram shown in Fig. 4.11 (Baraldi et al., 2005).

As depicted in Fig. 4.11, to begin with, DAMA employs a procedure called reference map generation by multiple clustering (RMC) to locate a set of unlabeled candidate representative raw areas from the input image, which are then organized separately by clustering algorithms, such as the K-means algo-

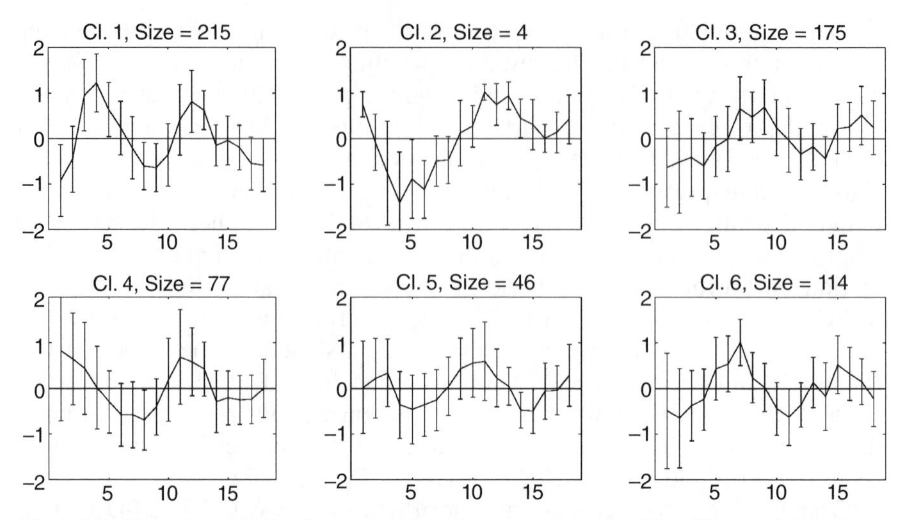

Fig. 4.10. Cluster analysis of yeast cell-cycle data using the CLICK algorithm. Each cluster is represented by the mean temporal profile with error bars displaying the standard deviation for the first 18 measured points. The sizes of the clusters are shown above each plot. (From Jiang, Tao, Ying Xu, and Michael Q. Zhang., *Current Topics in Computational Molecular Biology*, figure 11.13, © 2002 Massachusetts Institute of Technology, by permission of the MIT Press.)

rithm and the ELBG algorithm, to obtain reference cluster maps. The number of clusters in each cluster map is set as the number of label types in the test map. For each reference cluster map, the submaps of the test map are identified as the portions of the test map that overlap with the reference cluster map. The labeling fidelity of the explicit submap to the implicit reference cluster map and the segmentation fidelity of the explicit submap to the reference segmentation extracted for the reference cluster map can be calculated using the corresponding measures. The measures of fidelity are then combined based on certain image quality criteria.

The DAMA strategy is tested on a remote sensing satellite image, as shown in Fig. 4.12 (a). Three unobserved image areas are extracted from the raw image, one of which is given in Fig. 4.12 (b). The K-means algorithm is applied to the derived data, and one of the implicit reference cluster maps, which consists of 21 clusters, is illustrated in Fig. 4.12 (c). The three cluster maps, together with the computed labeling and spatial fidelity index, can further be used to compare and evaluate competing classifiers (Baraldi et al., 2005).

4.8.3 Detection of Multiple Sclerosis with Visual Evoked Potentials

Visual evoked potentials (VEPs) are the most commonly used test tool in the diagnosis of multiple sclerosis (MS) (Noseworthy et al., 2000). Based on the recorded brain waves using electrodes attached to the head, when the eyes are

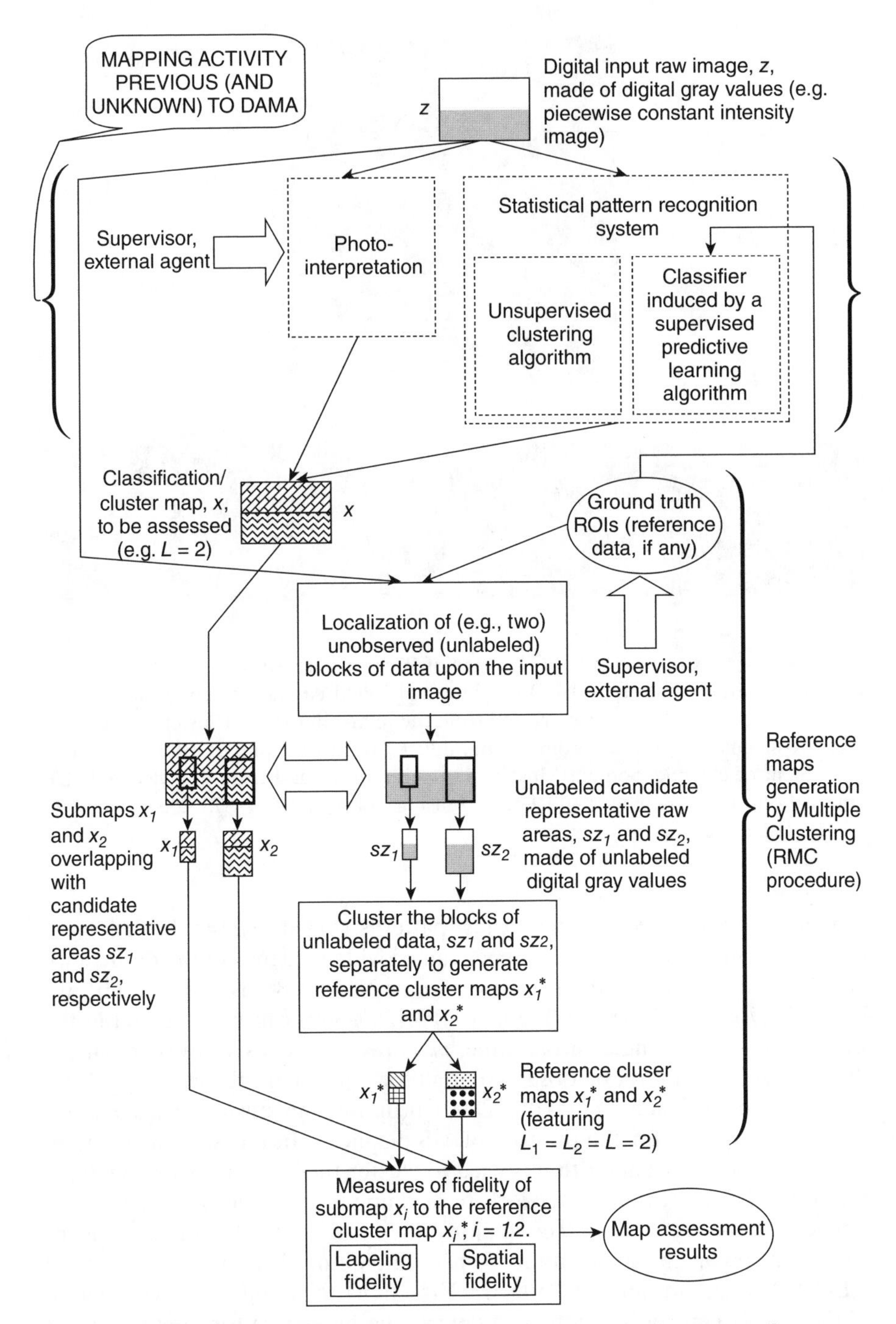

Fig. 4.11. Block diagram of the unsupervised DAMA strategy for the quality assessment of competing maps made from digital input images. (From A. Baraldi, L. Bruzzone, and P. Blonda. Quality assessment of classification and cluster maps without ground truth knowledge. IEEE Transactions on Geoscience and Remote Sensing, vol. 43, pp. 857–873, 2005. Copyright © 2005 IEEE.)

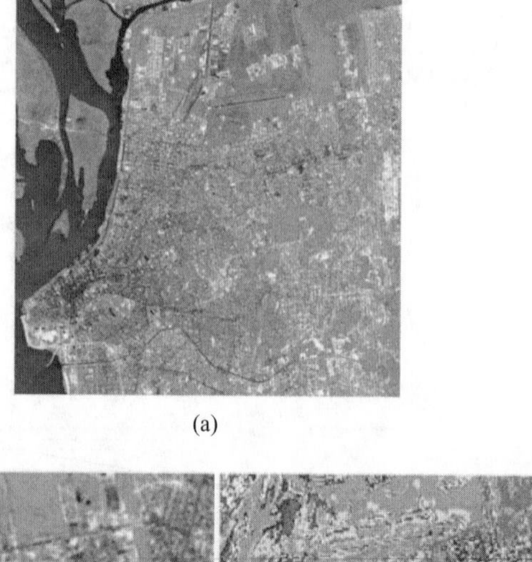

(a)

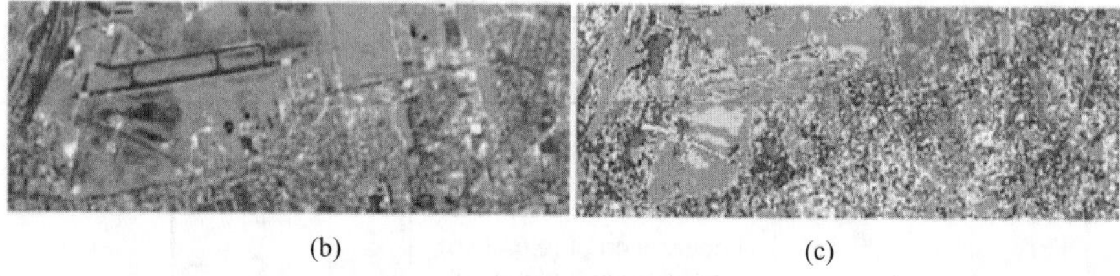

(b) (c)

Figure 4.12. (a) A three-band SPOT image of Porto Alegre, Brazil, 512 × 512 pixels in size, 20 m spatial resolution. (b) One of the unlabeled candidate representative raw areas, 100 × 300 pixels in size, extracted from the image in (a). (c) Cluster result of (b) using K-means algorithm. (From A. Baraldi, L. Bruzzone, and P. Blonda. Quality assessment of classification and cluster maps without ground truth knowledge. IEEE Transactions on Geoscience and Remote Sensing, vol. 43, pp. 857–873, 2005. Copyright © 2005 IEEE.)

stimulated by looking at different test patterns (Fig. 4.13), the distortions of the waveform from the normal ranges can provide evidence of the occurrence of MS. In normal cases, the peaks, known as N1 (60–80 ms), P1 (90–110 ms), N2 (135–170 ms), P2 (180–200 ms), and P3 (280–350) can be observed in the waveforms at the ranges in parentheses. However, peaks will occur outside these ranges or will not be observed at all for abnormal subjects.

Dasey and Micheli-Tzanakou (2000) proposed an unsupervised pattern recognition system for the purpose of MS diagnosis. In this system, a variant of the fuzzy c-means algorithm is used to cluster the patterns whose features are derived from a neural network performing a d-dimensional Karhunen-Loève feature extraction. The major modification of the fuzzy c-means algorithm lies in the application of the combinatorial optimization method ALOPEX (ALgorithms Of Pattern EXtraction) to optimize the calculation of the cluster centroids. Figure 4.14 depicts the histogram for the two-cluster result of a data set, including 10 normal subjects (NL), 5 definite MS (DMS), 3 probable MS (PMS), 1 neurosarcoidosis (NSD), and 1 subject who was normal but having difficulty controlling left eye movement (VIS), upon the

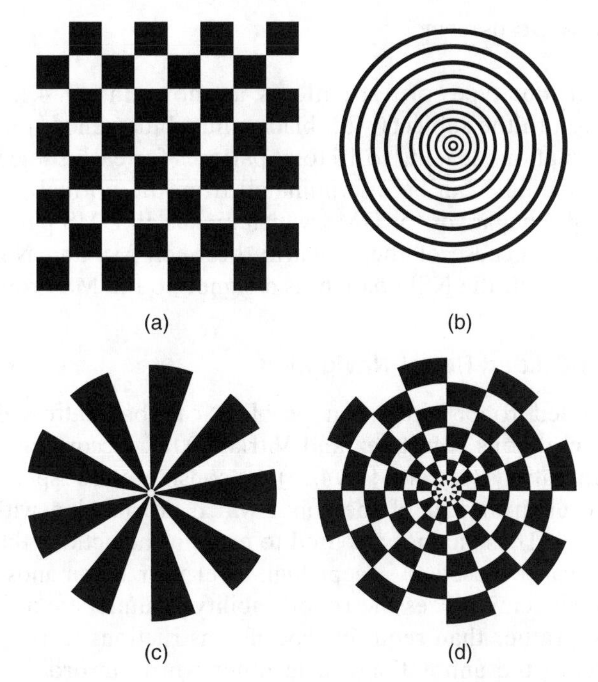

Fig. 4.13. Test patterns presented to subjects in order to collect VEPs. (a) Checkerboard. (b) Concentric circles. (c) Windmill. (d) Circular checkerboard. (From T. Dasey and E. Micheli-Tzanakou. Detection of multiple sclerosis with visual evoked potentials—an unsupervised computational intelligence system. IEEE Transactions on Information Technology in Biomedicine, vol. 4, pp. 216–224, 2000. Copyright © 2000 IEEE.)

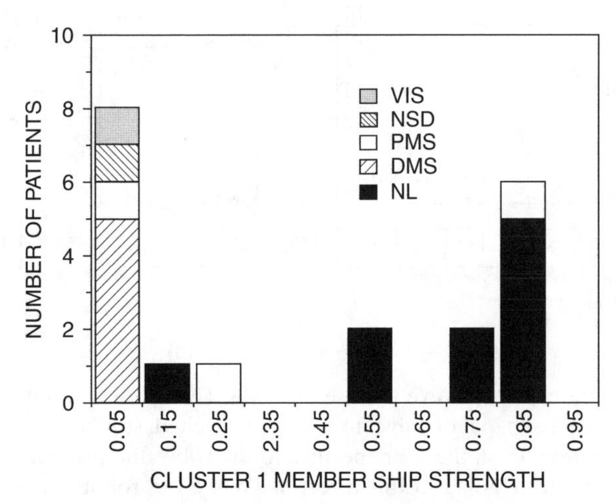

Fig. 4.14. Cluster membership histogram for a data set with 20 subjects using the cluster membership strengths of each of the 16 simulation patterns. (From T. Dasey and E. Micheli-Tzanakou. Detection of multiple sclerosis with visual evoked potentials—an unsupervised computational intelligence system. IEEE Transactions on Information Technology in Biomedicine, vol. 4, pp. 216–224, 2000. Copyright © 2000 IEEE.)

presentation of four types of test patterns, as shown in Fig. 4.13. These patterns were given to the subjects with black/white, blue/yellow, red/green, and red/blue combinations, leading to 16 total patterns for each subject. It is clear that most MS patients can be discriminated from the normal patients. Also, there are no false negatives for DMS, which means all DMS patients are correctly identified. Because of the similarity between MS and NSD, it is not surprising to see that the NSD patient is assigned to the MS group.

4.8.4 Vision-Guided Robot Navigation

The vision-guided robot navigation problem can be addressed using the concept of annotations (Martinez and Vitrià, 2001). According to Martinez and Vitrià, annotations (see also Fig. 4.15) are considered as "specific locations (areas) of the environment of the robot which are labeled with a specific (symbolic name). These names are used to explain the actions that the robot performs and can be used to accept high-level user commands." Here, the word "high-level" emphasizes the robot's ability to understand the goal of a particular task, rather than requiring specific instructions to reach that goal. In order to learn the annotations, or in other words, in order to equip the robot with the ability to identify its current location, the right moving direction, and the target point, a normal mixture models-based algorithm (NMM)

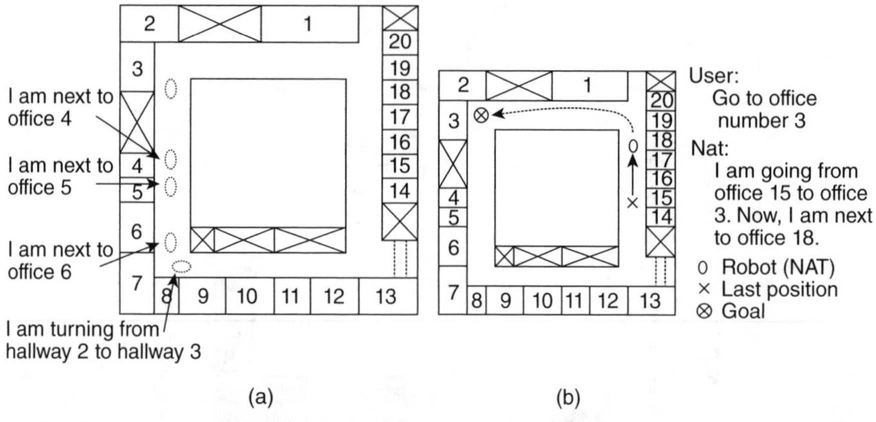

Fig. 4.15. (a) Annotations involve the use of labels. The robot annotates all actions it is performing once a subgoal or substate has been reached. (b) An example of navigation using annotation. First, the user specifies to the robot the place it has to go next. Second, the robot starts going in this direction. While the robot is moving, it makes annotations, i.e., it explains (using high-level descriptions) what it is doing and where it currently is within its environment. (From A. Martinez and J. Vitrià. Clustering in image space for place recognition and visual annotations for human-robot interaction. IEEE Transactions on Systems, Man, And Cybernetics—Part B: Cybernetics, vol. 31, pp. 669–682, 2001. Copyright © 2001 IEEE.)

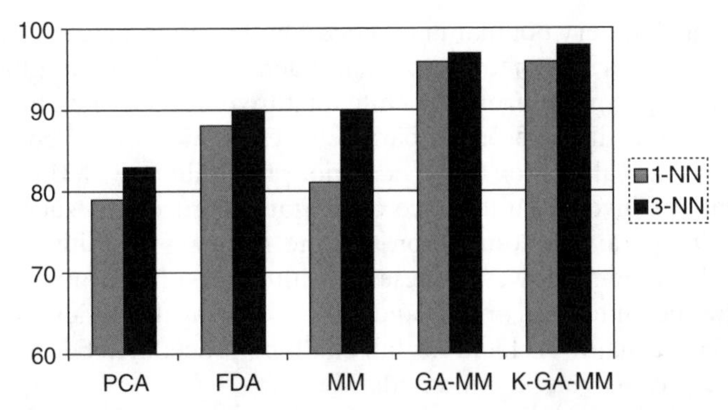

Fig. 4.16. Comparison of the performance of PCA, FDA, NMM, GA-NMM, and K-GA-NMM in terms of successful recognition rate. 1-NN means that the solution is within the first nearest neighbor, and 3-NN means that the solution is within the three nearest neighbors. (From A. Martinez and J. Vitrià. Clustering in image space for place recognition and visual annotations for human-robot interaction. IEEE Transactions on Systems, Man, And Cybernetics—Part B: Cybernetics, vol. 31, pp. 669–682, 2001. Copyright © 2001 IEEE.)

was developed to cluster the images of the working environment, which are represented as vectors in the space of pixel intensities and are further projected to a lower-dimensional space using principal component analysis (PCA) (see Chapter 9). Each model in the mixtures corresponds to a different area of the environment, and the position associated with the mean of the model is annotated by the user. A genetic algorithm is integrated with the EM algorithm to prevent the algorithm from being stuck to local optima, which leads to variants of NMM, called a GA-NMM algorithm and a K-GA-NMM algorithm. The latter also considers the quality of the mixture when performing selection operations in the evolutionary process. The successful rate results of the three mixture models, compared with PCA and Fisher discriminant analysis (FDA), on the environment illustrated in Fig. 4.15 (a) are summarized in Fig. 4.16. The working environment is one of the floors of a research institution and consists of 4 hallways with 20 offices. Two searching strategies were used; one method only examines the nearest neighbor while the other searches the three closest neighbors. As clearly indicated in the figure, the two GA-based versions of NMM have achieved the best performance in both cases, with a correct recognition rate over 95%.

4.9. SUMMARY

Partitional clustering directly produces a one-level partition of K clusters without the hierarchical structures. This can be achieved through an optimization process of a criterion function, for example, the sum-of-squared-error criterion, which is very popular in practice. On the other hand, cluster analysis can be viewed as a parametric probability density problem, and the data are assumed to be generated from the finite mixture densities. Thus, the clustering problem becomes the process of parameter estimation of the corresponding mixtures and the calculation of the posterior probabilities with Bayes theorem. Both situations involve an iterative optimization procedure, such as the EM algorithm for parameter estimation and the K-means algorithm for seeking the optimal solutions. However, these algorithms all suffer from the possibility of being trapped into local optima and therefore being dependent on initialization. This is the major problem for hill-climbing procedures because the solutions space is concave and multimodal.

Search techniques, such as evolutionary algorithms, provide a way to seek the global or approximate global optimum. However, the major drawback that plagues the application of search techniques in clustering is parameter selection. More often than not, search techniques introduce more parameters than other methods (like K-means). There are no theoretical guidelines to select appropriate and effective parameters. For example, Hall et al. (1999) provided some methods for setting parameters in their GA-based clustering framework, but most of these criteria are still obtained empirically. The same situation exists for TS and SA clustering (Al-Sultan, 1995; Selim and Alsultan, 1991). Another problem associated with search techniques is the computational complexity paid for the convergence to global optima. High computational requirements limit their applications in large-scale data sets.

Clusters are not always well-separated, and it is quite arbitrary to put the data point on the boundary of one cluster without considering the other cluster. Fuzzy clustering overcomes this limitation of hard or crisp clustering by allowing an object to be associated with all clusters. In this way, it is possible to find the second best or third best cluster of a certain object, which may provide more information about the data structure. For example, during gene clustering based on their functions, many genes are known to achieve more than one function.

In partitional clustering, it is also important to determine the number of clusters, which is usually assumed to be known and is required as a parameter from the users. However, this number unfortunately is unknown for many clustering practices, and estimating it becomes a major problem for cluster validation. We devote an entire chapter to this topic.

CHAPTER 5

NEURAL NETWORK–BASED CLUSTERING

5.1. INTRODUCTION

Neural networks have solved a wide range of problems and have good learning capabilities. Their strengths include adaptation, ease of implementation, parallelization, speed, and flexibility. Neural network–based clustering is closely related to the concept of competitive learning, which is traced back to the early works of Rosenblatt (1962), von der Malsburg (1973), Fukushima (1975), and Grossberg (1976a, b). According to Rumelhart and Zipser (1985), a competitive learning scheme consists of the following three basic components:

1. "Start with a set of units that are all the same except for some randomly distributed parameter which makes each of them respond slightly differently to a set of input patterns."
2. Limit the "strength" of each unit.
3. Allow the units to compete in some way for the right to respond to a given subset of inputs."

Specifically, a two-layer feedforward neural network that implements the idea of competitive learning is depicted in Fig. 5.1. The nodes in the input layer admit input patterns and are fully connected to the output nodes in the competitive layer. Each output node corresponds to a cluster and is associated with a prototype or weight vector $\mathbf{w}_j$, $j = 1, \ldots, K$, where K is the number of

Clustering, by Rui Xu and Donald C. Wunsch, II

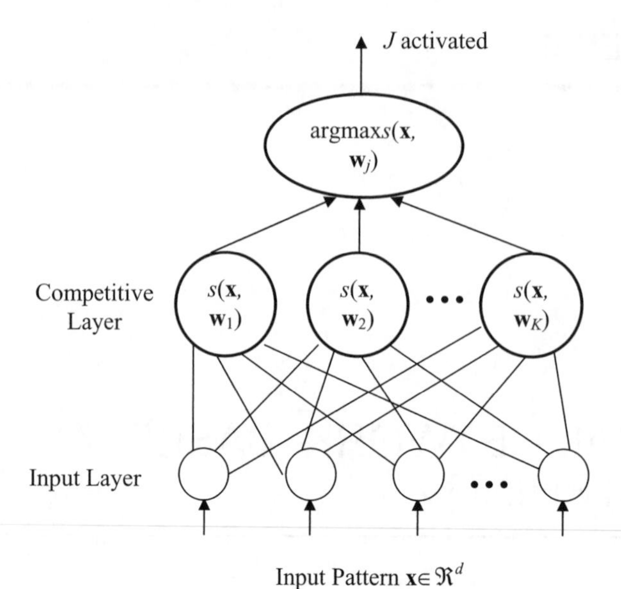

Fig. 5.1. A competitive learning network with excitatory connections from the input nodes to the output neurons. Each node in the competitive layer is associated with a weight vector $\mathbf{w}_j$. The neuron that is nearest to the input pattern, based on the prespecified similarity or distance function, is fired, and its prototype is adapted to the input pattern thereafter. However, updating will not occur for other losing neurons.

clusters, stored in terms of synaptic weights w_{ji}, $i = 1, \ldots, d$, representing the connection between input node i and output node j.

Given an input pattern $\mathbf{x} = \{x_1, x_2, \ldots x_d\}$ presented at iteration t, the similarity between the weight vector $\mathbf{w}_j$ of the randomly initialized cluster j, and $\mathbf{x}$ can be obtained by computing the net activation v_i,

$$s(\mathbf{x}, \mathbf{w}_j) = v_j = \mathbf{w}_j^T \mathbf{x} = \sum_{i=1}^{d} w_{ji} x_i. \tag{5.1}$$

The neurons in the competitive layer then compete with each other, and only the one with the largest net activation value becomes activated or fired, written as,

$$J = \arg\max_j s(\mathbf{x}, \mathbf{w}_j). \tag{5.2}$$

Note that if the distance function $D(\cdot)$ is used to measure dissimilarity, Eq. 5.2 becomes,

$$J = \arg\min_j D(\mathbf{x}, \mathbf{w}_j). \tag{5.3}$$

The weight vector of this winning neuron J is further moved towards the input pattern following the updating equation,

$$\mathbf{w}_J(t+1) = \mathbf{w}_J(t) + \eta(\mathbf{x}(t) - \mathbf{w}_J(t)), \tag{5.4}$$

where η is the learning rate. This learning rule is also known as the instar rule (Grossberg, 1976a, b). Note that by using the instar rule, the weight vector will also be normalized if the input pattern is normalized, such as $\|\mathbf{x}\| = 1$. This property avoids the possible problem of unlimited growth of the weight in magnitude when directly using Hebb's rule (Hebb, 1949),

$$\mathbf{w}_J(t+1) = \mathbf{w}_J(t) + \eta\mathbf{x}(t), \tag{5.5}$$

which may ultimately prevent the learning of other neurons. Both divisive enforcement and subtractive enforcement could be used to normalize the weights, and their difference in terms of input pattern representation has been further investigated by Goodhill and Barrow (1994).

The competitive learning paradigm described above only allows learning for a particular winning neuron that matches best with the given input pattern. Thus, it is also known as winner-take-all (WTA) or hard or crisp competitive learning (Baraldi and Blonda, 1999; Fritzke, 1997). On the other hand, learning can also occur in a cooperative way, which means that not just the winning neuron adjusts its prototype, but all other cluster prototypes have the opportunity to be adapted based on how proximate they are to the input pattern (Baraldi and Blonda, 1999). In this context, the learning scheme is called soft competitive learning or winner-take-most (WTM) (Baraldi and Blonda, 1999; Fritzke, 1997). Baraldi and Blonda (1999) also called algorithms that integrate soft competitive learning "fuzzy clustering algorithms." But here, we still use the term fuzzy clustering algorithms to refer to the algorithms that allow an input pattern to belong to all clusters with a degree of membership, based on the fuzzy set theory, as discussed in Chapter 4.

In the remaining sections of this chapter, we discuss neural network–based clustering algorithms according to whether a hard or soft competitive learning strategy is involved. In Section 5.2, we concentrate on the family of adaptive resonance theory (ART), together with the leader-follower algorithm. In Section 5.3, we introduce WTM-based algorithms, including learning vector quantization (LVQ), Self-Organizing Feature Maps (SOFM), neural gas, and growing neural gas. Applications of neural network–based clustering algorithms are illustrated in Section 5.4.

5.2. HARD COMPETITIVE LEARNING CLUSTERING

5.2.1 Online *K*-means Algorithm

As we saw in Chapter 4, the *K*-means or LBG algorithm (Forgy, 1965; Linde et al., 1980; MacQueen, 1967) iteratively assigns input patterns to their closest prototype vectors and then updates the prototype vectors as the means of their corresponding Voronoi sets. Clearly, the hard competitive learning paradigm is embedded in this algorithm. Particularly, when the input patterns are presented in an incremental mode rather than a batch mode, the online version of K-means algorithm proceeds in the following way:

1. Initialize K cluster prototype vectors, $\mathbf{m}_1, \ldots, \mathbf{m}_K \in \Re^d$, randomly;
2. Present a normalized input pattern $\mathbf{x} \in \Re^d$;
3. Choose the winner J that has the smallest Euclidean distance to $\mathbf{x}$,

$$J = \arg\min_j \|\mathbf{x} - \mathbf{m}_j\|; \tag{5.6}$$

4. Update the winning prototype vector towards $\mathbf{x}$,

$$\mathbf{m}_J(\text{new}) = \mathbf{m}_J(\text{old}) + \eta(\mathbf{x} - \mathbf{m}_J(\text{old})), \tag{5.7}$$

   where η is the learning rate;
5. Repeat steps 2–4 until the maximum number of steps is reached.

Apparently, the learning rate η determines the adaptation of the prototype vector towards the input pattern and is directly related to the convergence. If η equals zero, there is no learning. If η is set to one, it will result in fast learning, and the prototype vector is directly pointed to the input pattern. For the other choices of η, the new position of the prototype vector will be on the line between the old prototype vector and the input pattern. Generally, the learning rate could take a constant value or vary over time.

First, let η be a constant, i.e.,

$$\eta = \eta_c, 0 < \eta_c \leq 1. \tag{5.8}$$

If we assume that cluster J wins a sequence of presentation of t input patterns, $\mathbf{x}_1, \ldots, \mathbf{x}_t$, Eq. 5.7 can then be rewritten as,

$$\mathbf{m}_J(t) = (1-\eta_c)^t \mathbf{m}_J(0) + \eta_c \sum_{i=1}^{t} (1-\eta_c)^{t-i} \mathbf{x}_i, \tag{5.9}$$

where $\mathbf{m}_J(0)$ represents the random initial value for the prototype vector. Eq. 5.9 immediately brings two interesting observances: (1) the current input pattern always contributes a certain component (that may be large) to the prototype vector; and (2) the influence of the past input patterns on the prototype vector decays exponentially fast with the number of further input patterns (Fritzke, 1997). A direct consequence of such observances is that the prototypes may vary forever without convergence, although these properties may be useful for tracking slowly and gradually changing sequences of input patterns. A new input pattern may cause unstable learning and a considerable change of the winning prototype vector.

Instead of remaining constant, the learning rate η can also be dynamically decreased over time (Fritzke, 1997), for instance,

$$\eta(t) = \eta_0(\eta_1/\eta_0)^{t/t_1}, \tag{5.10}$$

where η_0 and η_1 are the initial and final values of the learning rate, respectively, and t_1 is the maximum number of iterations allowed. In this way, we prefer faster learning during the early phase while using smaller learning rates as the learning proceeds. The disadvantage of this method is that new patterns cannot be learned well in the late stage because the learning rate will be very small.

An adaptive learning rate strategy, which makes the learning rate exclusively dependent on the within-group variations without involving any user activities, was illustrated in Chinrungrueng and Séquin (1995). The key idea is that if we do not have a good partition of the input patterns, we choose a large learning rate to speed up the learning. In contrast, when we have clusters with good quality, we use a small learning rate to finely tune the positions of the prototype vectors. Specifically, the quality of a partition of the input patterns is measured in terms of the entropy of the normalized within-cluster variations, $v_1, \ldots, v_K$, for a K-partition, written as,

$$H(v_1, \mathrm{K}, v_K) = \sum_{i=1}^{K} -v_{i,norm} \ln(v_{i,norm}), \tag{5.11}$$

where

$$v_{i,norm} = \frac{v_i}{\sum_{j=1}^{K} v_j}. \tag{5.12}$$

Thus, the learning rate is defined as

$$\eta = \frac{\ln(K) - H(v_1, \mathrm{K}, v_K)}{\ln(K)}. \tag{5.13}$$

5.2.2 Leader-Follower Clustering Algorithm

As we saw in Chapter 4, one of the major disadvantages of the K-means algorithm is its requirement for determining the number of clusters K in advance. However, for many real problems, the number of clusters is actually seen as part of the unknown natural structure in the data and has to be estimated via the procedure of cluster analysis. An inappropriate selection of number of clusters may distort the real clustering structure and cause an ineffective clustering partition (Xu et al., 1993).

A possible approach to overcoming this obstacle is to produce the clusters in a constructive way, as the leader-follower algorithm does (Duda et al., 2001; Moore, 1989). The major difference between the leader-follower algorithm and the K-means algorithm lies in the former's introduction of an additional threshold parameter θ, which determines the level of match between the winning prototype and the input pattern. If the distance between the prototype and the input pattern is below the threshold, the learning occurs as discussed previously. Otherwise, a new cluster is created with its prototype vector taking the value of the input pattern. In this sense, the threshold θ functions as a cluster diameter. A small θ corresponds to a great number of small clusters, while a large θ requires a small number of clusters to represent the data. We describe the leader-follower algorithm using the following basic steps:

1. Initialize the first cluster prototype vector $\mathbf{m}_1$ with the first input pattern;
2. Present a normalized input pattern $\mathbf{x}$;
3. Choose the winner J that is closest to $\mathbf{x}$ based on the Euclidean distance,

$$J = \arg\min_j \|\mathbf{x} - \mathbf{m}_j\|; \tag{5.14}$$

4. If $\|\mathbf{x} - \mathbf{m}_J\| < \theta$, update the winning prototype vector,

$$\mathbf{m}_J(\text{new}) = \mathbf{m}_J(\text{old}) + \eta(\mathbf{x} - \mathbf{m}_J(\text{old})), \tag{5.15}$$

   where η is the learning rate. Otherwise, create a new cluster with the prototype vector equal to $\mathbf{x}$;
5. Repeat steps 2–4 until the maximum number of steps is reached.

5.2.3 Adaptive Resonance Theory

An important problem with competitive learning-based clustering is stability. Moore (1989) defines the stability of an incremental clustering algorithm in terms of two conditions: "(1) No prototype vector can cycle, or take on a value that it had at a previous time (provided it has changed in the meantime). (2) Only a finite number of clusters are formed with infinite presentation of the data." The first condition considers the stability of individual prototype vectors of the clusters, and the second one concentrates on the stability of all the cluster vectors. In this sense, the algorithms discussed before do not always produce stable clusters, as pointed out by Moore (1989) and Grossberg (1976a).

The reason for this instability lies in the algorithms' plasticity, which is required to adapt to important new patterns. However, this plasticity may cause the memories of prior learning to be lost, worn away by the recently-learned knowledge. Carpenter and Grossberg (1987a, 1988) refers to this problem as the stability-plasticity dilemma, i.e., how adaptable (plastic) should a learning system be so that it does not suffer from catastrophic forgetting of previously-learned rules (stability)?

Adaptive resonance theory (ART) was developed by Carpenter and Grossberg (1987a, 1988) as a solution to the stability-plasticity dilemma. ART can learn arbitrary input patterns in a stable, fast, and self-organizing way, thus overcoming the effect of learning instability that plagues many other competitive networks. ART is not, as is popularly imagined, a neural network architecture. It is a learning theory hypothesizing that resonance in neural circuits can trigger fast learning. As such, it subsumes a large family of current and future neural network architectures with many variants. ART1 is the first member, which only deals with binary input patterns (Carpenter and Grossberg, 1987a, 1988), although it can be extended to arbitrary input patterns by utilizing a variety of coding mechanisms. ART2 extends the applications to analog input patterns (Carpenter and Grossberg, 1987b), and ART3 introduces a mechanism originating from elaborate biological processes to achieve more efficient parallel searches in hierarchical structures (Carpenter and Grossberg, 1990). Fuzzy ART (FA) incorporates fuzzy set theory and ART and can work for all real data sets (Carpenter et al., 1991b). (It is typically regarded as a superior alternative to ART2.) Linares-Barranco et al. (1998) demonstrated the hardware implementations and very-large-scale integration (VLSI) design of ART systems. In Wunsch (1991) and Wunsch et al. (1993), the optical correlator-based ART implementation, instead of the implementation of ART in electronics, was also discussed.

5.2.3.1 ART1 As depicted in Fig. 5.2, the basic ART1 architecture consists of two-layer nodes or neurons, the feature representation field F_1, and the category representation field F_2, whose present state is known as short-term memory (STM). The neurons in layer F_1 are activated by the input pattern, while the prototypes of the formed clusters are stored in layer F_2. The neurons in layer F_2 that are already being used as representations of input patterns are said to be committed. Correspondingly, the uncommitted neuron encodes no input patterns. The two layers are connected via adaptive weights: a bottom-up weight matrix $\mathbf{W}^{12} = \{w_{ij}^{12}\}$, where the index represents the connection from the ith neuron in layer F_1 to the jth neuron in layer F_2, and a top-down weight matrix $\mathbf{W}^{21} = \{w_{ji}^{21}\}$, which is also called long-term memory (LTM). F_2 performs a winner-take-all competition, between a certain number of committed neurons and one uncommitted neuron. The winning neuron feeds back its template weights to layer F_1. This is known as top-down feedback expectancy. This template is compared with the input pattern. The prespecified vigilance parameter ρ $(0 \le \rho \le 1)$ determines whether the expectation and the input

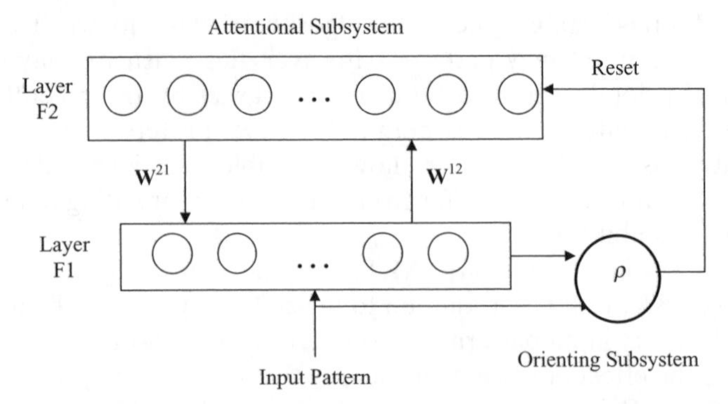

Fig. 5.2. ART1 architecture. Two layers are included in the attentional subsystem, connected via bottom-up and top-down adaptive weights. Their interactions are controlled by the orienting subsystem through a vigilance parameter.

pattern are closely matched. If the match meets the vigilance criterion, weight adaptation occurs, where both bottom-up and top-down weights are updated simultaneously. This procedure is called resonance, which suggests the name of ART. On the other hand, if the vigilance criterion is not met, a reset signal is sent back to layer F_2 to shut off the current winning neuron, which will remain disabled for the entire duration of the presentation of this input pattern, and a new competition is performed among the rest of the neurons. This new expectation is then projected into layer F_1, and this process repeats until the vigilance criterion is met. If an uncommitted neuron is selected for coding, a new uncommitted neuron is created to represent a potential new cluster. It is clear that the vigilance parameter ρ has a function similar to that of the threshold parameter θ of the leader-follower algorithm. The larger the value of ρ, the fewer mismatches will be tolerated; therefore, more clusters are likely to be generated.

At this point, we summarize the basic steps of ART1 as follows, which are also depicted in Fig. 5.3:

1. Initialize the weights as $w_{ij}^{12} = \xi/(\xi - 1 + d)$, where d is the dimensionality of the binary input pattern $\mathbf{x}$, ξ is a parameter that is larger than one, and $w_{ji}^{21} = 1$;
2. Present a new pattern $\mathbf{x}$ and calculate the input from layer F_1 to layer F_2 as

$$T_j = \sum_{i=1}^{d} w_{ij}^{12} x_i; \tag{5.16}$$

3. Activate layer F_2 by choosing neuron J with the winner-take-all rule,

$$T_J = \max_j \{T_j\}; \tag{5.17}$$

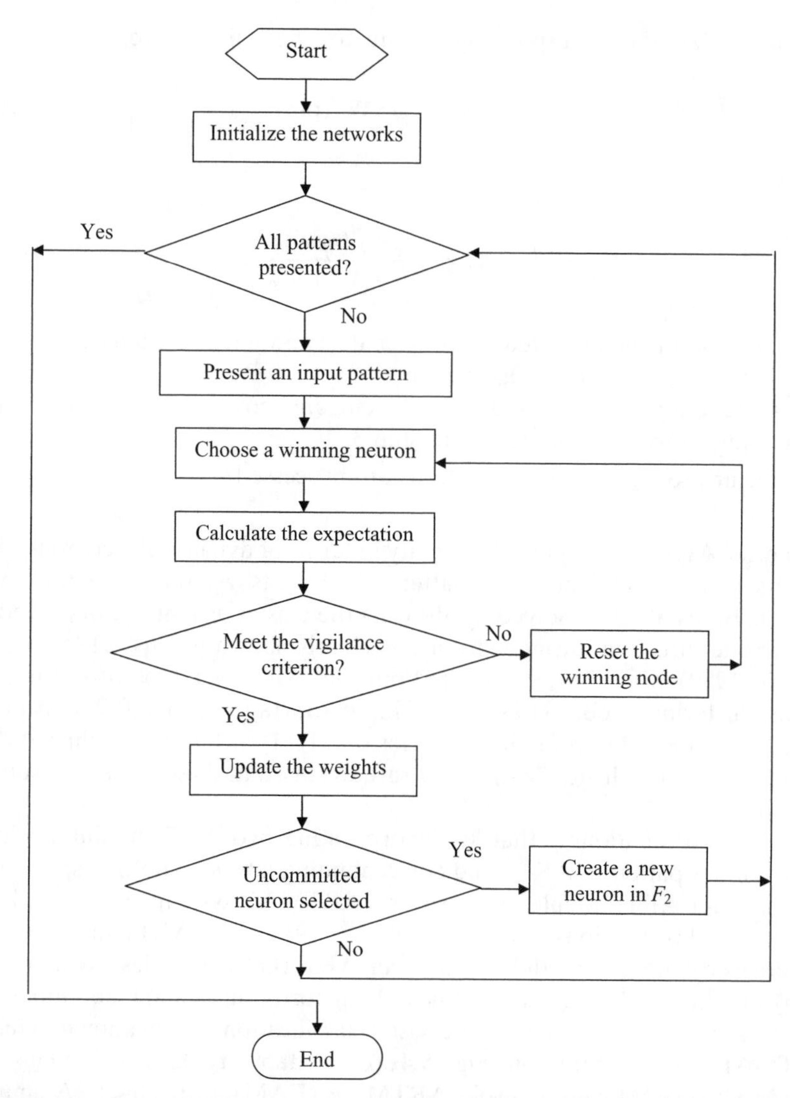

Fig. 5.3. Flowchart of ART1. Note that the learning occurs only when the vigilance test is passed. Otherwise, the current active neuron is suppressed by a reset signal and a new competition is performed among the remaining neurons.

4. Compare the expectation from layer F_2 with the input pattern. If

$$\rho \leq \frac{|\mathbf{x} \cap \mathbf{W}_J^{21}|}{|\mathbf{x}|}, \tag{5.18}$$

where $\cap$ represents the logic AND operation, go to step 5a; otherwise, go to step 5b.

5. a. Update the corresponding weights for the active neuron as

$$\mathbf{W}_J^{21}(\text{new}) = \mathbf{x} \cap \mathbf{W}_J^{21}(\text{old}), \tag{5.19}$$

and

$$\mathbf{W}_J^{12}(\text{new}) = \frac{\xi \mathbf{W}_J^{21}(\text{new})}{\xi - 1 + |\mathbf{W}_J^{21}(\text{new})|}. \tag{5.20}$$

If J is an uncommitted neuron, create a new uncommitted neuron with the initial values set as in Step 1;

b. Send a reset signal to disable the current active neuron by the orienting subsystem, and return to step 3;

6. Return to step 2 until all patterns are processed.

In brief, ART1 obtains its adaptability in terms of dynamically creating new clusters in order to learn new patterns or events. At the same time, the problem of instability is solved by allowing the cluster weight vectors to move only in one direction during learning, as clearly shown in Eqs. 5.19 and 5.20 (Moore, 1989). Moore (1989) pointed out that ART1 is a type of varying K-means clustering. Moore (1989) and Linares-Barranco et al. (1998) also discussed a number of important properties of ART1, such as self-scaling, direct access to a stored cluster, learning of rare events, and direct access to subset and superset.

It is worth mentioning that, by incorporating two ART1 modules, which receive input patterns (ART_a) and corresponding labels (ART_b), respectively, with an inter-ART module, the resulting ARTMAP system can be used for supervised classifications (Carpenter et al., 1991a). The ART1 modules can be replaced with FA modules, Gaussian ART (GA) modules (Williamson, 1996), or ellipsoid ART (EA) modules (Anagnostopoulos and Georgiopoulos, 2001), which correspond to the supervised classification system known as fuzzy ARTMAP, as illustrated in Fig. 5.4 (Carpenter et al., 1992), Gaussian ARTMAP (GAM), and ellipsoid ARTMAP (EAM), respectively. A similar idea, omitting the inter-ART module, is known as laterally primed adaptive resonance theory (LAPART) (Healy et al., 1993). Carpenter (2003) used a nested sequence to describe the relations among several variants of ARTMAP, as fuzzy ARTMAP $\subset$ default ARTMAP (Carpenter, 2003) $\subset$ ARTMAP-IC (Carpenter and Markuzon, 1998) $\subset$ distributed ARTMAP (Carpenter et al., 1998).

Wunsch (1991) and Wunsch et al. (1993) discussed the ease with which ART may be used for hierarchical clustering. The method, called ART tree, is a hierarchy in which the same input pattern is sent to every level. Which ART units in a given level get to look at the input are determined by the winning nodes of layer F_2 at a lower level. Thus, all nodes of layer F_2 in the

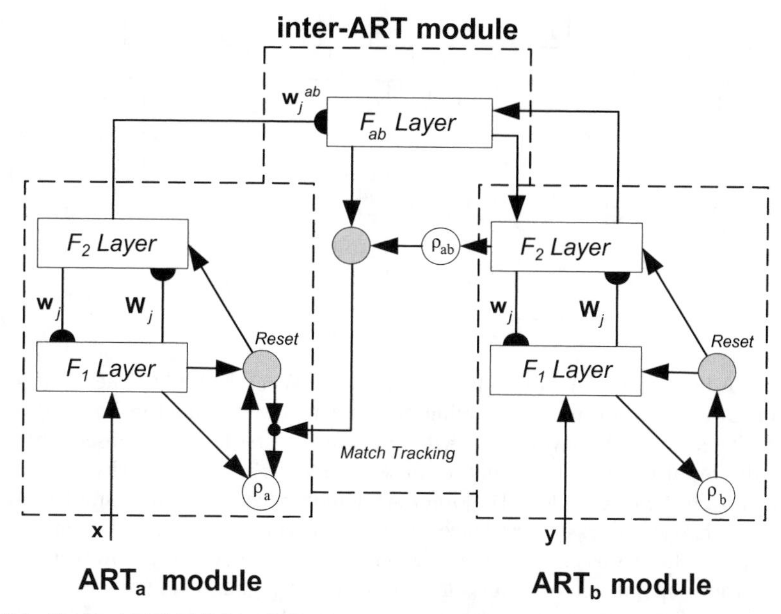

Fig. 5.4. Fuzzy ARTMAP block diagram. Fuzzy ARTMAP consists of two FA modules (ART_a and ART_b) interconnected via an inter-ART module. The ART_a module clusters patterns of the input domain, and ART_b the ones of the output domain. The match tracking strategy ensures the consistency of category prediction between two ART modules by dynamically adjusting the vigilance parameter of ART_a.

entire hierarchy see the same input pattern. This allows ART to perform hierarchical clustering in that the lower-level clusters will form perfect subsets of the higher-level clusters. An ART1 hierarchy with a total of 39 prototypes is illustrated in Fig. 5.5. Also, two ART-based approaches for hierarchical clustering were presented by Bartfai and White (1997), known as hierarchical ART with joining (HART-J) and hierarchical ART with splitting (HART-S).

5.2.3.2 Fuzzy ART Fuzzy ART extends the ART family by being capable of learning stable recognition clusters in response to both binary and real-valued input patterns with either fast or slow learning (Carpenter et al., 1991b). FA maintains architecture and operations similar to ART1 while using the fuzzy set operators to replace the binary operators so that it can work for all real data sets. We describe FA by emphasizing its main difference with ART1 in terms of the following five phases, known as preprocessing, initialization, category choice, category match, and learning.

- Preprocessing. Each component of a d-dimensional input pattern $\mathbf{x} = (x_1, \dots, x_d)$ must be in the interval [0,1].

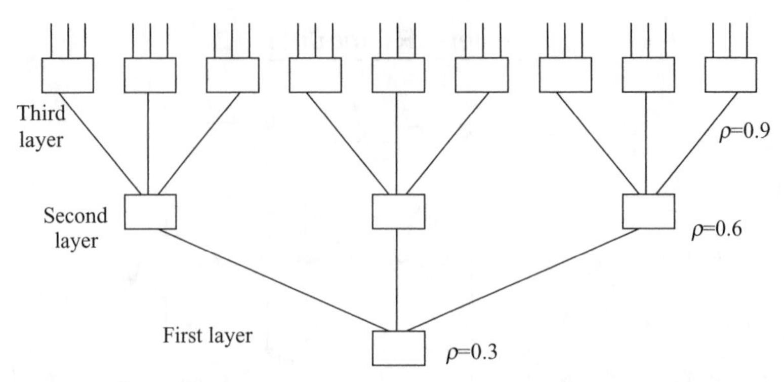

Fig. 5.5. A hierarchy of ART1 units (Adapted from Wunsch (1991).) The input pattern is fed in at the bottom, and the winning output is read out at the top. The entire input pattern is fed to the F2 layer of all ART units in the first layer. In subsequent layers, the entire input pattern is fed only to those ART units in the hierarchy of "winning" F2 units from the layer below. The other ART units receive no input at all. Note that each layer has an increasing vigilance threshold. This is done in order to more finely partition the clusters as classifications become more detailed. Note, however, that due to the order dependence of inputs, and the fact that higher-level units do not all see the same set of patterns, there is a possibility of interesting classifications being achieved even if vigilance is not adjusted in this manner. The study of vigilance selection in an ART Tree hierarchy still has some worthwhile open questions as of this writing.

- Initialization. The real-valued adaptive weights $\mathbf{W} = \{w_{ij}\}$, representing the connection from the ith neuron in layer F_2 to the jth neuron in layer F_1, include both the bottom-up and top-down weights of ART1. Initially, the weights of an uncommitted node are set to one. Larger values may also be used, however, this will bias the tendency of the system to select committed nodes (Carpenter et al., 1991b).
- Category choice. After an input pattern is presented, the nodes in layer F_2 compete by calculating the category choice function, defined as

$$T_j = \frac{|\mathbf{x} \wedge \mathbf{w}_j|}{\alpha + |\mathbf{w}_j|}, \tag{5.21}$$

where $\wedge$ is the fuzzy AND operator defined by

$$(\mathbf{x} \wedge \mathbf{y})_i = \min(x_i, y_i), \tag{5.22}$$

and $\alpha > 0$ is the choice parameter to break the tie when more than one prototype vector is a fuzzy subset of the input pattern. Particularly, the limit $\alpha \rightarrow 0$ is called the conservative limit (Carpenter et al., 1991b). Also, α is related to the vigilance parameter ρ; α should decrease as ρ decreases (Huang et al., 1995).

Similar to ART1, the neuron J becomes activated with the winner-take-all rule,

$$T_J = \max_j \{T_j\}. \tag{5.23}$$

- Category match. The category match function of the winning neuron is then tested with the vigilance criterion. If

$$\rho \leq \frac{|\mathbf{x} \wedge \mathbf{w}_J|}{|\mathbf{x}|}, \tag{5.24}$$

resonance occurs. Otherwise, the current winning neuron is disabled and a new neuron in layer F_2 is selected and examined with the vigilance criterion. This search process continues until Eq. 5.24 is satisfied.
- Learning. The weight vector of the winning neuron that passes the vigilance test at the same time is updated using the following learning rule,

$$\mathbf{w}_J(\text{new}) = \beta(\mathbf{x} \wedge \mathbf{w}_J(\text{old})) + (1-\beta)\mathbf{w}_J(\text{old}), \tag{5.25}$$

where $\beta \in [0, 1]$ is the learning rate parameter. Carpenter et al. (1991b) introduced a method, called fast-commit slow-recode, for achieving efficient coding of noisy input patterns. In this context, β is set to one when an uncommitted node is selected to represent the current input pattern. Correspondingly, Eq. 5.25 becomes

$$\mathbf{w}_J(\text{new}) = \mathbf{x}, \tag{5.26}$$

which indicates that the input pattern is directly copied as the prototype of the new cluster. On the other hand, committed prototypes are updated with a slow learning rate, $\beta < 1$, to prevent them from being corrupted by noise.

A practical problem in applying FA is the possibility of cluster proliferation, which occurs as a result of an arbitrarily small norm of input patterns (Carpenter et al., 1991b; Moore, 1989). Since the norm of weight vectors does not increase during learning, many low-valued prototypes may be generated without further access. The solution to the cluster proliferation problem is to normalize the input patterns (Carpenter et al., 1991b) so that,

$$|\mathbf{x}| = \varsigma, \quad \varsigma > 0. \tag{5.27}$$

This is an extended step of the preprocessing phase.

One way to normalize an input pattern $\mathbf{x}$ is to divide it by its norm, written as,

$$\mathbf{x}^* = \frac{\mathbf{x}}{|\mathbf{x}|}. \tag{5.28}$$

However, this method does not maintain the amplitude information of the input patterns. Alternately, Carpenter et al. (1991b) proposed a normalization rule, known as complement coding, to normalize input patterns without losing the amplitude information. Specifically, an input pattern d-dimensional $\mathbf{x} = (x_1, \ldots, x_d)$ is expanded as a $2d$-dimensional vector

$$\mathbf{x}^* = (\mathbf{x}, \mathbf{x}^c) = (x_1, \cdots, x_d, x_1^c, \cdots, x_d^c), \tag{5.29}$$

where $x_i^c = 1 - x_i$ for all i. A direct mathematical manipulation shows that input patterns in complement coding form are automatically normalized,

$$|\mathbf{x}^*| = |(\mathbf{x}, \mathbf{x}^c)| = \sum_{i=1}^{d} x_i + \sum_{i=1}^{d} x_i^c = \sum_{i=1}^{d} x_i + d - \sum_{i=1}^{d} x_i = d. \tag{5.30}$$

Corresponding to the expansion of the input patterns, now, the adaptive weight vectors $\mathbf{w}_j$ are also in the $2d$-dimensional form, represented as,

$$\mathbf{w}_j = (\mathbf{u}_j, \mathbf{v}_j^c). \tag{5.31}$$

Initially, $\mathbf{w}_j$ is still set to one, which causes $\mathbf{u}_j$ to be set to one and $\mathbf{v}_j$ to be set to zero. The adaptation of $\mathbf{w}_j$ also follows the same rule in Eq. 5.25.

A two-dimensional geometric interpretation of FA cluster update with complement coding and fast learning is illustrated in Fig. 5.6, where each category is represented as a rectangle. Note that complement coding has the advantage that one can inspect templates to determine how much learning has occurred. This, together with the normalization property, has popularized complement coding in ART1 as well. Another method that has hyper-rectangular representations of clusters is called fuzzy min-max clustering neural networks (Gabrys and Bargiela, 2000; Simpson, 1993).

As can be seen in Fig. 5.6, $\mathbf{u}_j$ and $\mathbf{v}_j$ in Eq. 5.31 are both two-dimensional vectors defining two corners of rectangle R_j, which is considered a geometric representation of cluster j. The size of R_j can be calculated using

$$|R_j| = |\mathbf{v}_j - \mathbf{u}_j|. \tag{5.32}$$

Note that when an uncommitted node j is eligible to encode an input pattern $\mathbf{x}^* = (\mathbf{x}, \mathbf{x}^c)$, the fast learning in Eq. 5.26 leads to

$$\mathbf{w}_j(\text{new}) = \mathbf{x}^* = (\mathbf{x}, \mathbf{x}^c), \tag{5.33}$$

which implies that both $\mathbf{u}_j$ and $\mathbf{v}_j$ are equal to $\mathbf{x}$. In this situation, rectangle R_j coincides with the point $\mathbf{x}$ with zero size.

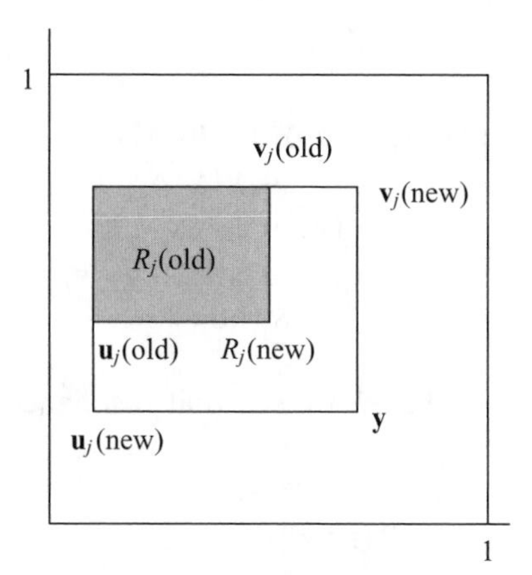

Fig. 5.6. Category update of FA with complement coding and fast learning. Each category j is geometrically represented as a rectangle R_j. The shaded rectangle expands to the smallest rectangle to incorporate the presented input pattern **x** into the cluster. For the reasons discussed in the text, complement coding has become popular in ART1 as well.

Suppose cluster j corresponding to the shaded area is now eligible to encode a new input pattern $\mathbf{y}^* = (\mathbf{y}, \mathbf{y}^c)$. Again, following the fast learning rule in Eq. 5.26, we have

$$\begin{aligned} \mathbf{w}_j(\text{new}) &= \mathbf{y}^* \wedge \mathbf{w}_j(\text{old}) \\ &= (\mathbf{y} \wedge \mathbf{u}_j(\text{old}), \mathbf{y}^c \wedge \mathbf{v}_j^c(\text{old})) \\ &= \left(\mathbf{y} \wedge \mathbf{u}_j(\text{old}), (\mathbf{y} \vee \mathbf{v}_j(\text{old}))^c\right)' \\ &= (\mathbf{u}_j(\text{new}), \mathbf{v}_j^c(\text{new})) \end{aligned} \tag{5.34}$$

where $\vee$ represents the fuzzy OR operator

$$(\mathbf{x} \vee \mathbf{y})_i = \max(x_i, y_i). \tag{5.35}$$

As can be seen from Eq. 5.34, the rectangle R_j expands with the smallest size to include both the previous representation region and the new input pattern. It is also interesting to see that if **y** is already inside R_j, there will be no change for the weight vector and, correspondingly, for the rectangle R_j.

Now we turn to examine the relation between the vigilance parameter ρ and the size of the rectangle R_j. we have already shown, learning will occur only if the winning cluster j meets the vigilance criterion in Eq. 5.24. Particularly, when the input pattern is two-dimensional and complement coding is used, we have $|\mathbf{y}^*| = 2$. Then, we can rewrite Eq. 5.24 as

$$2\rho \le |\mathbf{y}^* \wedge \mathbf{w}_j|. \tag{5.36}$$

By using Eq. 5.34, we have

$$\begin{aligned}|\mathbf{y}^* \wedge \mathbf{w}_j| &= |(\mathbf{y}, \mathbf{y}^c) \wedge (\mathbf{u}_j, \mathbf{v}_j^c)| \\ &= |(\mathbf{y} \wedge \mathbf{u}_j), (\mathbf{y}^c \wedge \mathbf{v}_j^c)| \\ &= |(\mathbf{y} \wedge \mathbf{u}_j), (\mathbf{y} \vee \mathbf{v}_j)^c| \\ &= |(\mathbf{y} \wedge \mathbf{u}_j)| + 2 - |\mathbf{y} \vee \mathbf{v}_j| \\ &= 2 - |R_j(\text{new})| \end{aligned} \quad (5.37)$$

By combining Eqs. 5.36 and 5.37, we see that resonance will occur when the expanded rectangle meets

$$|R_j(\text{new})| \leq 2(1-\rho). \quad (5.38)$$

Clearly, the closer the vigilance parameter ρ is to 1, the smaller the size of the rectangles will be, and correspondingly, the smaller the number of input patterns that are represented by the cluster prototypes, as discussed previously.

Similar manipulations can be applied to a more general situation with d-dimensional input patterns, which infers the size of a hyper-rectangle R_j,

$$|R_j| \leq d - |\mathbf{w}_j|, \quad (5.39)$$

together with the constraint on its maximum size,

$$|R_j| \leq d(1-\rho). \quad (5.40)$$

The discussions above can be summarized with the stable category learning theorem (Carpenter et al., 1991b):

> "In response to an arbitrary sequence of analog or binary input vectors, a Fuzzy ART system with complement coding and fast learning forms stable hyper-rectangular categories R_j, which grow during learning to a maximum size $|R_j| \leq d(1-\rho)$ as $|\mathbf{w}_j|$ monotonically decreases. In the conservative limit, one-pass learning obtains such that no reset or additional learning occurs on subsequent presentations of any input. Similar properties hold for the fast-learn slow-recode case, except that repeated presentations of an input may be needed before stabilization occurs."

As we have already seen, FA exhibits many desirable characteristics, such as fast and stable learning, a transparent learning paradigm, and atypical pattern detection. Huang et al. (1995) investigated and discussed more properties of FA in terms of prototype, access, reset, and the number of learning epochs required for weight stabilization. A comparison of the performance of FA and ART2 was presented by Frank et al. (1998).

5.2.3.3 Other ART Networks FA produces a hyper-rectangular representation of clusters in the feature space, which is more suitable for representing data that are uniformly distributed within hyper-rectangles (Williamson, 1996). When this assumption does not hold, the fuzzy categories may become an inefficient geometrical representation for exploring the potential data structures (Anagnostopoulos and Georgiopoulos, 2001; Williamson, 1996). Moreover, FA is sensitive to noise and has a problem of category proliferation in noisy data (Baraldi and Alpaydin, 2002; Baraldi and Blonda, 1999; Williamson, 1996). Williamson (1996) pointed out two possible causes of the category proliferation problem: (1) both the category choice and category match functions are flat within a cluster's hyper-rectangle and (2) fast learning is performed. As a solution to this problem, Williamson (1996) further suggested the Gaussian-defined category choice and match functions, which monotonically increase toward the center of a cluster, to replace those of FA. The obtained new ART module, in which each cluster is represented as a hyper-ellipsoid geometrically, is called Gaussian ART.

In the context of Gaussian distributions, each GA cluster j, representing d-dimensional input patterns, is described by a $(2d + 1)$-dimensional prototype vector $\mathbf{w}_j$ consisting of three components: $\boldsymbol{\mu}_j$ is the d-dimensional mean vector, $\boldsymbol{\sigma}_j$ is the d-dimensional standard deviation vector, and N_j is a scalar recording the number of patterns cluster j has encoded. Correspondingly, the category choice function is defined as a discriminant function examining the posteriori probability of cluster j given an input pattern $\mathbf{x}$,

$$T_j = -\frac{1}{2}\sum_{i=1}^{d}\left(\frac{\mu_{ji} - x_i}{\sigma_{ji}}\right)^2 - \log\left(\prod_{i=1}^{d}\sigma_{ji}\right) + \log P(j), \tag{5.41}$$

where the priori probability of cluster j is calculated as

$$P(j) = \frac{N_j}{\sum_{i=1}^{C} N_i}, \tag{5.42}$$

with C being the number of clusters.

After the cluster J with the maximum discriminant function is activated, the vigilance test is performed via the calculation of the value of the category match function, written as,

$$\rho_J = -\frac{1}{2}\sum_{i=1}^{d}\left(\frac{\mu_{Ji} - x_i}{\sigma_{Ji}}\right)^2, \tag{5.43}$$

which determines how well **x** matches with J in terms of the measurement of its distance to the mean of J, relative to the standard deviation.

The learning of the GA winning cluster J includes the update of the three elements of the prototype vector, given as follows,

$$N_J = N_J + 1, \tag{5.44}$$

$$\mu_{Ji}(\text{new}) = \left(1 - \frac{1}{N_J}\right)\mu_{Ji}(\text{old}) + \frac{1}{N_J}x_i, \tag{5.45}$$

$$\sigma_{Ji}(\text{new}) = \begin{cases} \sqrt{\left(1 - \frac{1}{N_J}\right)\sigma_{Ji}^2(\text{old}) + \frac{1}{N_J}(x_i - \mu_{Ji}(\text{new}))^2}, & \text{if } N_J > 1 \\ \gamma, & \text{otherwise} \end{cases}, \tag{5.46}$$

where γ is the initial standard deviation.

GA is designed as a class of probability density function estimators for Gaussian mixtures, using the maximum likelihood (ML) method (Baraldi and Alpaydin, 2002). The relation between GA and the expectation-maximization (EM) approach for mixture modeling was discussed by Williamson (1997). It was pointed out that the EM algorithm in the framework of Gaussian mixtures is essentially the same as the GA learning for modeling the Gaussian density of the input space. Furthermore, Baraldi and Alpaydin (2002) generalized GA in their defined constructive incremental clustering framework, called simplified ART (SART), which includes two other ART networks, known as symmetric fuzzy ART (SFART) and fully self-organizing SART (FOSART) networks. It is interesting to point out that FOSART uses a "soft-to-hard competitive model transition" to minimize the distortion error (Baraldi and Alpaydin, 2002), which displaces it from the category of hard competitive learning to which other ART networks belong.

Noticing the disadvantage of the lack of the fast learning law in GA, Anagnostopoulos and Georgiopoulos (2001) proposed ellipsoid ART, which evolved as a generalization of an early ART network, called hyper-sphere ART (HA) for hyper-spherical clusters (Anagnostopoulos and Georgiopoulos, 2000), to explore a hyper-ellipsoidal representation of EA clusters while following the same learning and functional principles of FA. A typical example of such a cluster representation, when the input space is two dimensional, is depicted in Fig. 5.7, where each category j is described by a center location vector $\mathbf{m}_j$, orientation vector $\mathbf{d}_j$, and Mahalanobis radius M_j, which are collected as the prototype vector $\mathbf{w}_j = [\mathbf{m}_j, \mathbf{d}_j, M_j]$. The orientation vector will be constant once it is set. If we define the distance between an input pattern **x** and a category j as

$$D(\mathbf{x}, \mathbf{w}_j) = \max\left\{\|\mathbf{x} - \mathbf{m}_j\|_{\mathbf{S}_j}, M_j\right\} - M_j, \tag{5.47}$$

$$\|\mathbf{x} - \mathbf{m}_j\|_{\mathbf{S}_j} = \sqrt{(\mathbf{x} - \mathbf{m}_j)^T \mathbf{S}_j (\mathbf{x} - \mathbf{m}_j)}, \tag{5.48}$$

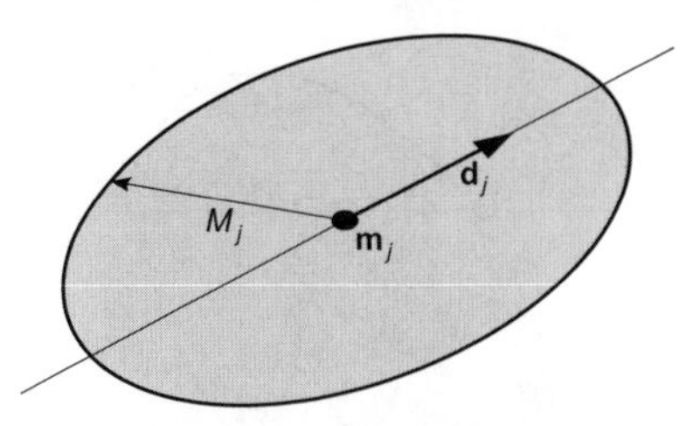

Fig. 5.7. Example of the geometric representation of an EA cluster j in a two dimensional feature space.

where $\mathbf{S}_j$ is the cluster's shape matrix, defined as

$$\mathbf{S}_j = 1/\mu^2(\mathbf{I} - (1-\mu^2)\mathbf{d}_j\mathbf{d}_j^T), \tag{5.49}$$

and $\mu \in (0, 1]$ is a constant ratio between the length of the hyper-ellipsoid's minor axes (with equal length) and major axis (for $\mu = 1$, the geometric representations become hyper-spheres, in which case the network is called HA), then the representation region of j, which is the shaded area in Fig. 5.7, can be defined as a set of points in the input space, satisfying the condition

$$D(\mathbf{x}, \mathbf{w}_j) = 0 \Rightarrow \|\mathbf{x} - \mathbf{m}_j\|_{\mathbf{S}_j} \leq M_j. \tag{5.50}$$

Similar to FA, the competition is performed via the category choice function, defined as

$$T_j = \frac{D_{\max} - 2M_j - D(\mathbf{x}, \mathbf{w}_j)}{D_{\max} - 2M_j + a}, \tag{5.51}$$

where $a > 0$ is the choice parameter, $D_{\max}$ is a parameter also greater than 0, and the match between the input pattern and the winning category's representation region is examined through the category match function

$$\rho_j = \frac{D_{\max} - 2M_j - D(\mathbf{x}, \mathbf{w}_j)}{D_{\max}}. \tag{5.52}$$

When it has been decided that a category j must be updated by a training pattern $\mathbf{x}$, its representation region expands so that it becomes the minimum-volume hyper-ellipsoid that contains the entire, original representation region and the new pattern. An example of this process for a two dimensional feature space is shown in Fig. 5.8, where the original representation region E_j expands to become E'_j. More specifically, the center location vector, orientation vector, and Mahalanobis radius are updated with the following equations:

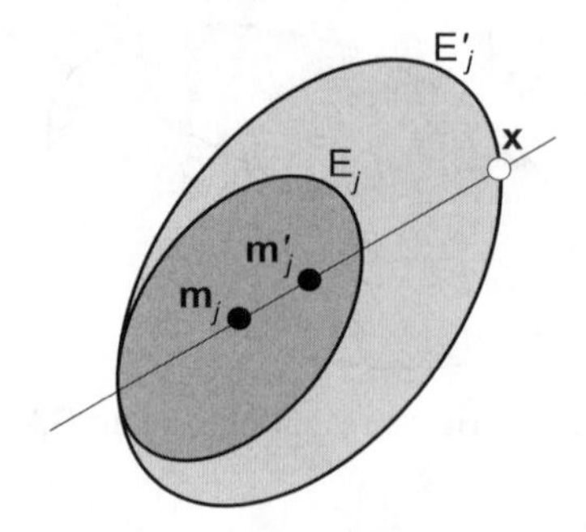

Fig. 5.8. Update of an EAM category j due to a training pattern $\mathbf{x}$ when the feature space is two dimensional. The representation region expands to contain the original region and the new pattern.

$$\mathbf{m}_j(\text{new}) = \mathbf{m}_j(\text{old}) + \frac{\eta}{2}\left(1 - \frac{\min\left(M_j(\text{old}), \|\mathbf{x}-\mathbf{m}_j(\text{old})\|_{\mathbf{S}_j(\text{old})}\right)}{\|\mathbf{x}-\mathbf{m}_j(\text{old})\|_{\mathbf{S}_j(\text{old})}}\right)(\mathbf{x}-\mathbf{m}_j(\text{old})), \tag{5.53}$$

where η is the learning rate,

$$\mathbf{d}_j = \frac{\mathbf{x}^{(2)} - \mathbf{m}_j}{\|\mathbf{x}^{(2)} - \mathbf{m}_j\|_2}, \quad \mathbf{x}^{(2)} \neq \mathbf{m}_j, \tag{5.54}$$

where $\mathbf{x}^{(2)}$ represents the second pattern encoded by cluster j, and

$$M_j(\text{new}) = M_j(\text{old}) + \frac{\eta}{2}\left(\max\left(M_j(\text{old}), \|\mathbf{x}-\mathbf{m}_j(\text{old})\|_{\mathbf{S}_j(\text{old})}\right) - M_j(\text{old})\right). \tag{5.55}$$

Notice that if $\mathbf{x}$ falls inside the representation region of j, no update occurs because j has already taken into account the presence of $\mathbf{x}$.

5.3. SOFT COMPETITIVE LEARNING CLUSTERING

5.3.1 Leaky Learning, Conscience Mechanism, and Rival Penalized Competitive Learning

One of the major problems with hard competitive learning is the underutilized or dead neuron problem (Grossberg, 1987; Hecht-Nielsen, 1987; Rumelhart and Zipser, 1985), which refers to the possibility that the weight vector of a neuron is initialized farther away from any input patterns than other weight vectors so that it has no opportunity to ever win the competition and, therefore, no opportunity to be trained. One solution to addressing this problem is to allow both winning and losing neurons to move towards the presented input pattern, but with different learning rates. The winning neuron moves much faster than the other neurons. This paradigm is called the leaky learning model (Rumelhart and Zipser, 1985), which is a special case of the partial contrast model described in Grossberg (1976a), and the learning equation is written as

$$\mathbf{w}_j(t) = \begin{cases} \mathbf{w}_j(t-1) + \eta_w(\mathbf{x}(t) - \mathbf{w}_j(t-1)), & \text{if } \mathbf{w}_j \text{ wins} \\ \mathbf{w}_j(t-1) + \eta_l(\mathbf{x}(t) - \mathbf{w}_j(t-1)), & \text{if } \mathbf{w}_j \text{ loses} \end{cases}, \tag{5.56}$$

where η_w and η_l are the learning rates for the winning and losing neurons, respectively, and $\eta_w >> \eta_l$. In this way, even if a neuron is initially located far away from the input patterns, it will gradually and slowly move towards the region that the input patterns reside in and have the chance to learn to represent certain patterns. However, "this scheme cannot be fully effective without having catastrophic results on code stability," as shown by Grossberg (1987).

Alternatively, another common strategy that can deal effectively with the dead neuron problem is to add a so-called conscience to hard competitive learning (Desieno, 1988; Grossberg, 1976a, b, 1987; Rumelhart and Zipser, 1985), which penalizes the neurons that win very often and provides chances for other neurons to learn. The basic theory of conscience was first introduced by Grossberg (1976a, 1987) in his variable-threshold model, where a history-dependent threshold is adjusted to decrease the sensitivity of the frequently winning neurons while making the losing neurons more sensitive. In addition, the learning rate also varies as a result of the competition: the more winning, the smaller the learning rate and vice versa.

Naturally, in order to admit the conscience mechanism into the competition process, we need to modify the distance definition described in Eq. 5.3. Desieno (1988) adds a bias term b_j to the squared Euclidean distance, represented as,

$$D^*(\mathbf{x}, \mathbf{w}_j) = \|\mathbf{w}_j - \mathbf{x}\|^2 - b_j. \tag{5.57}$$

The bias term is defined as,

$$b_j = C(1/K - p_j), \tag{5.58}$$

where K is the number of neurons in the competitive layer, C is a constant bias factor, and p_j is the fraction of time that a neuron j wins and is updated as

$$p_j(\text{new}) = p_j(\text{old}) + B(o_j - p_j(\text{old})), \tag{5.59}$$

where B is a constant satisfying $0 < B << 1$, and o_j takes the value of one when j wins or zero when j loses. In frequency sensitive competitive learning (FSCL), Ahalt et al. (1990) obtained the new distance measurement by multiplying the number of times the neuron j wins, c_j, to the original distance function,

$$D^*(\mathbf{x}, \mathbf{w}_j) = D(\mathbf{x}, \mathbf{w}_j)c_j. \tag{5.60}$$

In this way, the neurons that win often in the competitions will have a large value of the counter, which discourages the corresponding neurons in the competition. A fuzzy extension of FSCL is further discussed by Chung and Lee (1994), where the counter is calculated as the accumulation of the neuron's win membership in the past competitions. Other methods that implement a similar strategy can be found in Chen and Chang (1994), Butler and Jiang (1996), and Choy and Siu (1997).

The conscience mechanism provides an effective way to address the dead neuron problem. However, it is observed that the performance of methods like FSCL, which uses the conscience strategy, deteriorate rapidly when the number of clusters K in the competitive layer is not selected appropriately (Xu et al., 1993). In particular, if the chosen K is larger than the real number in the data, then the conscience strategy also moves the extra prototype vectors into the region where the input patterns reside. As a result of such undesired movement, some prototype vectors will be located at "some boundary points between different clusters or at points biased from some cluster centers" (Xu et al., 1993), instead of the real centers we expect.

Rival penalized competitive learning (RPCL) (Xu et al., 1993) solves the above problem by introducing an additional rival penalized scheme into FSCL. Given an input pattern $\mathbf{x}$ from a data set, the K neurons with associated weight vectors $\mathbf{w}_j, j = 1, \ldots, K$ compete with each other based on the modified squared Euclidean distance, which is integrated with a conscience mechanism,

$$D^*(\mathbf{x}, \mathbf{w}_j) = \lambda_j \|\mathbf{w}_j - \mathbf{x}\|^2, \tag{5.61}$$

where

$$\lambda_j = \frac{c_j}{\sum_{i=1}^{K} c_i}, \tag{5.62}$$

and c_j is the cumulative number of times that $\mathbf{w}_j$ won in the past.

In contrast to FSCL, during the learning of RPCL, not only is the winning neuron J with the minimum distance to $\mathbf{x}$ updated, but the weight vector of the second winner R, called the rival of the winner, is "de-learned" to push it away from $\mathbf{x}$. Specifically, we write the update equations of the winner and rival neuron as,

$$\mathbf{w}_J(\text{new}) = \mathbf{w}_J(\text{old}) + \eta_J(\mathbf{x} - \mathbf{w}_J(\text{old})), \tag{5.63}$$

$$\mathbf{w}_R(\text{new}) = \mathbf{w}_R(\text{old}) - \eta_R(\mathbf{x} - \mathbf{w}_R(\text{old})), \tag{5.64}$$

where η_J and η_R are the learning and de-learning rates for the winner and rival, respectively, with η_J generally much larger than η_R. These learning rules

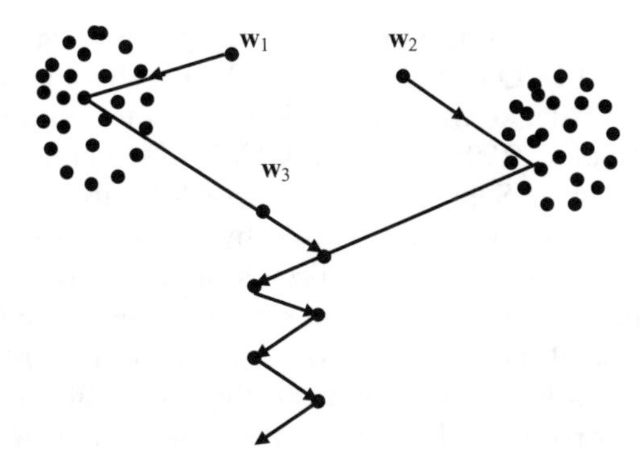

Fig. 5.9. Rival penalized competitive learning. The rival $\mathbf{w}_3$ of both $\mathbf{w}_1$ and $\mathbf{w}_2$ is driven out along a zigzag path. (From L. Xu, A. Krzyżak, and E. Oja. Rival penalized competitive learning for clustering analysis, RBF net, and curve detection. IEEE Transactions on Neural Networks, vol. 4, pp. 636–649, 1993. Copyright © 1993 IEEE.)

are also depicted in Fig. 5.9. In Cheung (2005), an additional step is proposed to dynamically adjust η_R by relating it to the competition strength. The closer the winner is to the rival than to the input pattern, the more penalties should be on the rival. This idea leads to the following method of determining η_R,

$$\eta_R = \eta_J \frac{\min(\|\mathbf{w}_R - \mathbf{w}_J\|, \|\mathbf{x} - \mathbf{w}_J\|)}{\|\mathbf{w}_R - \mathbf{w}_J\|}. \tag{5.65}$$

From Fig. 5.9, it is clear that the de-learning of the rival $\mathbf{w}_3$ keeps pushing it away from the two clusters that $\mathbf{w}_1$ and $\mathbf{w}_2$ win to represent. The real clusters will only have one prototype vector associated with them, thus avoiding the confusion caused by redundant weight vectors. In this sense, RPCL achieves an automatic estimation of the number of potential clusters in the data. The extra weight vectors due to the over-estimation of the number of clusters are driven away from the regions where the data are dense. The weight vectors are adjusted to learn the input data in a balance of the forces from both the conscience and the rival penalized mechanism (Xu et al., 1993). The former encourages the sharing of a cluster among weight vectors, while the latter prefers an exclusive prototype vector for representing a cluster. Furthermore, RPCL is analyzed as a special case of Bayesian Ying-Yang harmony learning and has been extended by combining finite mixture modeling and multi-sets modeling (Xu, 1998, 2001, 2002). The convergence behavior of RPCL under a general form, called distance-sensitive RPCL, was investigated by Ma and Wang (2006) in terms of the construction of a cost function. Exention of RPCL in sequential data clustering was in Law and Kwok (2000).

5.3.2 Learning Vector Quantization

In the literature, learning vector quantization (LVQ) refers to two families of models dealing with labeled and unlabeled data (Karayiannis and Bezdek,

1997). LVQ1, LVQ2, and LVQ3 perform supervised learning (Kohonen, 2001). Unsupervised LVQ (which is called LVQ in the literature, but we add "unsupervised" in order to avoid possible confusion) (Kohonen, 1989), generalized LVQ (Pal et al., 1993), and Fuzzy LVQ (Tsao et al., 1994) all achieve unsupervised clustering. Supervised LVQ classifiers are equipped with competitive mechanisms similar to those used by the methods we discuss in the chapter. For instance, when learning occurs in LVQ2, two winners that are closest to the input pattern are updated simultaneously. The one corresponding to the correct label is reinforced towards the input pattern, while the other with the wrong label is driven away from the input pattern. Clearly, RPCL, which uses a similar rival penalized mechanism and is discussed in the previous section, is an unsupervised counterpart of LVQ2.

The basic architecture of unsupervised LVQ is similar to the one described in Fig. 5.1, in the context of the competitive network, except that the Euclidean metric is used to measure the distance between the K prototype vectors of the neurons in the competitive layer $\mathbf{W} = \{\mathbf{w}_1, \mathbf{w}_2, \ldots, \mathbf{w}_K\} \subset \Re^d$ and the input pattern $\mathbf{x} \subset \Re^d$. Thus, it belongs to the category of hard competitive learning. A set of input patterns $\mathbf{X} = \{\mathbf{x}_1, \mathbf{x}_2, \ldots, \mathbf{x}_N\}$ is repeatedly presented to the LVQ network until the termination is met. However, because unsupervised LVQ lacks a definite clustering object, the resulting prototypes are not guaranteed to be a good representation of the clusters in the data (Pal et al., 1993). Fig. 5.10 provides a description of the basic steps for unsupervised LVQ.

In order to overcome the limitations of unsupervised LVQ, such as sensitivity to initiation, which is associated with hard competitive learning, a generalized LVQ (GLVQ) algorithm for clustering was proposed (Pal et al., 1993) by explicitly inferring the learning rules from the optimization of a cost function. Soft competitive learning is adopted for the neurons in the competitive layer so that every prototype vector could be updated during the presentation of an input pattern. Supposing that J is the neuron that wins the competition upon the presentation of the input pattern $\mathbf{x}$, the cost function $J(\mathbf{W}, \mathbf{x})$ is then defined on the locally weighted error between $\mathbf{x}$ and J,

$$J(\mathbf{W}, \mathbf{x}) = \sum_{i=1}^{K} v_{Ji} \|\mathbf{x} - \mathbf{w}_i\|^2, \tag{5.66}$$

where

$$v_{Ji} = \begin{cases} 1, & \text{if } i = J \\ \dfrac{1}{\sum_{i=1}^{K} \|\mathbf{x} - \mathbf{w}_i\|^2}, & \text{otherwise.} \end{cases} \tag{5.67}$$

Formally, for a finite set of input patterns $\mathbf{X} = \{\mathbf{x}_1, \mathbf{x}_2, \ldots, \mathbf{x}_N\}$, we could construct the clustering problem as seeking a set of prototype vectors $\mathbf{W} = \{\mathbf{w}_1, \mathbf{w}_2, \ldots, \mathbf{w}_K\}$ in order to minimize the expectation function

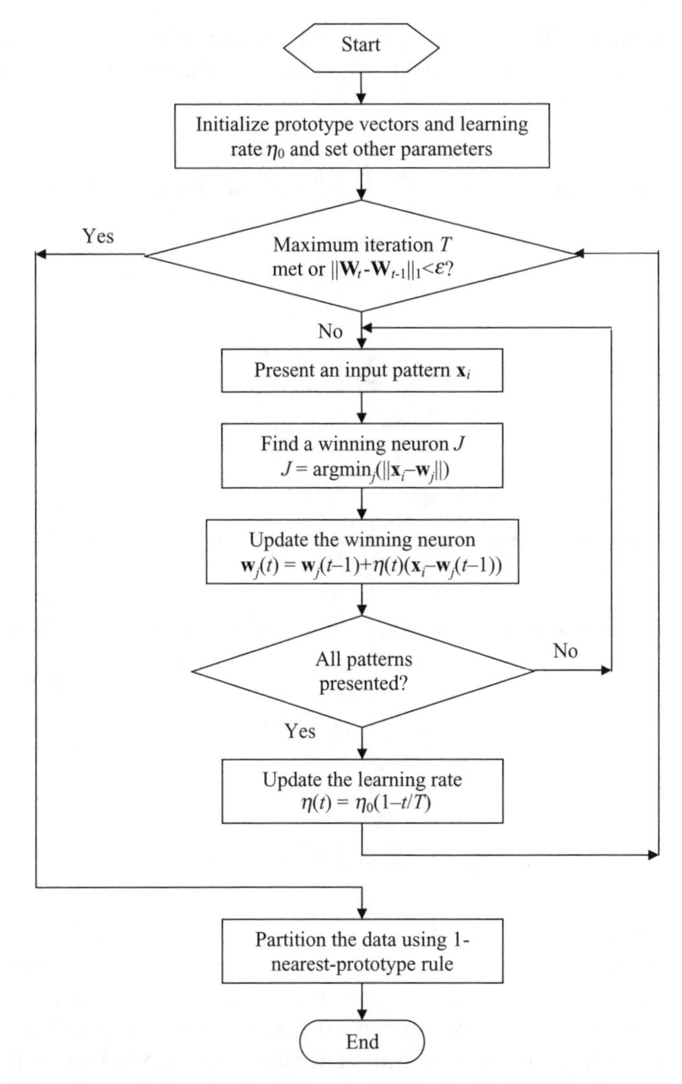

Fig. 5.10. Flowchart of the standard form of unsupervised LVQ. Input patterns are repeatedly presented to the LVQ network until the maximum iteration is met or the change of prototype vectors is below a prespecified threshold.

$$\Gamma(\mathbf{W}) = \frac{\sum_{l=1}^{N}\sum_{i=1}^{K} v_{Ji}\|\mathbf{x}_l - \mathbf{w}_i\|^2}{N} = \frac{\sum_{l=1}^{N} J(\mathbf{W}, \mathbf{x}_l)}{N} \tag{5.68}$$

It is shown that $\Gamma(\mathbf{W})$ can be minimized via the steepest gradient descent search of $J(\mathbf{W}, \mathbf{x}_l)$ (Pal et al., 1993), which leads to the following learning rules at iteration t

$$\mathbf{w}_J(t+1) = \mathbf{w}_J(t) + \eta(t)\frac{D^2 - D + \|\mathbf{x} - \mathbf{w}_J(t)\|^2}{D^2}(\mathbf{x} - \mathbf{w}_J(t)), \text{ for winner } J, \tag{5.69}$$

where

$$D = \sum_{i=1}^{K} \|\mathbf{x} - \mathbf{w}_i\|^2, \tag{5.70}$$

and

$$\mathbf{w}_j(t+1) = \mathbf{w}_j(t) + \eta(t)\frac{\|\mathbf{x} - \mathbf{w}_J(t)\|^2}{D^2}(\mathbf{x} - \mathbf{w}_j(t)), \quad \text{for } j \neq J. \tag{5.71}$$

In particular, the learning rate $\eta(t)$ should meet two conditions, illustrated as

$$t \to \infty, \eta(t) \to 0, \tag{5.72}$$

and

$$\sum_t \eta(t) \to \infty, \tag{5.73}$$

in order to prevent the prototype vectors from oscillating (Pal et al., 1993). The learning rate series used in LVQ provides such a potential choice. From Eqs. 5.69 and 5.71, we can also observe that GLVQ is identical to LVQ when the input pattern matches exactly with the winning prototype. Otherwise, the influence of the input pattern on the other neurons is dependent on the degree of its match with the winning neuron, which exerts a change to the learning rate. Gonzalez et al. (1995) and Karayiannis et al. (1996) further identified several situations in which the behaviors of GLVQ become inconsistent. Particularly, the fuzzy memberships of input patterns are used to replace the weighted factors in the definition of cost function in Eq. 5.66 in order to make the algorithm invariant to data scaling (Karayiannis et al., 1996). The resulting learning rules lead to an infinite family of competitive learning paradigms, called GLVQ-F.

Another member of LVQ that incorporates the fuzzy membership function with the learning rule is the fuzzy LVQ (FLVQ), originally known as a fuzzy Kohonen clustering network (FKCN) (Tsao et al., 1994). Different from the unsupervised LVQ or GLVQ-F, FLVQ is a batch clustering algorithm, where learning occurs only after all the input patterns are presented. Owing to the integration of the fuzzy memberships, FLVQ can automatically determine the size of the update neighborhood without the requirement of defining the neighborhood, as used in SOFM discussed in the next section.

Using the fuzzy membership coefficient u_{ji} ($j = 1, \ldots, K, i = 1, \ldots, N$) of $\mathbf{x}_i$ in cluster j defined in fuzzy c-means algorithm (FCM) in Chapter 4, we have

$$u_{ji} = \frac{1}{\sum_{l=1}^{K} \left(\|\mathbf{x}_i - \mathbf{w}_j\|^2 / \|\mathbf{x}_i - \mathbf{w}_l\|^2 \right)^{\frac{1}{m-1}}} \forall j, i, \quad (5.74)$$

where m is the weighting exponent. The key idea of FLVQ is to obtain the learning rate η_{ji} at iteration t via u_{ji}, written as

$$\begin{aligned} \eta_{ji}(t) &= (u_{ji}(t))^{m(t)} \\ &= \left(\frac{1}{\sum_{l=1}^{K} \left(\|\mathbf{x}_i - \mathbf{w}_j(t)\|^2 / \|\mathbf{x}_i - \mathbf{w}_l(t)\|^2 \right)^{\frac{1}{m(t)-1}}} \right)^{m(t)} \quad \forall j, i. \end{aligned} \quad (5.75)$$

Note that the weighting exponent m is no longer fixed and varies with the iteration,

$$m(t) = m_0 + t((m_f - m_0)/T), \quad (5.76)$$

where both final and initial parameters, m_f and m_0, are greater than 1, and T is the maximum number of iterations. By choosing different values of m_f and m_0, three families of FLVQ can be derived (Bezdek and Pal, 1995):

- Descending FLVQ, where $m_0 > m_f \Rightarrow \{m(t)\} \downarrow m_f$;
- Ascending FLVQ, where $m_f > m_0 \Rightarrow \{m(t)\} \uparrow m_f$;
- FLVQ ≡ FCM, where $m_f = m_0 \Rightarrow \{m(t)\} = m_f = m_0$.

In practice, it is preferred to use descending FLVQ and the values of m_f and m_0 were recommended to be restricted as $1.1 < m_f < m_0 < 7$ (Bezdek and Pal, 1995). The asymptotic behaviors of FLVQ when $m(t)$ approaches either of two extremes (1 or ∞) were further investigated (Baraldi et al., 1998). The relations of FLVQ with another learning algorithm, called the soft competition scheme (SCS) (Yair et al., 1992) were also discussed in terms of the properties of their learning rates (Bezdek and Pal, 1995).

At this point, we can write the learning rule of FLVQ as

$$\mathbf{w}_j(t+1) = \mathbf{w}_j(t) + \sum_{i=1}^{N} \eta_{ji}(t)(\mathbf{x}_i - \mathbf{w}_j(t)) \Big/ \sum_{l=1}^{N} \eta_{jl}(t). \quad (5.77)$$

Again, we can see that this rule is equivalent to the update equation of prototype vectors used in FCM when $m(t)$ is fixed. Moreover, a family of batch LVQ algorithms was introduced, known as an extended FLVQ family (EFLVQ-F), by explicitly inferring the learning rules from the minimization of the proposed cost function with gradient descent search (Karayiannis and Bezdek, 1997). The cost function is defined as the average generalized mean between the prototypes and the input pattern, which can also be constructed via a broad class of aggregation operators, such as ordered weighted operators (Karayiannis, 2000). EFLVQ-F generalizes the learning rule of FLVQ and FCM that have a restricted weighting exponent under certain conditions. A class of incremental fuzzy algorithms for LVQ (FALVQ) was also developed, including the FALVQ1, FALVQ2, and FALVQ3 family of algorithms (Karayiannis, 1997; Karayiannis and Pai, 1996).

5.3.3 Self-Organizing Feature Maps

SOFMs developed from the work of von der Malsburg (1973), Grossberg (1976 a, b, 1978), and Kohonen (1989, 1990). In his study of the visual cortex, von der Malsburg (1973) first introduced a self-organizing model with the corresponding learning rule. The learning law requires an additional step to normalize the weights associated with the neurons after each update. Grossberg (1976 a, b, 1978) further extended and developed the self-organizing model by using the instar learning rule, defining the correct recurrent competitive dynamics, and proving a number of the main properties in his stable sparse learning theorem. The instar rule guarantees the automatic normalization of the weights as long as the input patterns are normalized. SOFM's learning rule, suggested by Kohonen (1989, 1990), is related to the instar rule by allowing the reinforcement of all neurons in the neighborhood of the winning neuron.

Basically, the objective of SOFM is to represent high-dimensional input patterns with prototype vectors that can be visualized in, usually, a two-dimensional lattice structure, or sometimes a one-dimensional linear structure, while preserving the proximity relationships of the original data as much as possible (Kohonen, 1990, 2001). Each unit in the lattice is called a neuron, and the input patterns are fully connected to all neurons via adaptable weights, as depicted in Fig. 5.11. During training, neighboring input patterns are projected into the lattice, corresponding to adjacent neurons. These adjacent neurons are connected to each other, giving a clear topology of how the network fits into the input space. Therefore, the regions with a high probability of occurrence of sampled patterns will be represented by larger areas in the feature map (Haykin, 1999). In this sense, some authors prefer to think of SOFM as a method of displaying latent data structures in a visual way rather than through a clustering approach (Pal et al., 1993). However, SOFM could be integrated with other clustering approaches, such as K-means or hierarchical clustering, to reduce the computational cost and provide fast clustering (Vesanto and Alhoniemi, 2000).

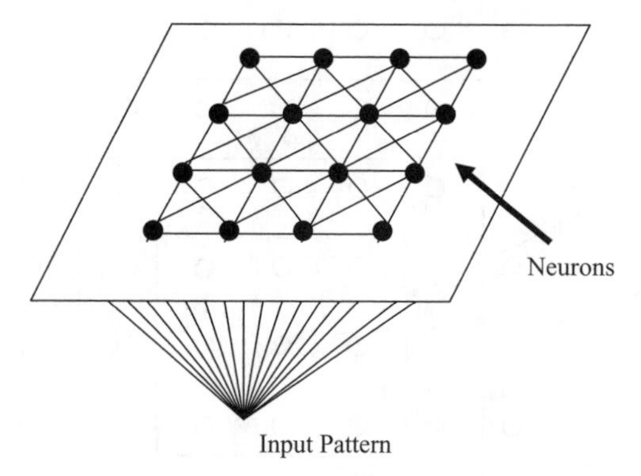

Fig. 5.11. Basic architecture of SOFM. The input pattern is fully connected to all neurons placed on a two-dimensional lattice. The size of the network (4×4 here) must be determined in advance.

After the competition among the neurons is complete, SOFM updates a set of weight vectors within the neighborhood of the winning neuron. Learning will not occur for the neurons lying outside the neighborhood. The neighborhood is determined in a topological sense, making it different from FLVQ, which uses metric neighbors in the input space (Baraldi et al., 1998). Moreover, the size of the neighborhood is designed to decrease monotonically with time, as shown in Fig. 5.12 (Kohonen, 2001). Given a winning neuron J upon the presentation of an input pattern $\mathbf{x}$, its updating neighborhood Ω_J starts with a wide field and gradually shrinks with time until there are no other neurons inside, i.e., $\Omega_J = \emptyset$. Correspondingly, the learning paradigm transits from soft competitive learning, which updates a neighborhood of neurons, to hard competitive learning, which only updates the winner. More specifically, we can write the updating equation for a neuron j at iteration t as,

$$\mathbf{w}_j(t+1) = \begin{cases} \mathbf{w}_j(t) + \eta(t)(\mathbf{x} - \mathbf{w}_j(t)), & \text{if } j \in \Omega_J(t) \\ \mathbf{w}_j(t), & \text{if } j \notin \Omega_J(t) \end{cases}, \tag{5.78}$$

where $\eta(t)$ is the monotonically decreasing learning rate. Alternately, by using the neighborhood function $h_{Jj}(t)$, Eq. 5.78 could be rewritten as,

$$\mathbf{w}_j(t+1) = \mathbf{w}_j(t) + h_{Jj}(t)(\mathbf{x} - \mathbf{w}_j(t)). \tag{5.79}$$

Here, the neighborhood function is defined as

$$h_{Jj}(t) = \begin{cases} \eta(t), & \text{if } j \in \Omega_J(t) \\ 0, & \text{if } j \notin \Omega_J(t) \end{cases}. \tag{5.80}$$

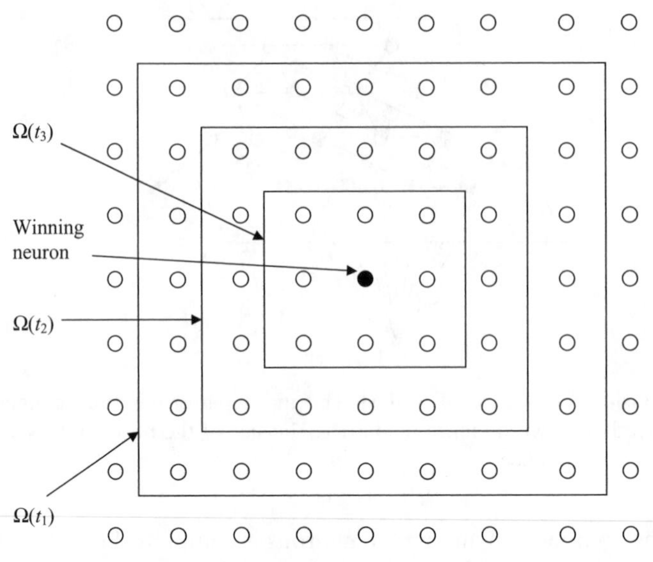

Fig. 5.12. Rectangular topological neighborhood Ω of SOFM. Ω decreases monotonically with time until it is empty. Hexagonal or other irregular topological neighborhoods could also be used.

More often, the neighborhood function takes the form of a Gaussian function that is appropriate for representing the biological lateral interaction, shaped as a bell curve (Kohonen, 1990),

$$h_{Jj}(t) = \eta(t)\exp\left(\frac{-\|\mathbf{r}_J - \mathbf{r}_j\|^2}{2\sigma^2(t)}\right), \tag{5.81}$$

where $\mathbf{r}_J$ and $\mathbf{r}_j$ represent the positions of the corresponding neurons on the lattice and $\sigma(t)$ is the monotonically decreasing kernel width function. A possible choice of $\sigma(t)$ is

$$\sigma(t) = \sigma_0 \exp(-t/\tau), \tag{5.82}$$

where σ_0 is the initial value and τ is the time constant (Haykin, 1999).

We now summarize the basic steps of SOFM, whose flowchart is depicted in Fig. 5.13.

1. Determine the topology of the SOFM. Initialize the weight vectors $\mathbf{w}_j(0)$ for $j = 1, \dots, K$, randomly;
2. Present an input pattern $\mathbf{x}$ to the network. Choose the winning node J that has the minimum Euclidean distance to $\mathbf{x}$, i.e.

$$J = \arg\min_j (\|\mathbf{x} - \mathbf{w}_j\|) \tag{5.83};$$

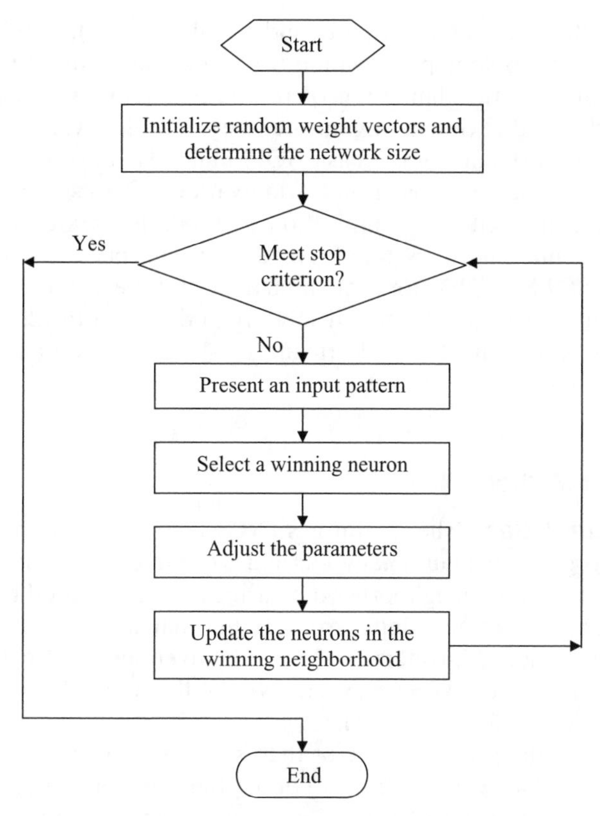

Fig. 5.13. Flowchart of SOFM. The algorithm repeatedly adapts the weight vectors until the change of neuron position is below a prespecified threshold.

3. Calculate the current learning rate and size of the neighborhood;
4. Update the weight vectors of all the neurons in the neighborhood of J using Eq. 5.79;
5. Repeat steps 2 to 4 until the change of neuron position is below a prespecified small positive number.

While SOFM enjoys the merits of input space density approximation and independence of the order of input patterns in a batch mode (Kohonen, 2001), the requirement for determining the size of the lattice, i.e., the number of clusters, in advance and the inefficiency of handling noise and outliers may limit its applications in some real-world problems (Baraldi and Blonda, 1999). As pointed out by Haykin (1999), trained SOFM may suffer from input space density misrepresentation, where areas of low pattern density may be over-represented and areas of high density under-represented. Moreover, because SOFM only aims to preserve the topology of the inverse mapping from the lattice to the input manifold, but not necessarily of the mapping from the input manifold to the network (Martinetz and Schulten, 1994), SOFM no longer provides perfect topology preservation when the dimension of the input patterns is larger than the dimension of the output network. Kohonen (2001) provided a detailed review of a great number of SOFM variants in order to improve the performance of the basic SOFM and also to broaden its applications. These methods are based on a wide variety of considerations, such as different matching criteria, optimization methods for search improvement, dynamically defined network topology, adaptive-subspace SOFM, evolutionary-learning SOFM, SOFM for sequential data, and speed-up of SOFM. Some recent advances can also be found in Berglund and Sitte (2006), Hammer et al. (2004), Kohonen (2006), Porrmann et al. (2003), Seo and Obermayer (2004), and Su and Chang (2000).

5.3.4 Other Methods

5.3.4.1 Neural Gas The neural gas (NG) algorithm also belongs to the class of self-organizing neural networks and is capable of adaptively determining the updating of the neighborhood by using a neighborhood ranking of the prototype vectors within the input space, rather than a neighborhood function in the output lattice (Martinetz et al., 1993). Given an input pattern $\mathbf{x} \in \Re^d$, the K prototype vectors $\mathbf{W} = \{\mathbf{w}_1, \mathbf{w}_2, \ldots, \mathbf{w}_K\}$ of the network are sorted based on their Euclidean distance to $\mathbf{x}$, i.e., $(j_0, j_1, \ldots, j_{K-1})$ is a sequence of indices such that $\mathbf{w}_{j0}$ is the prototype vector that is closest to $\mathbf{x}$ and $\mathbf{w}_{jK-1}$ is the one that is farthest from $\mathbf{x}$. Let $k_j(\mathbf{x}, \mathbf{W})$ denote the number k associated with a prototype vector $\mathbf{w}_j$, where $k = 0, \ldots K - 1$ is the prototype vector for which there exist k vectors $\mathbf{w}_i$. With

$$\|\mathbf{x} - \mathbf{w}_i\| < \|\mathbf{x} - \mathbf{w}_{jk}\|, \tag{5.84}$$

we can derive the learning rule of the prototype vectors via the optimization of a global cost function (Martinetz et al., 1993), defined as,

$$J(\mathbf{W}, \lambda) = \frac{1}{\sum_{k=0}^{K-1} h_\lambda(k)} \sum_{j=1}^{K} \int h_\lambda(k_j(\mathbf{x}, \mathbf{W}))(\mathbf{x} - \mathbf{w}_j)^2 P(\mathbf{x}) d^d\mathbf{x}, \tag{5.85}$$

where $P(\mathbf{x})$ represents the probability distribution of the data points, λ is a characteristic decay constant, and $h_\lambda(k_j(\mathbf{x}, \mathbf{W}))$ is a bell-shaped curve, written as

$$h_\lambda(k_j(\mathbf{x}, \mathbf{W})) = \exp(-k_j(\mathbf{x}, \mathbf{W})/\lambda). \tag{5.86}$$

Note that the cost function is closely related to the framework of fuzzy clustering (Bezdek, 1981) discussed in Chapter 4. The value

$h_\lambda(k_j(\mathbf{x}, \mathbf{W})) \Big/ \sum_{j=1}^{K} h_\lambda(k_j)$ is regarded as the degree of membership that the input pattern belongs to all the prototype vectors. It has the same function as the fuzzy membership u_{ij} defined in the fuzzy c-means algorithm.

By using the gradient descent method on the cost function, the prototype vectors are updated as

$$\mathbf{w}_j(t+1) = \mathbf{w}_j(t) + \eta(t) h_\lambda(k_j(\mathbf{x}, \mathbf{W}))(\mathbf{x} - \mathbf{w}_j(t)). \tag{5.87}$$

A batch version of NG, equal to the minimization of the cost function with the Newton method, was also proposed (Cottrell et al., 2006). Typically, both learning rate η and characteristic decay constant λ monotonically decreases with time according to a cooling scheme, such as

$$\eta(t) = \eta_0 (\eta_f / \eta_0)^{t/T}, \tag{5.88}$$

where $\eta_0 > \eta_f > 0$ are the initial and final learning rates and T is the maximum number of iterations, and

$$\lambda(t) = \lambda_0 (\lambda_f / \lambda_0)^{t/T}, \tag{5.89}$$

where $\lambda_0 > \lambda_f$ are the initial and final decay constants. The designs, together with the definition of the h function, assure a gradual decrease of the number of updating neurons and the adjusting strength. As λ approximates zero, the learning rule in Eq. 5.87 becomes equivalent to that of hard competitive learning.

The major process of the NG algorithm is as follows:

1. Initialize a set of prototype vectors $\mathbf{W} = \{\mathbf{w}_1, \mathbf{w}_2, \ldots, \mathbf{w}_K\}$ randomly;
2. Present an input pattern $\mathbf{x}$ to the network. Sort the index list in order from the prototype vector with the smallest Euclidean distance from $\mathbf{x}$ to the one with the greatest distance from $\mathbf{x}$;
3. Calculate the current learning rate and $h_\lambda(k_j(\mathbf{x}, \mathbf{W}))$ using Eqs. 5.88, 5.89, and 5.86. Adjust the prototype vectors using the learning rule in Eq. 5.87;
4. Repeat steps 2 and 3 until the maximum number of iterations is reached.

For NG, the number of prototype vectors still must be decided in advance, which leads to the development of a dynamic NG algorithm, called plastic NG, whose converge properties were also investigated (Ridella et al., 1998). The NG algorithm can also be combined with the competitive Hebbian rule to construct models of topology-preserving maps (Martinetz and Schulten, 1994).

5.3.4.2 Growing Neural Gas Growing neural gas (GNG) (Fritzke, 1995, 1997) combines the growth mechanism inherited from growing cell structures (GCS) (Fritzke, 1994) with the topology generation rules of competitive Hebbian learning (Martinetz and Schulten, 1994). Each prototype vector $\mathbf{w}_j$ is connected with its topological neighborhood via a set of edges to form an induced Delaunay triangulation (Martinetz and Schulten, 1994). Upon the presentation of an input pattern $\mathbf{x}$, an edge between the two closest prototype vectors with respect to $\mathbf{x}$ is created. An edge removal mechanism, called the edge aging scheme, is designed to discard the edges that are no longer valid in the subgraph due to the adaptation of the corresponding neurons. When prototype learning occurs, not only is the prototype vector of the winning neuron J_1 updated towards $\mathbf{x}$, but the prototypes within its topological neighborhood N_{J_1} are also adapted, although with a smaller updating strength. Different from NG, GCS, or SOFM, which require a fixed network dimensionality *a priori*, GNG is developed as a self-organizing network that can dynamically increase (usually) and remove the number of neurons in the network. A succession of new neurons is inserted into the network every λ iterations near the neuron with the maximum accumulated error. At the same time, a neuron removal rule could also be used to eliminate the neurons featuring the lowest utility for error reduction (Baraldi and Blonda, 1999). This utility measures the increase in overall distortion error caused by the removal of the neuron of interest. Moreover, the edge aging scheme also provides a way to detect and remove the neurons that are inactive over a predefined number of iterations.

The complete GNG algorithm proceeds with the following steps:

1. Initialize a set of prototype vectors (typically 2) $\mathbf{W} = \{\mathbf{w}_1, \mathbf{w}_2\}$ randomly and a connection set $\mathbf{C}$ to empty, $\mathbf{C} \subset \mathbf{W} \times \mathbf{W}$ and $\mathbf{C} = \varnothing$;
2. Present an input pattern $\mathbf{x}$ to the network. Choose the winning neuron J_1 and the second winning neuron J_2 according to the Euclidean distance to $\mathbf{x}$

$$J_1 = \arg\min_{\mathbf{w}_j \in \mathbf{W}} \|\mathbf{x} - \mathbf{w}_j\|, \quad (5.90)$$

$$J_2 = \arg\min_{\mathbf{w}_j \in \mathbf{W} \setminus \{\mathbf{w}_{J_2}\}} \|\mathbf{x} - \mathbf{w}_j\|; \quad (5.91)$$

3. Create and add a connection between J_1 and J_2 into $\mathbf{C}$ if it does not already exist,

$$\mathbf{C} = \mathbf{C} \cup \{(J_1, J_2)\}. \quad (5.92)$$

Set the age of the connection between J_1 and J_2 to zero,

$$\text{age}(J_1, J_2) = 0; \quad (5.93)$$

4. Update the local error of the winning neuron J_1,

$$E_{J_1}(t) = E_{J_1}(t-1) + \|\mathbf{x} - \mathbf{w}_{J_1}(t)\|^2; \tag{5.94}$$

5. Update the prototype vectors of the winning neuron J_1 and its direct topological neighbors with the following rules:

$$\mathbf{w}_{J_1}(t+1) = \mathbf{w}_{J_1}(t) + \eta_w(\mathbf{x} - \mathbf{w}_{J_1}(t)), \tag{5.95}$$

   where η_w is the learning rate for the winning neuron,
   and,

$$\mathbf{w}_j(t+1) = \mathbf{w}_j(t) + \eta_n(\mathbf{x} - \mathbf{w}_j(t)), \forall j \in N(J_1), \tag{5.96}$$

   where $N(J_1)$ is the set of direct topological neighbors of J_1, and η_n is the corresponding learning rate;
6. Increase the age of all edges emanating from J_1 by one,

$$\text{age}(J_1, j) = \text{age}(J_1, j) + 1, \forall j \in N(J_1); \tag{5.97}$$

7. Remove the edges whose age values are larger than a prespecified threshold α_{max}. Remove also the neurons that no longer have emanating edges as a result of the previous operation;
8. If the number of iterations is an integer multiple of the prespecified parameter λ, insert a new neuron in the network,
   a. Choose the neuron *Je* with the maximum accumulated error,

$$J_e = \arg\max_{\mathbf{w}_j \in \mathbf{W}} E_j; \tag{5.98}$$

   b. Choose the neuron J_f in the neighborhood of J_e with the maximum accumulated error,

$$J_f = \arg\max_{j \in N(J_e)} E_j; \tag{5.99}$$

   c. Insert a new neuron J_n into the network,

$$\mathbf{W} = \mathbf{W} \cup \{\mathbf{w}_{J_n}\}, \tag{5.100}$$

   with the corresponding prototype vector initialized as

$$\mathbf{w}_{J_n} = \frac{\mathbf{w}_{J_e} + \mathbf{w}_{J_f}}{2}; \tag{5.101}$$

d. Insert edges that connect J_n with J_e and J_f, and remove the edge between J_e and J_f,

$$\mathbf{C} = (\mathbf{C} \backslash \{(J_e, J_f)\}) \cup \{(J_n, J_e), (J_n, J_f)\}; \tag{5.102}$$

e. Adjust the accumulated error of J_e and J_f,

$$E_{J_e} = E_{J_e} - \alpha E_{J_e}; \tag{5.103}$$

$$E_{J_f} = E_{J_f} - \alpha E_{J_f}; \tag{5.104}$$

where α is a small constant. Set the initial error of J_n,

$$E_{J_n} = \frac{E_{J_e} + E_{J_f}}{2}; \tag{5.105}$$

9. Decrease the accumulated error of all neurons,

$$E_j = E_j - \beta E_j, \forall \mathbf{w}_j \in \mathbf{W}; \tag{5.106}$$

where β is a constant.

10. Repeat steps 2 through 9 until the allowed maximum network size is reached or the mean accumulated error is smaller than a pre-specified threshold.

An extension of GNG for achieving better robustness is possible by incorporating the outlier resistant strategy, adaptive learning rates, and the cluster repulsion scheme into the original algorithm (Qin and Suganthan, 2004). The grow when required (GWR) network inserts new neurons into the network whenever the presented input pattern is not represented well by any existing neurons, instead of maintaining a fixed number of iterations as in GNG (Marsland et al., 2002). It is also interesting to note that both GNG and GCS can be modified for supervised classification (Fritzke, 1994; Heinke and Hamker, 1998).

5.4. APPLICATIONS

5.4.1 Neural Information Retrieval System for Group Technology

In the manufacturing industry, it is important to avoid unnecessary redesigning of parts, which can be costly in both time and money. Group technology refers to the study and implementation of information retrieval systems that can

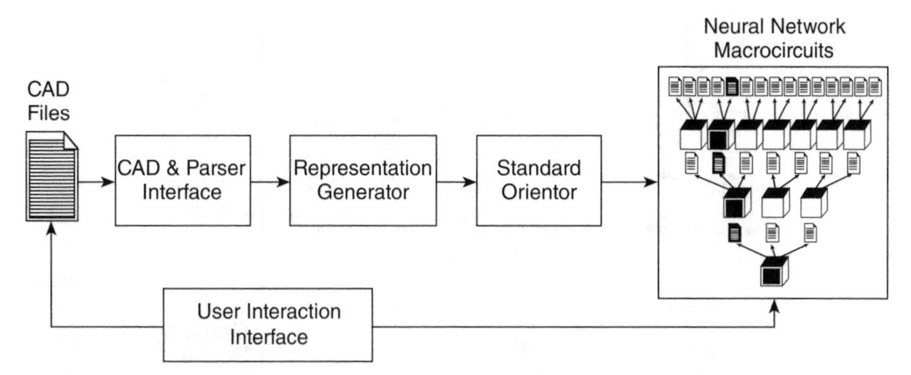

Fig. 5.14. A schematic of the components of a neural information retrieval system for design retrieval applications. The user interaction interface provides control of the abstraction level and discrimination features within the network during retrieval. (Reprinted from Neural Networks, vol. 7, T. Caudell, S. Smith, R. Escobedo, and M. Anderson. NIRS: Large scale ART-1 neural architectures for engineering design retrieval, pp. 1339–1350, Copyright © 1994, with permission from Elsevier.)

retrieve, store, and compare designs. A neural information retrieval system using ART1 networks was developed for application to this problem of group technology (Caudell et al., 1991, 1994). Two- or three-dimensional representations of engineering designs are input to ART1 to produce groups or families of similar parts. A new part design is then queried and compared with these families to prevent duplication of design efforts.

Figure 5.14 describes the generic system architecture for group technology applications, which includes five basic components:

1. CAD system interface and parser;
2. Feature representation generator;
3. Standard orientor;
4. Neural network macrocircuits; and
5. User interaction interface.

After the design of a part has been completed, a list of instructions on how to draw and annotate a diagram of the part is stored in a file. The parser extracts the salient information from the CAD system interface, which the representation generator then converts and compresses into a form usable by ART1. A modified version of ART1 is used in the system in order to provide direct operations on compressed codes of the input vectors and the memory prototypes (Caudell et al., 1991). The orientor assures that similar parts are represented at similar locations and orientations within the graphics viewport.

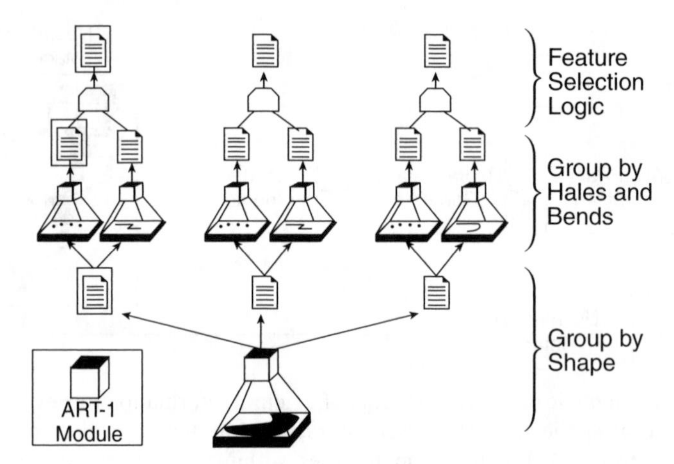

Fig. 5.15. The macrocircuit of ART1 modules that implement a feature selection option for the sheet metal floor stiffener design retrieval system. (Reprinted from Neural Networks, vol. 7, T. Caudell, S. Smith, R. Escobedo, and M. Anderson. NIRS: Large scale ART-1 neural architectures for engineering design retrieval, pp. 1339–1350, Copyright © 1994, with permission from Elsevier.)

The structure of ART1 macrocircuits provides an implementation of feature selection, as illustrated in Fig. 5.15. The detailed structure of this macrocircuit evolves during training, during which a training set of part designs is repetitively presented to the networks. Within this macrocircuit, shape representation is considered first by the lowest "shape" ART1 module. For each cluster formed by this module, an additional pair of ART1 modules, known as the "holes" and "bends" modules, is spawned for secondary training. After learning has stabilized in the shape module, those parts from the training set assigned to a shape cluster are used to separately train the pair of holes and bends ART1 modules associated with the cluster. The logic at the very top of the figure intersects the names on the lists. This architecture makes it possible to choose to discriminate based on shape alone, shape and holes, shape and bends, or shape, bends, and holes.

Another user requirement is the ability to specify on-line among degrees of family similarity. This ability can be implemented with a hierarchical abstraction tree of macrocircuit modules, as shown in Fig. 5.16. Each module in the tree is trained separately with the input patterns associated only with that branch cluster, and each module receives the complete set of feature representations, depicted in Fig. 5.15. The modules at the top of the tree have the greatest discrimination, while the one at the bottom has the least. When a query occurs, the lowest module places the design into one of its families or clusters. Families at this level represent the most general abstraction of the possible set of designs stored in the system. When a winning cluster is selected at the first level, the appropriate module within the next level is activated. This module

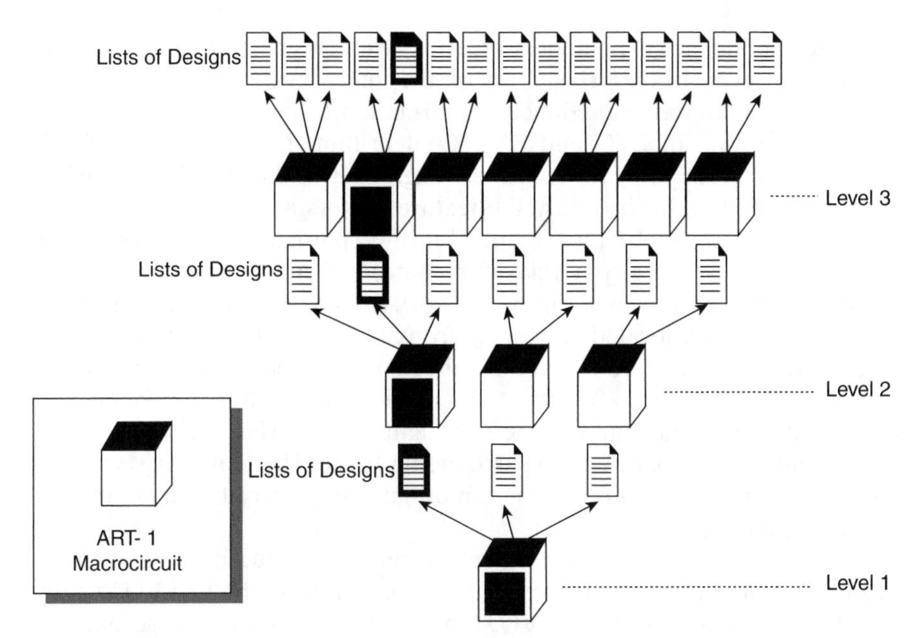

Fig. 5.16. An example of an ART tree database structure. Each cube represents a macrocircuit of ART1 neural networks to provide the feature selection option to query. (Reprinted from Neural Networks, vol. 7, T. Caudell, S. Smith, R. Escobedo, and M. Anderson. NIRS: Large scale ART-1 neural architectures for engineering design retrieval, pp. 1339–1350, Copyright © 1994, with permission from Elsevier.)

places the design into one of its clusters, and the process repeats. The user selects the level of abstraction at retrieval time according to the requirements of the current design.

5.4.2 Magnetic Resonance Imaging Segmentation

Magnetic resonance imaging (MRI) provides a visualization of the internal tissues and organs in the living organism, which is valuable in its applications in disease diagnosis (such as cancer and heart and vascular disease), treatment and surgical planning, image registration, location of pathology, and so on. As an important image-processing step, MRI segmentation aims to partition an input image into significant anatomical areas, each of which is uniform and homogeneous according to certain image properties (Bezdek et al., 1993; Pham et al., 2000). MRI segmentation can be formulated as a clustering problem in which a set of feature vectors, which are obtained through transforming image measurements and positions, is grouped into a relatively small number of clusters (Jain et al., 1999; Karayiannis, 1997). Thus, each cluster, represented with a prototype vector in the feature space, corresponds to an image segment. Since the number of clusters is much smaller than the number of intensity levels in the original image, the unsupervised clustering process provides a way to remove redundant information from the MR images.

An application of LVQ and FALVQ algorithms in the segment of MR images of the brain of a patient with meningioma was illustrated in Karayiannis and Pai (1999). In this study, the feature vectors at every image location are composed of pixel values of the T1 (the spin-lattice relaxation time)-weighted, T2 (the spin-spin relaxation time)-weighted, and SD (the spin density) images, as shown in Fig. 5.17(a)–(c), respectively. According to these images, the tumor is located in the right frontal lobe, i.e., the upper-left quarter of the MR images. More specifically, the tumor appears bright on the T2-weighted image and dark on the T1-weighted image. After the patient was given Gadolinium, the tumor on the T1-weighted image (Fig. 5.17(d)) becomes very bright and is isolated from surrounding tissue. There also exists a large amount of edema surrounding the tumor, appearing very bright on the T2-weighted image.

Figure 5.18(a)–(d) show the segmented images produced using LVQ, the algorithm from the FALVQ1 family, the algorithm from the FALVQ2 family, and the algorithm from the FALVQ3 family, respectively. In all these analyses, the number of clusters is set as 8, which leads to 8 different segments. From the results, one can see that although LVQ can identify the edema, it is unsuccessful in discriminating the tumor from the surrounding tissue. For all FALVQ families of algorithms, the performance is more promising as both the tumor and the edema can be identified successfully. More segmented results with different parameter selections of LVQ and FALVQ algorithms are given in Karayiannis (1997) and Karayiannis and Pai (1999).

5.4.3 Condition Monitoring of 3G Cellular Networks

The 3G mobile networks combine new technologies such as WCDMA and UMTS and provide users with a wide range of multimedia services and applications with higher data rates (Laiho et al., 2005). At the same time, emerging new requirements make it more important to monitor the states and conditions of 3G cellular networks. Specifically, in order to detect abnormal behaviors in 3G cellular systems, four competitive learning neural networks, LVQ, FSCL, SOFM (see another application of SOFM in WCDMA network analysis in Laiho et al. (2005)), and NG, were applied to generate abstractions or clustering prototypes of the input vectors under normal conditions, which are further used for network behavior prediction (Barreto et al., 2005). The input pattern vectors are composed of a set of selected key performance indicators (KPIs), which refer to the essential measurements that can be used to represent and summarize the behaviors of cellular networks, such as number of users, noise rise, downlink throughput, and other-cells interference, used in the study. Upon presentation of a new state input vector, the behavior of the cellular system is evaluated in terms of a hypothesis test based on the distribution

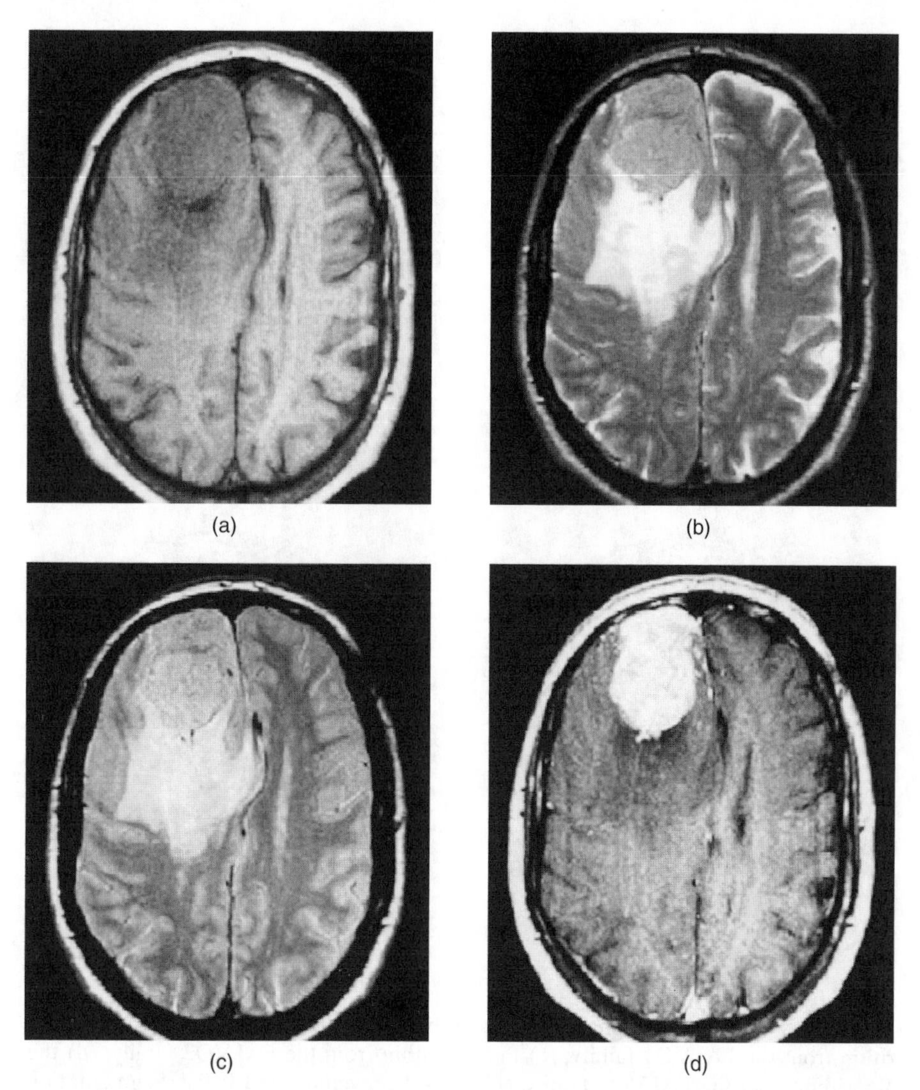

Fig. 5.17. Magnetic resonance image of the brain of an individual suffering from meningioma. (a) T1-weighted image; (b) T2-weighted image; (c) Spin-density image; (d) T1-weighted image after the patient was given Gadolinium. (From N. Karayiannis and P. Pai. Segmentation of magnetic resonance images using fuzzy algorithms for learning vector quantization. IEEE Transactions on Medical Imaging, vol. 18, pp. 172–180, 1999. Copyright © 1999 IEEE.)

of quantization errors of the normal state vectors, called global normality profiles. Supposing $\mathbf{x}_1, \ldots, \mathbf{x}_N$ is a set of state vectors, the quantization errors E_i, $i = 1, \ldots, N$ are calculated as

$$E_i = \|\mathbf{x}_i - \mathbf{w}_J\|, \tag{5.107}$$

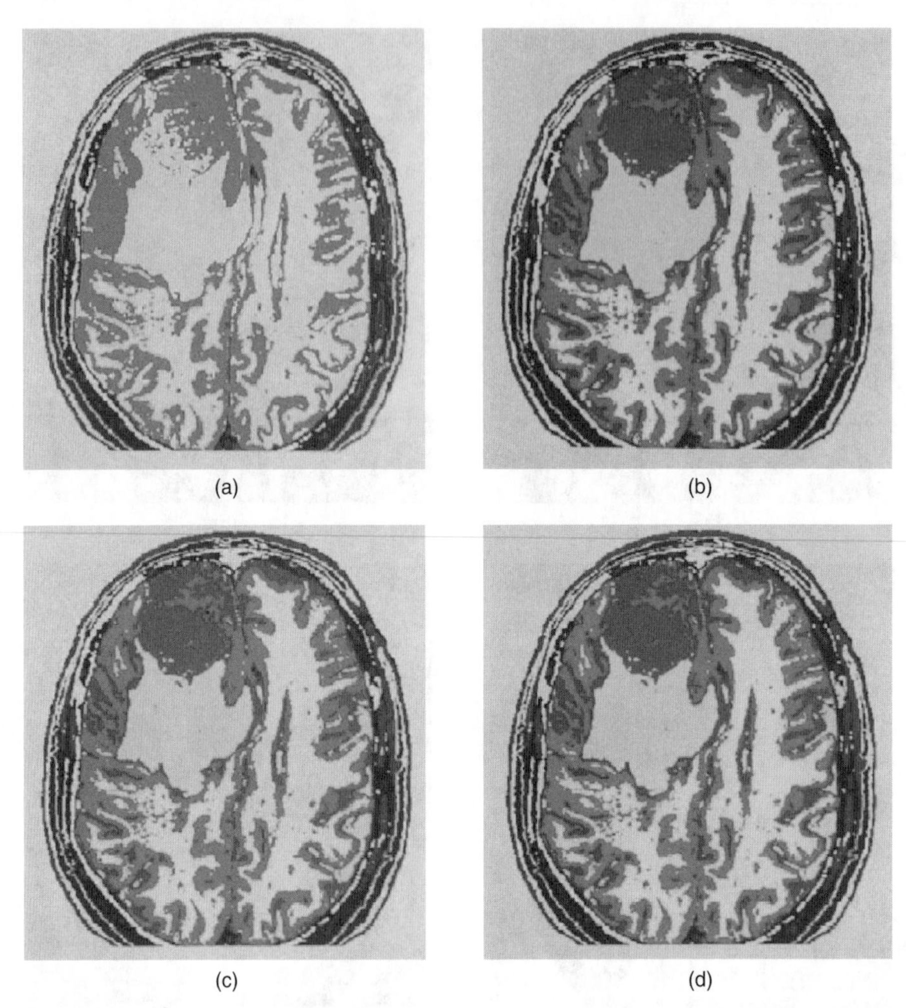

Fig. 5.18. Segmented MR images produced by (a) the LVQ algorithm; (b) the algorithm from the FALVQ1 family; (c) the algorithm from the FALVQ2 family; (d) the algorithm from the FALVQ3 family. (From N. Karayiannis and P. Pai. Segmentation of magnetic resonance images using fuzzy algorithms for learning vector quantization. IEEE Transactions on Medical Imaging, vol. 18, pp. 172–180, 1999. Copyright © 1999 IEEE.)

where $\mathbf{w}_J$ is the prototype vector associated with the winning neuron. By defining the upper limit E_{p}^{+} as the $100(1 + p)/2$ th percentile and the lower limit E_p^{-} as the $100(1 + p)/2^{\mathrm{th}}$ percentile, where p is a percentage of normal values of the variable found within the decision interval, the hypothesis for a new state vector $\mathbf{x}^*$ can be written as,

TABLE 5.1 Typical false alarm rates and intervals of normality for four competitive neural networks.

Model	CI, FA (95%)	CI, FA (99%)
LVQ	[0.366, 1.534], 12.43	[0.074, 1.836], 5.41
FSCL	[0.214, 1.923], 10.20	[0.136, 4.584], 1.80
SOM	[0.361, 1.815], 8.75	[0.187, 2.710], 1.43
NG	[0.277, 1.944], 9.50	[0.165, 4.218], 2.10

(From G. Barreto, J. Mota, L. Souza, R. Frota, and L. Aguayo. Condition monitoring of 3G cellular networks through competitive neural models. IEEE Transactions on Neural Networks, vol. 16, pp. 1064–1075, 2005. Copyright © 2005 IEEE.)

$$\begin{matrix} H_0 : \mathbf{x}^* \text{ is normal} \\ H_1 : \mathbf{x}^* \text{ is abnormal} \end{matrix}, \tag{5.108}$$

and the rejection rule for the above test is

$$\text{Reject } H_0 \quad \text{if } E^* \notin [E_p^-, E_p^+]. \tag{5.109}$$

Similarly, the abnormality of the components of the state vector that has abnormal behavior can also be evaluated.

Table 5.1 summarizes the simulation results for a network scenario of 100 mobile stations initially trying to connect to seven base stations, with the intervals of normality for 95% and 99% confidence levels. The false alarm refers to the incorrect rejection of the null hypothesis when it is true. The clustering prototypes provide a good summary of the normal behaviors of the cellular networks, which can then be used to detect abnormalities. Moreover, in general, SOFM and NG can achieve better performance than LVQ and FSCL.

5.4.4 Document Clustering

WEBSOM was developed as an information retrieval system based on SOFM (Honkela et al., 1997; Kaski et al., 1998). The object of WEBSOM is to meaningfully and automatically organize document collections and databases so that users can have an easier experience browsing and exploring them. In this system, documents are mapped onto a two-dimensional lattice of neurons, called a document map, and those that share similar contents will be represented by the same or neighboring neurons on the map. Links are created for each neuron to provide access to the corresponding documents.

Figure 5.19 summarizes the basic architecture of the WEBSOM system. The documents are first encoded as word histograms, which are then compressed via self-organizing semantic maps or word category maps to generate

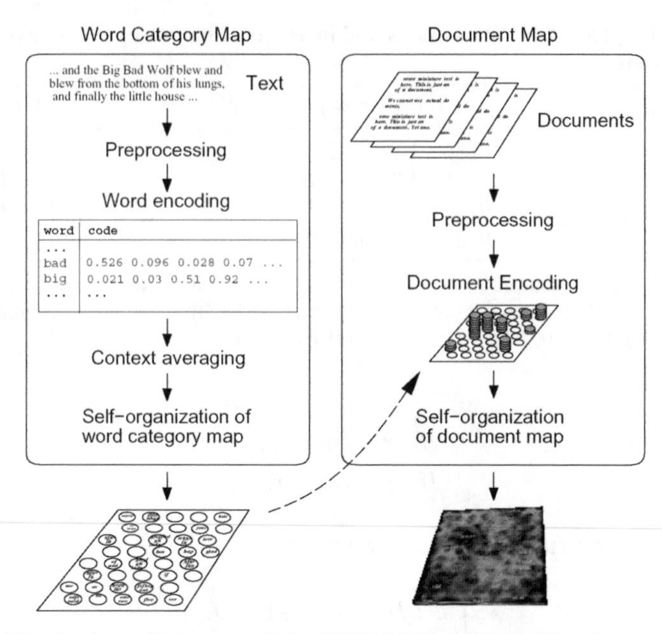

Fig. 5.19. The basic architecture of the WEBSOM method. The document map is organized based on documents encoded with the word category map. Both maps are produced with the SOFM algorithm. When the maps have been constructed, the processing of new documents is much faster. (From T. Honkela, S. Kaski, K. Lagus, and T. Kohonen, WEBSOM—Self-organizing maps of document collections, In Proceedings of Workshop on Self-Organizing Maps (WSOM'97), pp. 310–315, Helsinki University of Technology, Neural Networks Research Center, Espoo, Finland, by permission of T. Honkela.)

word clusters. The document map is finally formed with the reduced histograms as fingerprints of the documents. The modified version of WEBSOM, called WEBSOM2 (Fig. 5.20) (Kohonen et al., 2000), uses a random projection method to encode word histograms instead of word category maps. WEBSOM2 combines several speed-up methods, such as the initialization of large SOFM based on carefully constructed smaller ones and the parallelized batch map algorithm, in order to scale to large-scale document databases. In comparison with the original SOFM, the shortcut methods achieve a nine-tenths decrease of the computational time with comparable maps (Kohonen et al., 2000).

Figure 5.21 illustrates the results of a keyword search case of the application of WEBSOM in a database with 6,840,568 patent abstracts, written in English. The number of neurons on the map is up to 1,002,240, which is the largest WEBSOM map known so far. Given the keyword of interest, speech recognition in this case, the matching nodes are found from the built index, and the best matches are displayed as circles on the map, which are distributed into a

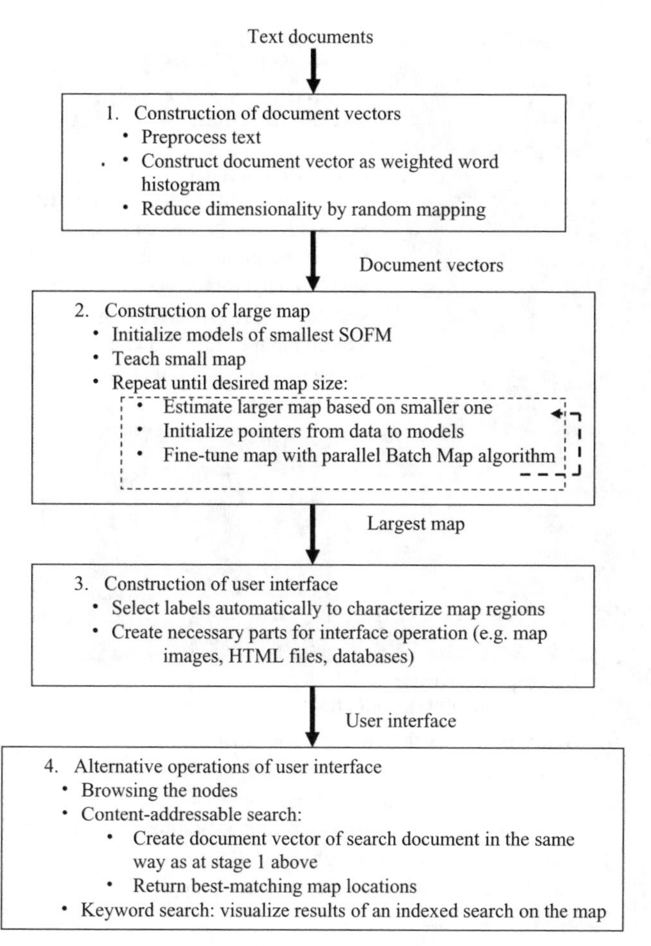

Fig. 5.20. Overview of the construction and operation of the WEBSOM2 system. (From T. Kohonen, S. Kaski, K. Lagus, J. Salojärvi, J. Honkela, V. Paatero, and A. Saarela. Self organization of a massive document collection. IEEE Transactions on Neural Networks, vol. 11, pp. 574–585, 2000. Copyright © 2000 IEEE.)

set of clusters, such as the two shown in Fig. 5.21. It is also interesting to see that the generated clusters could provide users with different aspects of the search keywords and unveil more relevant information with regard to the query. For instance, one cluster in Fig. 5.21 focuses on the signal aspect of speech, while the other shifts the emphasis to speech input devices as well as recognition methods. Another application using the Encyclopedia Britannica, consisting of 68,000 articles and 43,000 additional summaries, updates, and other miscellaneous material, was shown in Lagus et al. (2004). The size of the document map is 72×168, resulting in 12,096 nodes. An example of the usefulness of ordering the map in querying a topic is illustrated in

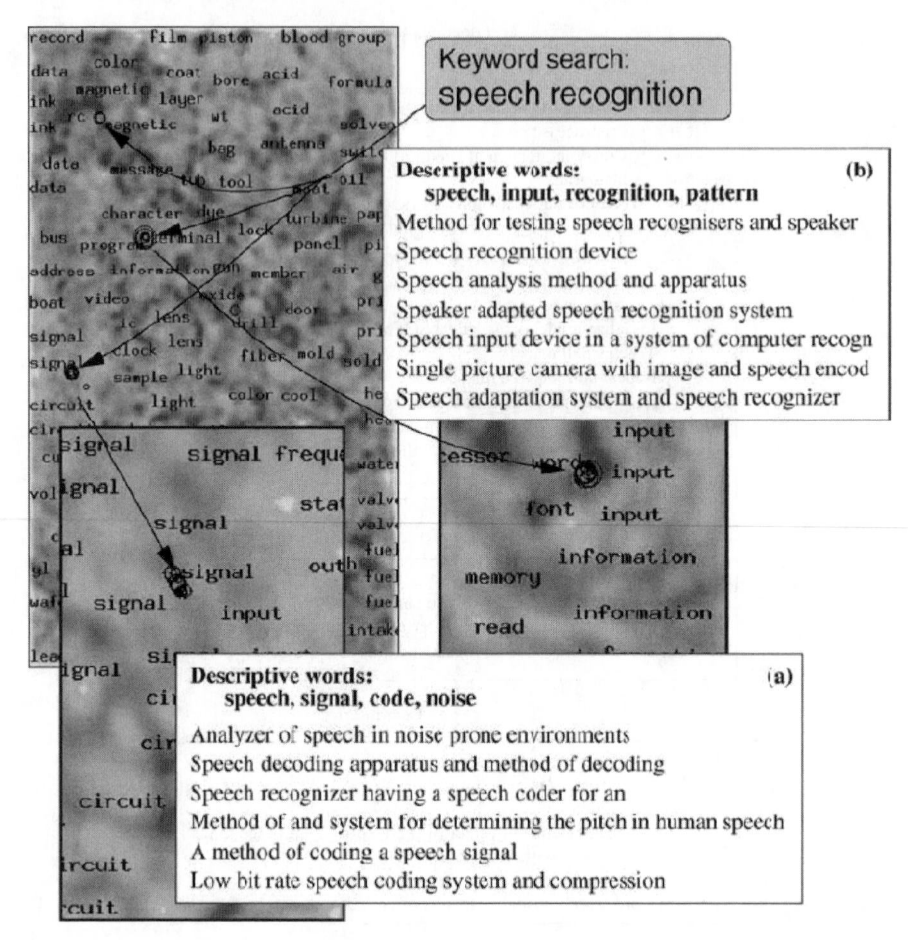

Fig. 5.21. A map of 7 million patent abstracts. The results reflect the search for the words "speech recognition." The best matches are marked with circles on the map display. A detailed view of an area appears upon clicking the map display. Two of the generated clusters are further examined, with their partial contents shown in the insets. Note that in the inset (a) the emphasis is on the signal aspect of speech, whereas in (b) there are patents concerning speech input devices as well as recognition methods. (Reprinted from Information Sciences, vol. 163, K. Lagus, S. Kaski, and T. Kohonen, Mining massive document collections by the WEBSOM method, pp. 135–156, Copyright © 2004, with permission from Elsevier.)

Fig. 5.22. The recent advances of WEBSOM and several demonstrations of WEBSOM in news bulletins and Usenet newsgroups can be accessed at http://websom.hut.fi/websom/.

5.4.5 Gene Expression Data Analysis

SOFM has been used to represent the gene expression space using a two-dimensional lattice of neurons (Tamayo et al., 1999; Törönen et al., 1999). One such SOFM application is on human hematopoietic differentiation modeled with four cell lines (HL-60, U937, Jurkat, and NB4 cells), where SOFM is implemented in a software package called GENECLUSTER (Tomayo et al., 1999). The process of hematopoietic differentiation is largely controlled at the transcriptional level, and blocks in the developmental program likely underlie the pathogenesis of leukemia. Fig. 5.23 shows the result when a 6×4 SOFM is applied to a data set with 1,036 genes that are involved with at least one of four cell lines. The time courses for four cell lines are shown as HL-60+PMA, U937+PMA, NB4+ATRA, and Jurkat+PMA, from left to right. Biologically interesting insights are observed from the formed clusters. For instance, cluster 21 consists of 21 genes induced in the closely related cell lines HL-60 and U937, while its neighboring clusters, 17 and 20 include genes induced in only one of the two cell lines. This suggests that although HL-60 and U937 have similar macrophage maturation responses to PMA simulation, they can still be distinguished by some transcriptional responses (Tomayo et al., 1999).

In another study on gene expression data analysis, FA was applied to the data during sporulation of *Saccharomyces cerevisiae* (Tomida et al., 2002). The data were obtained from a previous study by Chu et al. (1998) and consist of the expression levels of around 6,100 genes measured at 7 time points. Forty-five genes that are involved in meiosis and sporulation were selected for further analysis. Figure 5.24 shows a partition of the 45 genes into 5 clusters, where 14 characterized early genes, 2 mid-late genes, and 3 late genes are correctly organized into the corresponding clusters, while 2 middle genes are assigned to incorrect groups. Accordingly, the prototype vector of each generated cluster, which represents the profiles of the genes within the cluster, is depicted in Fig. 5.25. It is clearly observed that genes in cluster 1 are induced at the early phase of sporulation and maintain high expression levels throughout the entire process. The expression levels of genes in cluster 2 continuously increase for 7 hours until they start to drop. Cluster 3 consists of genes without an apparent induction phase. The expression levels of genes in cluster 4 rapidly increase during the mid-late phase of sporulation. Genes in cluster 5 also display high expression levels during the late phase.

SOFM and ART can also be used to discriminate different types of cancers, such as identifying two types of human acute leukemias: acute myeloid

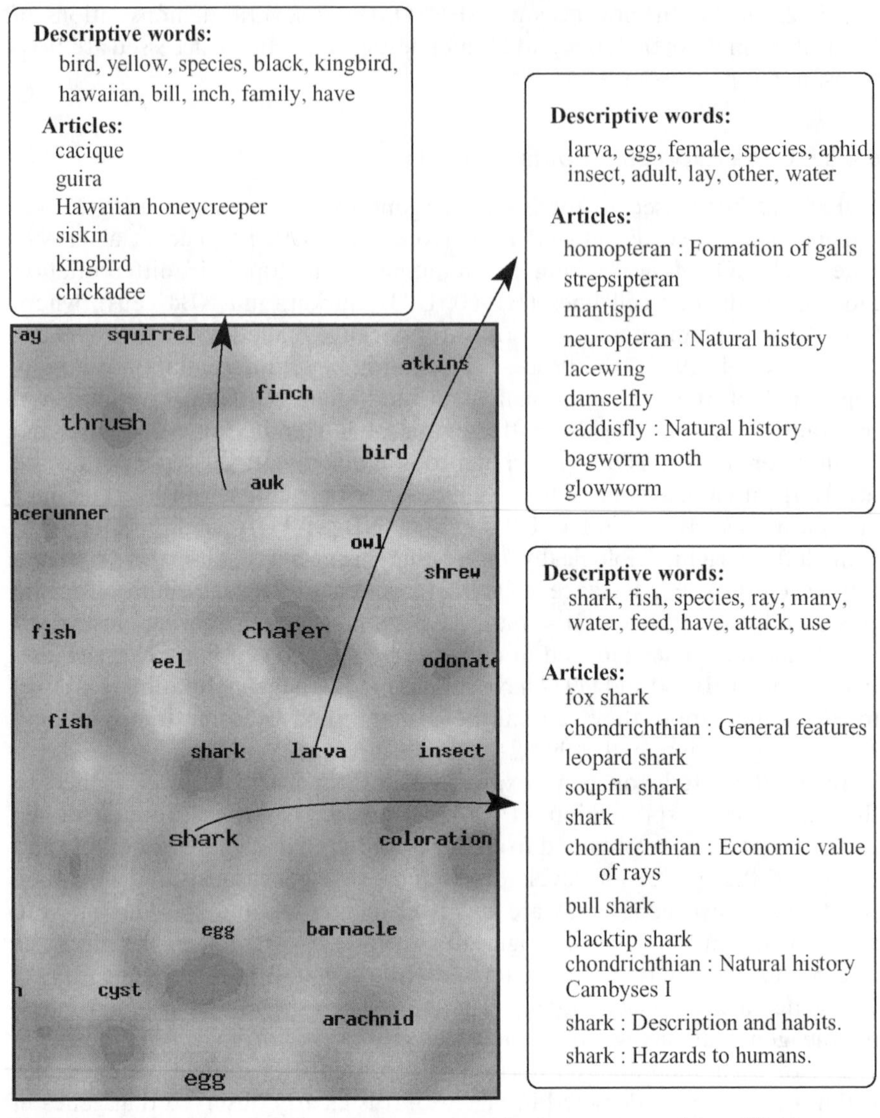

Fig. 5.22. A close-up of the map of Encyclopedia Britannica articles. Clicking on a map region labeled "shark" leads to a view of a section of the map with articles on sharks; various species of fishes and eel (in the middle and left); insects and larvae (lower right corner); various species of birds (upper right corner), and so on. Thus, a topic of interest is displayed in a context of related topics. The three insets depict the contents of three map nodes. (Reprinted from Information Sciences, vol. 163, K. Lagus, S. Kaski, and T. Kohonen, Mining massive document collections by the WEBSOM method, pp. 135–156, Copyright © 2004, with permission from Elsevier.)

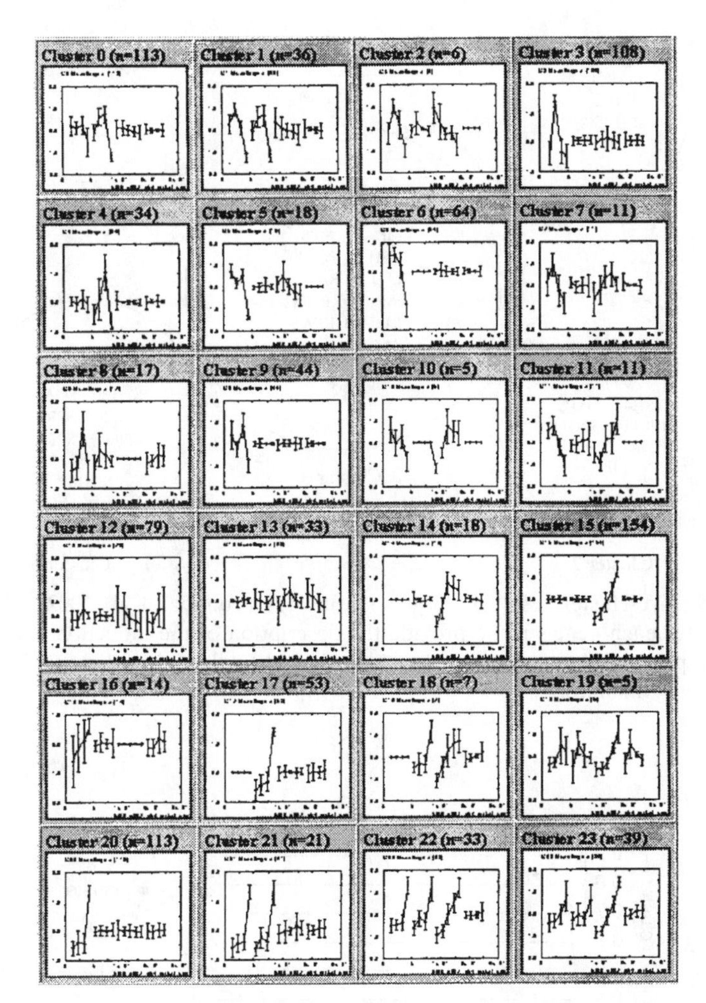

Fig. 5.23. Hematopoietic Differentiation SOFM. Each cluster is visualized using the average expression pattern of the genes in the cluster, together with the error bars indicating the standard deviation at each condition. The number *n* indicates the number of genes within each cluster. (From P. Tamayo, D. Slonim, J. Mesirov, Q. Zhu, S. Kitareewan, E. Dmitrovsky, E. Lander, and T. Golub. Interpreting patterns of gene expression with self-organizing maps: Methods and application to hematopoietic differentiation. Proceedings of National Academy of Sciences, vol. 96, pp. 2907–2912. Copyright © 1999 National Academy of Sciences, U.S.A. Reprinted with permission.)

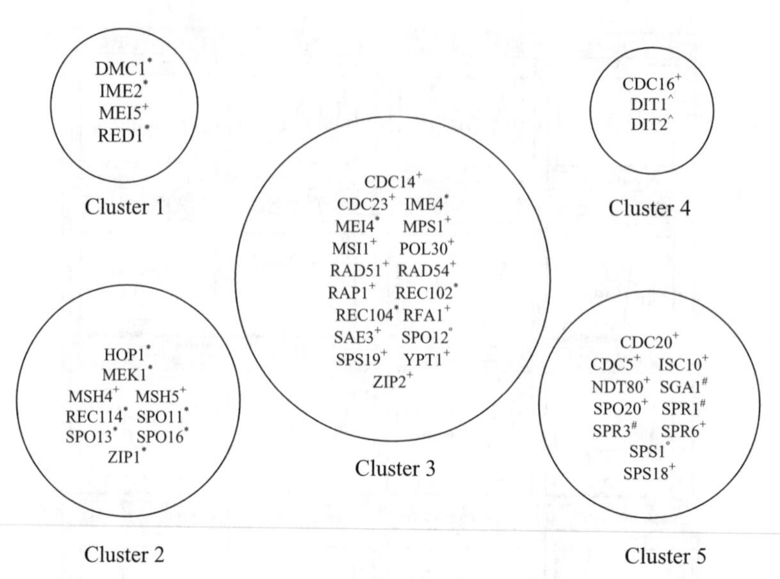

Fig. 5.24. Five-cluster result of 45 genes using FA (Tomida et al., 2002). The superscripts after each gene's name provide the description of the induction period based on some previously reported results (*: Early, °: Middle, ^: Mid-Late, #: Late, +: no description available).

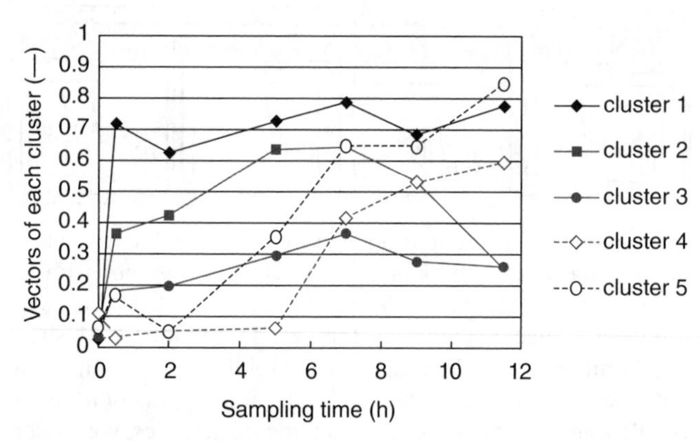

Fig. 5.25. Weight vector **W** of five clusters generated by Fuzzy ART. (From S. Tomida, T. Hanai, H. Honda, and T. Kobayashi, Analysis of expression profile using fuzzy adaptive resonance theory, Bioinformatics, vol. 18, no. 8, pp. 1073–1083, 2002, by permission of Oxford University Press.)

leukemia (AML) and acute lymphoblastic leukemia (ALL) (Golub et al., 1999; Xu et al., 2002). In particular, ALL can be classified further as T-lineage ALL and B-lineage ALL. The entire data set consists of 6,817 human genes plus 312 control genes with their expression levels measured across 72 samples, including bone marrow samples, peripheral blood samples, and childhood AML cases. Fig. 5.26 (a) and (b) show the clustering result when a 2×1 SOFM and a 4×1 SOFM were applied to a subset of 38 leukemia samples using the expression profiles of 6,817 genes (Golub et al., 1999). According to the results,

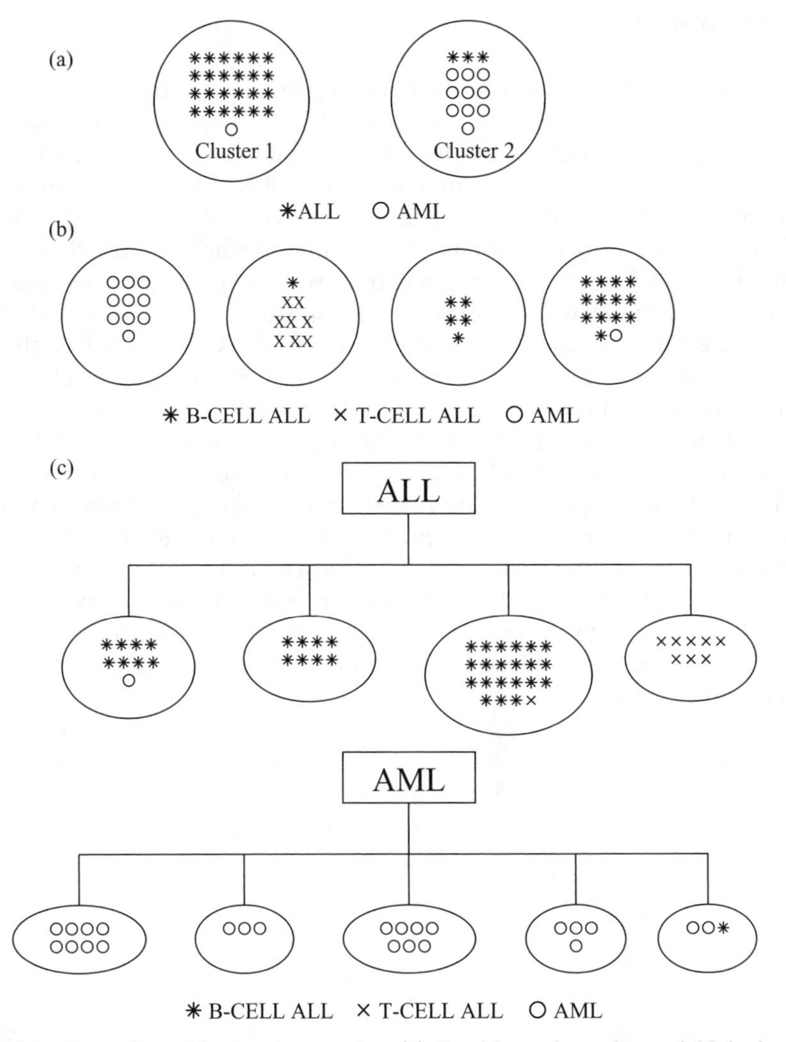

Fig. 5.26. Clustering of leukemia samples. (a) Partition of a subset of 38 leukemia samples with a 2×1 SOFM; (b) Partition of a subset of 38 leukemia samples with a 4×1 SOFM (Golub et al., 1999); (c) Partition of all 72 leukemia samples with EA.

not only can ALL and AML samples be discriminated using their gene profiles, but the subclass of ALL can also be exposed with few classification errors. Results with EA on all 72 samples are shown in Fig. 5.26 (c) (Xu et al., 2002). It can be seen that the ALL samples are grouped into 4 clusters (3 for B-lineage ALL and 1 for T-lineage ALL), while the AML samples form 5 clusters. Only 3 errors occurred: 1 AML sample (AML66) and 1 ALL sample (ALL6) were assigned to the wrong clusters, and a T-lineage ALL sample (ALL11) was not distinguished from other B-lineage ALL samples.

5.5. SUMMARY

Neural network–based clustering is tightly related to the concept of competitive learning. Prototype vectors, associated with a set of neurons in the network and representing clusters in the feature or output space, compete with each other upon the presentation of an input pattern. The active neuron or winner reinforces itself (hard competitive learning) or its neighborhood within certain regions (soft competitive learning). More often, the neighborhood decreases monotonically with time, i.e., changing from soft competitive learning to hard competitive learning, which ensures the independence of initialization at the early phase while emphasizing the optimization of distortion error in the late stage. In addition, this strategy also prevents the occurrence of dead neurons in the network, and the prototype vectors are less likely to be stuck to local optima. One important problem that learning algorithms need to deal with is the stability and plasticity dilemma. A system should have the capability of learning new and important patterns while maintaining stable cluster structures in response to irrelevant inputs. The design of the ART family provides a solution to this problem. Additionally, as most neural network–based clustering algorithms are based on incremental or online learning, they are order dependent. The weight of prototype vectors will be affected by the order in which a sequence of input patterns is presented. Users should pay extra attention to this problem.

CHAPTER 6

KERNEL-BASED CLUSTERING

6.1. INTRODUCTION

Since the 1990s, kernel-based learning algorithms have become increasingly important in pattern recognition and machine learning, particularly in supervised classification and regression analysis, with the introduction of support vector machines (Burges, 1998; Haykin, 1999; Müller et al., 2001; Schölkopf et al., 1999, Schölkopf and Smola, 2002; Vapnik, 1998, 2000). According to Cover's theorem (Cover, 1965), by nonlinearly transforming a set of complex and nonlinearly separable patterns into a higher-dimensional feature space, it is more likely to obtain a linear separation of these patterns. In other words, given a nonlinear map $\Phi: \Re^d \rightarrow F$, where F represents a feature space with arbitrarily high dimensionality, a set of patterns $\mathbf{x}_j \in \Re^d, j = 1, \ldots, N$ is mapped into the feature space F so that a linear algorithm can be performed. The idea is illustrated in Fig. 6.1.

The difficulty of the curse of dimensionality (Bellman, 1957), which describes the exponential growth in computational complexity as a result of high dimensionality in the problem space, can be overcome by the kernel trick, arising from Mercer's theorem (Mercer, 1909):

If $k(\mathbf{x}, \mathbf{z}): C \times C \rightarrow \Re$ is a continuous symmetric kernel, where $C \subset \Re^d$ is a compact set, the necessary and sufficient condition for expanding the kernel $k(\mathbf{x}, \mathbf{z})$ in a uniformly convergent series

Clustering, by Rui Xu and Donald C. Wunsch, II

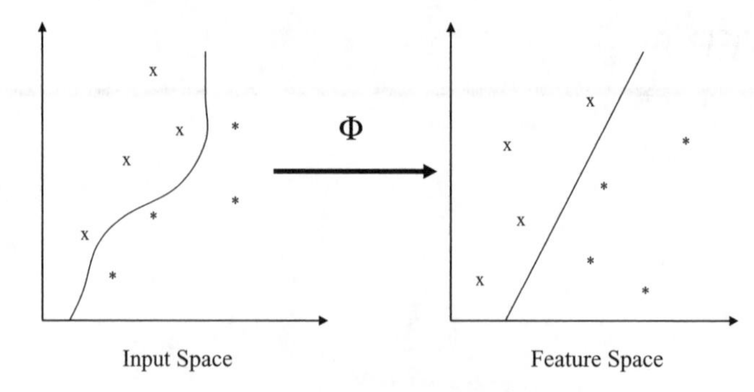

Fig. 6.1. Map input patterns into a higher-dimensional feature space. The two categories with a nonlinear boundary in the original space can be linearly separated in the transformed feature space.

$$k(\mathbf{x}, \mathbf{z}) = \sum_{j=1}^{\infty} \lambda_j \psi_j(\mathbf{x}) \psi_j(\mathbf{z}), \tag{6.1}$$

where $\forall \lambda_j > 0$, is that

$$\int_{C \times C} k(\mathbf{x}, \mathbf{z}) \psi_j(\mathbf{x}) \psi_j(\mathbf{z}) d\mathbf{x} d\mathbf{z} \geq 0 \tag{6.2}$$

holds for all $\psi(\cdot)$ for which

$$\int_C \psi^2(\mathbf{x}) d\mathbf{x} < \infty. \tag{6.3}$$

By designing and calculating an inner-product kernel, we can avoid the time-consuming, sometimes even infeasible process of explicitly describing the nonlinear mapping and computing the corresponding points in the transformed space. The commonly used kernel functions include polynomial kernels,

$$k(\mathbf{x}, \mathbf{z}) = (\mathbf{x} \cdot \mathbf{z} + 1)^p, \tag{6.4}$$

where p is the polynomial degree; Gaussian radial basis function (RBF) kernels,

$$k(\mathbf{x}, \mathbf{z}) = \exp\left(-\frac{1}{2\sigma^2} \|\mathbf{x} - \mathbf{z}\|^2\right), \tag{6.5}$$

where σ is the RBF width parameter; and sigmoid kernels,

$$k(\mathbf{x}, \mathbf{z}) = \tanh(\theta_0(\mathbf{x} \cdot \mathbf{z}) + \theta_1), \tag{6.6}$$

where θ_0 and θ_1 are user-specified parameters. While the polynomial and RBF kernels completely satisfy Mercer's conditions, Mercer's theorem only holds for certain parameters of θ_0 and θ_1 for the sigmoid kernel (Haykin, 1999). Different kernel functions lead to different nonlinear decision surfaces. However, the selection of an appropriate kernel is still an open problem and currently is determined through experiments.

Similarly, the kernel trick can be used in clustering to explore the potentially nonlinear structure in the data. Section 6.2 introduces a nonlinear version of principal component analysis (PCA), i.e., kernel PCA. Section 6.3 focuses on the squared-error-based clustering algorithms. Section 6.4 discusses the support vector clustering, which finds the smallest hypersphere in the feature that can enclose most of the data points. Section 6.5 illustrates two applications in the feature extraction in regression tasks and prototype identification within handwritten digits patterns. Section 6.6 concludes the chapter with some general discussion.

6.2. KERNEL PRINCIPAL COMPONENT ANALYSIS

Principal component analysis (PCA) projects high-dimensional data onto a lower-dimensional subspace by seeking a linear combination of a set of projection vectors that can best describe the variance of data in a sum of squared-error sense, as introduced in Chapter 9. Kernel PCA extends the capability of linear PCA by capturing nonlinear structure in the data, since a linear PCA performance in the feature space corresponds to a nonlinear projection in the original data space (Müller et al., 2001; Schölkopf et al., 1998).

For a set of data points $\mathbf{x}_j \in \Re^d, j = 1, \ldots, N$, we map them into an arbitrary high-dimensional feature space with the nonlinear function $\Phi: \Re^d \rightarrow F$. The transformed data are centered, i.e., the mean is 0. This can be achieved by using the substitute kernel matrix (Schölkopf et al., 1998),

$$\mathbf{k} = \mathbf{k} - 1_N\mathbf{k} - \mathbf{k}1_N + 1_N\mathbf{k}1_N, \tag{6.7}$$

where $\mathbf{k} = \{k(\mathbf{x}_i, \mathbf{x}_j)\}$ is the kernel matrix and $(1_N)_{ij} = 1/N$.

Similar to linear PCA, the principal components are obtained by calculating the eigenvectors $\mathbf{e}$ and eigenvalues $\lambda > 0$ of the covariance matrix $\sum^{\Phi} = \frac{1}{N}\sum_{j=1}^{N}\Phi(\mathbf{x}_j)\Phi(\mathbf{x}_j)^T$,

$$\lambda\mathbf{e} = \sum^{\Phi}\mathbf{e}. \tag{6.8}$$

By multiplying with $\Phi(\mathbf{x}_i)$ from the left and noticing that $\mathbf{e} = \sum_{l=1}^{N}\alpha_l\Phi(\mathbf{x}_l)$, a straightforward manipulation of Eq. 6.8 yields

$$\lambda \sum_{l=1}^{N} \alpha_l (\Phi(\mathbf{x}_l) \cdot \Phi(\mathbf{x}_l)) = \frac{1}{N} \sum_{l=1}^{N} \alpha_l \left(\Phi(\mathbf{x}_l) \cdot \sum_{j=1}^{N} \Phi(\mathbf{x}_j) \right) (\Phi(\mathbf{x}_j) \cdot \Phi(\mathbf{x}_l)). \quad \text{for all } l = 1, \ldots, N \tag{6.9}$$

Using the kernel function, Eq. 6.9 can be written as

$$\lambda \boldsymbol{\alpha} = \mathbf{k} \boldsymbol{\alpha}, \tag{6.10}$$

where $\boldsymbol{\alpha} = (\alpha_1, \ldots, \alpha_N)^T$. The achieved solutions $(\lambda_i, \boldsymbol{\alpha}_i)$ must be normalized following the condition,

$$\lambda_i (\boldsymbol{\alpha}_i \cdot \boldsymbol{\alpha}_i) = 1. \tag{6.11}$$

Given a new data point, its projection can be calculated as

$$(\boldsymbol{\alpha}_i \cdot \Phi(\mathbf{x})) = \sum_{l=1}^{N} \alpha_{il} k(\mathbf{x}_l, \mathbf{x}). \tag{6.12}$$

The basic steps for kernel PCA are summarized in Fig. 6.2. A discussion of kernel PCA can also be found in Schölkopf et al. (1999) and Tipping (2000).

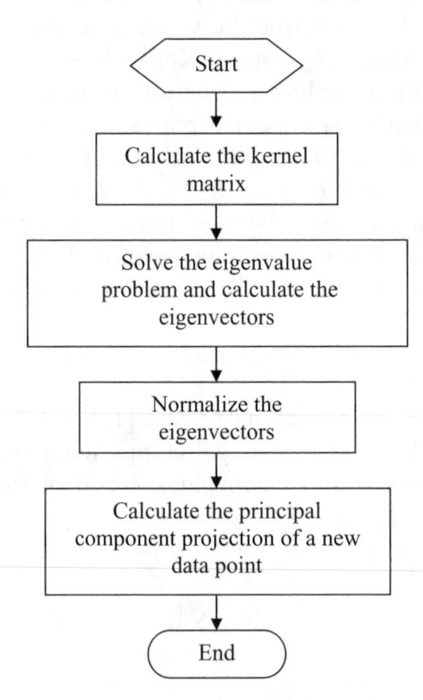

Fig. 6.2. Flowchart of the kernel PCA algorithms. Note that the nonlinearity is introduced by mapping the input data from the input space to a high-dimensional feature space.

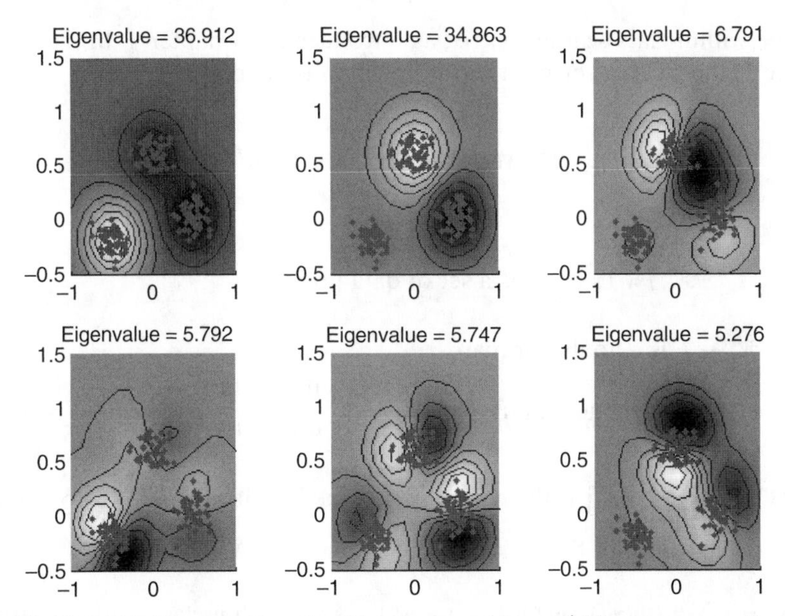

Fig. 6.3. Kernel PCA on a Gaussian distributions data set.[1] The contours reflect the equivalent projection on the corresponding principal components. The number of principal components can be much larger than the dimensionality of the data.

Jenssen et al. (2006b) proposed a similar algorithm for data transformation, called kernel MaxEnt, which is based on the principle of maximum entropy preservation. Kernel MaxEnt requires the kernel to be a density, while kernel PCA works on Mercer kernels that are not necessarily densities.

Figure 6.3 depicts the results of kernel PCA with an RBF kernel to a data set, which consists of three clusters following Gaussian distributions. Each cluster includes 50 data points. The first six principal components were extracted. It is obvious that the two largest principal components provide an effective separation of the data. However, this is not always the case, since the primary object of kernel PCA is to seek the maximal variance in the data structure (Müller et al., 2001).

6.3. SQUARED-ERROR-BASED CLUSTERING WITH KERNEL FUNCTIONS

The squared-error-based clustering algorithms seek a cluster partition that minimizes the criterion function, which is equivalent to minimizing the trace

[1] The figure is obtained using the Matlab program written by Schölkopf, and can be downloaded at http://www.kernel-machines.org.

of the within-class scatter matrix, as introduced in Chapter 4. In the feature space F, the squared-error criterion function is defined as

$$J(\mathbf{\Gamma}, \mathbf{M}^{\Phi}) = \sum_{i=1}^{K}\sum_{j=1}^{N}\gamma_{ij}\left\|\Phi(\mathbf{x}_j)-\mathbf{m}_i^{\Phi}\right\|^2, \\ = Tr(S_W^{\Phi}) \tag{6.13}$$

where $\mathbf{x}_j \in \Re^d$, $j = 1, \ldots, N$ is a set of data points;

$\Phi: \Re^d \rightarrow F$ is a nonlinear map;

$\mathbf{\Gamma} = \{\gamma_{ij}\}$ is a partition matrix and $\gamma_{ij} = \begin{cases} 1 & \text{if } \mathbf{x}_j \in \text{cluster } C_i \\ 0 & \text{otherwise} \end{cases}$ with $\sum_{i=1}^{K}\gamma_{ij} = 1 \forall j$;

$\mathbf{M}^{\Phi} = [\mathbf{m}_1^{\Phi}, \ldots, \mathbf{m}_K^{\Phi}]$ is the cluster prototype matrix and $\mathbf{m}_i^{\Phi} = \frac{1}{N_i}\sum_{j=1}^{N}\gamma_{ij}\Phi(\mathbf{x}_j)$ is the center of the cluster C_i, which includes N_i data points.

$S_W^{\Phi} = \sum_{i=1}^{K}\sum_{j=1}^{N}\gamma_{ij}(\Phi(\mathbf{x}_j)-\mathbf{m}_i^{\Phi})(\Phi(\mathbf{x}_j)-\mathbf{m}_i^{\Phi})^T$ is the within-class scatter matrix.

Now, the problem is formulated to determine an optimal partition $\mathbf{\Gamma}$ to minimize the trace of the within-class scatter matrix (Girolami, 2002),

$$\begin{aligned}
\mathbf{\Gamma} &= \arg\min_{\mathbf{\Gamma}} Tr(S_W^{\Phi}) \\
&= \arg\min_{\mathbf{\Gamma}} Tr\left\{\frac{1}{N}\sum_{i=1}^{K}\sum_{j=1}^{N}\gamma_{ij}(\Phi(\mathbf{x}_j)-\mathbf{m}_i^{\Phi})(\Phi(\mathbf{x}_j)-\mathbf{m}_i^{\Phi})^T\right\} \\
&= \arg\min_{\mathbf{\Gamma}} \sum_{i=1}^{K}\sum_{j=1}^{N}\gamma_{ij}(\Phi(\mathbf{x}_j)-\mathbf{m}_i^{\Phi})^T(\Phi(\mathbf{x}_j)-\mathbf{m}_i^{\Phi}) \\
&= \arg\min_{\mathbf{\Gamma}} \sum_{i=1}^{K}\sum_{j=1}^{N}\gamma_{ij}\left(\Phi(\mathbf{x}_j)-\frac{1}{N}\sum_{j=1}^{N}\gamma_{ij}\Phi(\mathbf{x}_j)^T\right)\left(\Phi(\mathbf{x}_j)-\frac{1}{N_i}\sum_{j=1}^{N}\gamma_{ij}\Phi(\mathbf{x}_j)\right) \\
&= \arg\min_{\mathbf{\Gamma}} \sum_{i=1}^{K}\sum_{j=1}^{N}\gamma_{ij}\left(\Phi(\mathbf{x}_j)\cdot\Phi(\mathbf{x}_j)-\frac{2}{N_i}\sum_{l=1}^{N}\gamma_{il}\Phi(\mathbf{x}_j)\cdot\Phi(\mathbf{x}_l)\right. \\
&\quad \left. +\frac{2}{N_i^2}\sum_{l=1}^{N}\sum_{m=1}^{N}\gamma_{il}\gamma_{im}\Phi(\mathbf{x}_l)\cdot\Phi(\mathbf{x}_m)\right).
\end{aligned} \tag{6.14}$$

Using the kernel function $k(\mathbf{x}_i, \mathbf{x}_j) = \Phi(\mathbf{x}_i) \cdot \Phi(\mathbf{x}_j)$, we obtain

$$\begin{aligned}
\mathbf{\Gamma} = \arg\min_{\mathbf{\Gamma}} \sum_{i=1}^{K}\sum_{j=1}^{N}\gamma_{ij}\left(k(\mathbf{x}_j, \mathbf{x}_j)-\frac{2}{N_i}\sum_{l=1}^{N}\gamma_{il}k(\mathbf{x}_j, \mathbf{x}_l)\right. \\
\left. +\frac{2}{N_i^2}\sum_{l=1}^{N}\sum_{m=1}^{N}\gamma_{il}\gamma_{im}k(\mathbf{x}_l, \mathbf{x}_m)\right).
\end{aligned} \tag{6.15}$$

If we let

$$R(\mathbf{x}|C_i) = \frac{1}{N_i^2} \sum_{l=1}^{N} \sum_{m=1}^{N} \gamma_{il} \gamma_{im} k(\mathbf{x}_l, \mathbf{x}_m), \quad (6.16)$$

Eq. 6.15 is then written as

$$\mathbf{\Gamma} = \arg\min_{\mathbf{\Gamma}} \sum_{i=1}^{K} \sum_{j=1}^{N} \gamma_{ij} k(\mathbf{x}_j, \mathbf{x}_j) - \sum_{i=1}^{N} N_i R(\mathbf{x}|C_i). \quad (6.17)$$

If an RBF kernel is used, Eq. 6.17 can be reduced further to

$$\mathbf{\Gamma} = \arg\min_{\mathbf{\Gamma}} \left(1 - \sum_{i=1}^{N} N_i R(\mathbf{x}|C_i)\right) = \arg\max_{\mathbf{\Gamma}} \sum_{i=1}^{N} N_i R(\mathbf{x}|C_i). \quad (6.18)$$

This nonlinear optimization problem can be solved with an iterative procedure (Girolami, 2002):

$$\gamma_{ij} = \frac{\alpha_i \exp(-2\beta D_{ij}^{new})}{\sum_{l=1}^{K} \alpha_l \exp(-2\beta D_{lj}^{new})}, \quad (6.19)$$

where $D_{ij}^{new} = 1 - \sum_{l=1}^{N} \sum_{m=1}^{N} \frac{\gamma_{im}}{\gamma_{il}} k(\mathbf{x}_j, \mathbf{x}_m)$, $\alpha_i = \exp(-\beta R(\mathbf{x}(C_i))$ and β is the parameter controlling the assignment softness. When β approaches infinity, this becomes a batch-mode RBF kernel K-means algorithm.

Herein, the kernel-K-means algorithm can be summarized as below:

1. Initialize K cluster centers $\mathbf{m}_l^{\Phi}$ randomly or based on some prior knowledge;
2. Assign each data point $\mathbf{x}_j$, $j = 1, \ldots, N$ to the nearest cluster C_i, $i = 1, \ldots, K$, i.e.

$$\mathbf{x}_j \in C_i, \quad \text{if } \|\Phi(\mathbf{x}_j) - \mathbf{m}_i^{\Phi}\|^2 < \|\Phi(\mathbf{x}_j) - \mathbf{m}_l^{\Phi}\|^2, \quad i \neq 1;$$

3. Recalculate the cluster centers based on the current partition;
4. Repeat steps 2 and 3 until there is no change for each cluster;
5. Finally, the data point whose image is closest to the center is selected as the representative of the corresponding cluster.

An incremental kernel-K-means algorithm was described by Schölkopf et al. (1998) and Corchado and Fyfe (2000), who also introduced two variants, motivated by SOFM and ART networks. These variants consider effects of neighborhood relations and use a vigilance parameter to control the process of producing mean vectors. Discussions on Kernel K-means in large-scale data clustering can be found in Zhang and Rudnicky (2002).

It is also worthwhile to point out that kernel K-means can be interpreted in terms of information theoretic quantities such as Renyi quadratic entropy and integrated squared error (Jenssen et al., 2006a). In this context, the kernel K-means algorithm corresponds to the maximization of an integrated squared error divergence measure between the Parzen window estimated cluster probability density functions (Jenssen and Eltoft, 2006). Similarly, kernel independent component analysis (Bach and Jordan, 2002) can also be analyzed in an information theoretic framework, i.e., it is equivalent to Cauchy-Schwartz independence measure, estimated via a weighted Parzen windowing procedure (Xu et al., 2005). Further discussions on the connection between information theoretic learning and the kernel methods can be found in Jenssen et al. (2005 and 2006).

6.4. SUPPORT VECTOR CLUSTERING

The support vector clustering (SVC) algorithm is inspired by the support vector machines and solves a global optimization problem by turning the Lagrangian into the dual quadratic form (Ben-Hur et al., 2000, 2001). The object is to find the smallest enclosing hypersphere in the transformed high-dimensional feature space that contains most of the data points. The hypersphere is then mapped back to the original data space to form a set of contours, which are regarded as the cluster boundaries in the original data space.

The SVC algorithm includes two key steps: SVM training and cluster labeling, which are summarized in Fig. 6.4 (Ben-Hur et al., 2001; Lee and Lee, 2005; Yang et al., 2002). The former determines the hypersphere construction and the distance definition from a point's image in the feature space to the hypersphere center. The latter aims to assign each data point to its corresponding cluster. Given a set of data points $\mathbf{x}_j \in \Re^d$, $j = 1, \ldots, N$ and a nonlinear map $\Phi: \Re^d \rightarrow F$, the object is to find a hypershpere with the minimal radius R, such as

$$\|\Phi(\mathbf{x}_j) - \boldsymbol{\alpha}\|^2 \leq R^2 + \xi_j, \tag{6.20}$$

where $\boldsymbol{\alpha}$ is the center of the hypersphere and $\xi_j \geq 0$ are the slack variables allowing soft constraints. The primal problem is solved in its dual form by introducing the Lagrangian

$$L = R^2 - \sum_j \left(R^2 + \xi_j - \|\Phi(\mathbf{x}_j) - \boldsymbol{\alpha}\|^2\right)\beta_j - \sum_j \xi_j \mu_j + C\sum_j \xi_j, \tag{6.21}$$

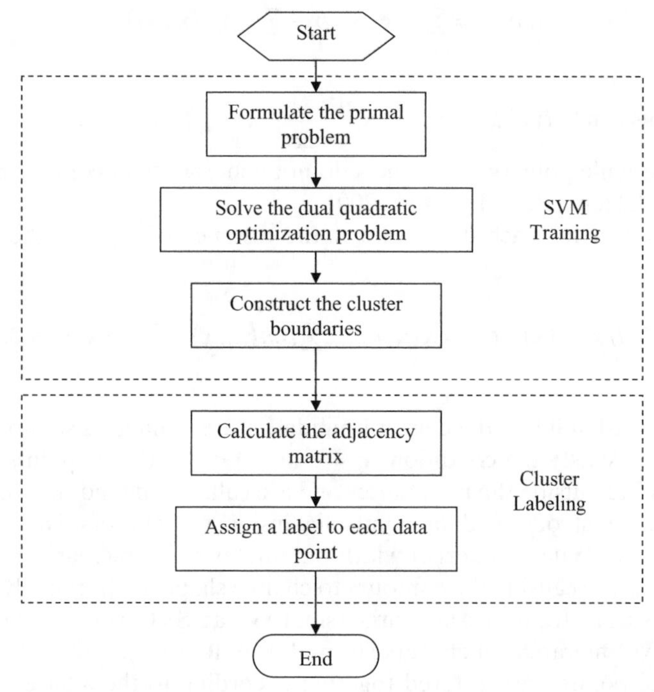

Fig. 6.4. Flowchart of SVC algorithm. The SVC algorithm consists of two main phases: SVM training for generating the cluster boundaries and cluster labeling for determining the cluster membership of each data point.

where $\beta_j \geq 0$ and $\mu_j \geq 0$ are Lagrange multipliers and $C\sum_j \xi_j$ is a penalty term with C as a regularization constant. The dual form of the constrained optimization is constructed as

$$\max_{\beta_j} W = \sum_j \Phi(\mathbf{x}_j)^2 \beta_j - \sum_{i,j} \beta_i \beta_j \Phi(\mathbf{x}_i) \cdot \Phi(\mathbf{x}_j), \tag{6.22}$$

subject to the constraints:

$$(1) \quad 0 \leq \beta_j \leq C, \tag{6.23}$$

$$(2) \quad \sum_j \beta_j = 1 \quad \text{for } j = 1, 2, \ldots, N \tag{6.24}$$

Using the kernel representation $k(\mathbf{x}_i, \mathbf{x}_j) = \Phi(\mathbf{x}_i) \cdot \Phi(\mathbf{x}_j)$, Eq. 6.22 is rewritten as

$$\max_{\beta_j} W = \sum_j k(\mathbf{x}_j, \mathbf{x}_j)\beta_j - \sum_{i,j} \beta_i \beta_j k(\mathbf{x}_i, \mathbf{x}_j). \tag{6.25}$$

The Gaussian kernel $k(\mathbf{x}_i, \mathbf{x}_j) = \exp\left(-\frac{1}{2\sigma^2}\|\mathbf{x}_i - \mathbf{x}_j\|^2\right)$ is usually used for SVC algorithms, while polynomial kernels do not generate tight contour representations of clusters (Ben-Hur et al., 2001).

Furthermore, for each data point $\mathbf{x}$, the distance of $\Phi(\mathbf{x})$ to the center is calculated as

$$R^2(\mathbf{x}) = \|\Phi(\mathbf{x}) - \boldsymbol{\alpha}\|^2 = k(\mathbf{x}, \mathbf{x}) - 2\sum_j \beta_j k(\mathbf{x}_j, \mathbf{x}) - \sum_{i,j} \beta_i \beta_j k(\mathbf{x}_i, \mathbf{x}_j). \tag{6.26}$$

The points that lie on the cluster boundaries are defined as support vectors (SVs), which satisfy the conditions $\xi_j = 0$ and $0 < \beta_j < C$. The points with $\xi_j > 0$ and $\beta_j = C$ lie outside the boundaries and are called bounded support vectors (BSVs). The rest of the data points lie inside the clusters. Note that the increase of the Gaussian kernel width parameter σ can increase the number of SVs, therefore causing the contours to change shape. By iteratively decreasing (increasing) σ from a certain large (small) value, SVC can form agglomerative (divisive) hierarchical clusters (Ben-Hur et al., 2000).

The data points are clustered together according to the adjacency matrix $\mathbf{A}$, which is based on the observation that any corresponding path in the feature space, which connects a pair of data points belonging to different clusters, must exit from the hypersphere. Given each pair of $\mathbf{x}_i$ and $\mathbf{x}_j$, their adjacency value is defined as

$$A_{ij} = \begin{cases} 1, & \text{if } R(\mathbf{x}_i + \gamma(\mathbf{x}_j - \mathbf{x}_i)) \le R, \gamma \in [0, 1] \\ 0, & \text{otherwise.} \end{cases} \tag{6.27}$$

The number of sampling points for each edge between two data points is usually around 20 (Ben-Hur et al., 2001). The overall computational complexity of the labeling step is $O(N^2)$, which becomes a critical issue for large-scale data sets. Ben-Hur et al. (2001) suggested a heuristic that only calculates the adjacency value between support vectors to lower the time complexity to $O((N - N_{bsv})N_{sv}^2)$, where N_{bsv} is the number of BSVs and N_{sv} is the number of SVs. However, this heuristic still has quadratic computational complexity when N_{sv} is greater than $0.05N - 0.1N$, as pointed out by Yang et al. (2002), who proposed a proximity graph strategy to achieve $O(N\log N)$ time complexity. Furthermore, Lee and Lee (2005) presented a two-phase cluster labeling algorithm that has almost linear time complexity. The algorithm only needs to label the selected stable equilibrium points at which all the other points converge. Nath and Shevade (2006) proposed an additional preprocessing step to remove the data points that are not important to the clustering, i.e., nonsupport vectors. The chance for each point to be a nonsupport vector is determined with the nearest neighbor queries, implemented with an R*–tree data structure.

Chiang and Hao (2003) extended the idea of SVC by representing each cluster with a hypersphere, instead of using just one hypersphere in SVC overall. The proposed algorithm is then called multiplesphere support vector clustering (MSVC) and has the ability to identify the cluster prototypes. A mechanism similar to ART's orienting subsystem (Carpenter and Grossberg, 1987; Grossberg, 1976) is adopted to dynamically generate clusters. When a data point **x** is presented, clusters compete with each other based on the distance function, defined as,

$$D(\mathbf{x}, \mathbf{m}_l) = \left(k(\mathbf{x}, \mathbf{x}) - 2 \sum_{\mathbf{x}_j \in S_l} \beta_j k(\mathbf{x}_j, \mathbf{x}) - \sum_{\mathbf{x}_j, \mathbf{x}_i \in S_l} \beta_i \beta_j k(\mathbf{x}_i, \mathbf{x}_j) \right)^{1/2}, \qquad (6.28)$$

where $\mathbf{m}_l$ is the center of the sphere S_l. A validation test is performed to ensure the eligibility of the winning cluster to represent the input pattern. A new cluster is created as a result of the failure of all clusters available to the vigilance test. Furthermore, the radius of the hypersphere and the distance between the input pattern and the cluster center provide a way to calculate the fuzzy membership function. Specifically, the degree of a point **x** in a cluster C_i is calculated as,

$$\mu_i(\mathbf{x}) = \begin{cases} \frac{1}{2} \times \left(\frac{1 - \frac{D(\mathbf{x}, \mathbf{m}_i)}{R_i}}{1 + \lambda_1 \frac{D(\mathbf{x}, \mathbf{m}_i)}{R_i}} + 1 \right), & \text{if } D(\mathbf{x}, \mathbf{m}_i) \le R_i \\ \frac{1}{2} \times \left(\frac{1}{1 + \lambda_2 (D(\mathbf{x}, \mathbf{m}_i) - R_i)} \right), & \text{otherwise} \end{cases}, \qquad (6.29)$$

where R_i is the radius of the hypersphere representing the cluster C_i and λ_1 and λ_2 are parameters satisfying the restriction in order to assure the differentiability of $\mu_i(\mathbf{x})$ when $D(\mathbf{x}, \mathbf{m}_i) = R_i$,

$$\lambda_2 = \frac{1}{R_i(1 + \lambda_1)}. \qquad (6.30)$$

The basic steps of MSVC are described in Fig. 6.5.

Asharaf et al. (2005) introduced the concept of rough set (Pawlak, 1992) into the support vector clustering. In this context, the rough hypersphere has both an inner radius representing its lower approximation and an outer radius representing its upper approximation. If the mapping points in the feature space is within the lower approximation, they are considered to belong to one cluster exclusively. However, if the data points are in the upper approximation but not in lower approximation, they are regarded to be associated with more

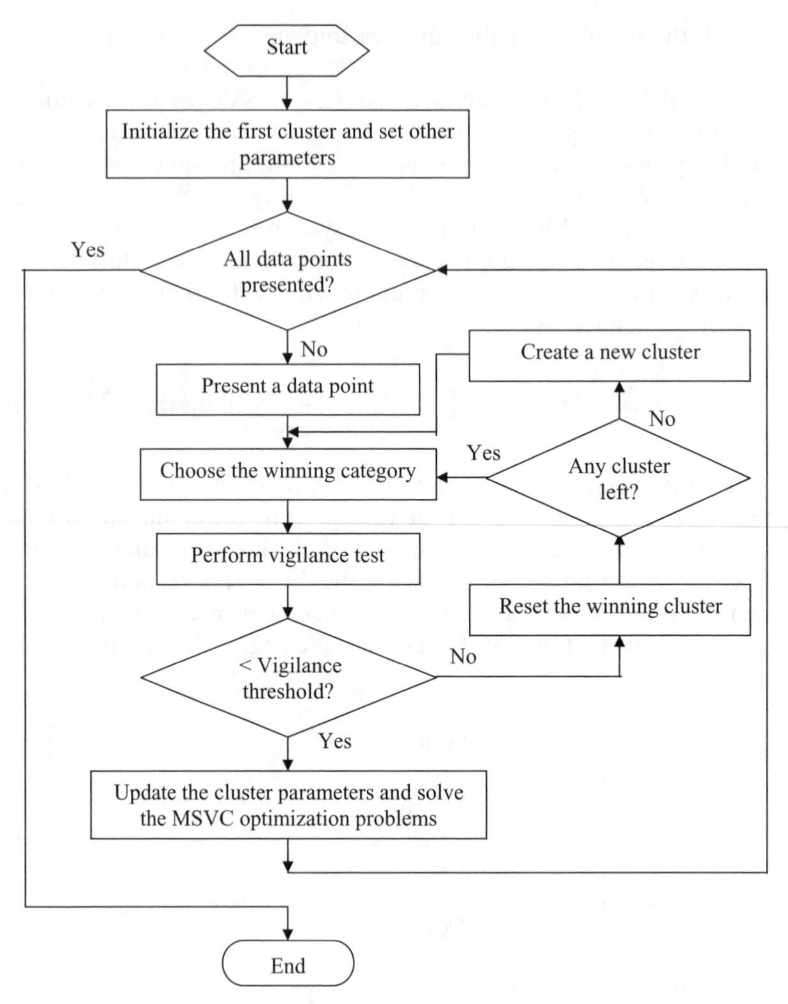

Fig. 6.5. Flowchart of MSSVC algorithm. Note that clusters are dynamically constructed as the introduction of the validity test mechanism.

than one cluster, which achieves the soft clustering. Correspondingly, the goal of the problem becomes to finding a smallest rough hypershpere with the inner radius R_i and outer radius R_o, such as

$$\|\Phi(\mathbf{x}_j)-\boldsymbol{\alpha}\|^2 \le R_i^2 + \xi_j + \xi_j', \tag{6.31}$$

$$0 \le \xi_j \le R_o^2 - R_i^2, \tag{6.32}$$

where $\xi_j' \ge 0$ are the slack variables. Again, this problem is solved with the construction of the Wolfe dual form.

6.5. APPLICATIONS

6.5.1 Feature Extraction in Regression Problems

Feature extraction is an important preprocessing step for regression and classification tasks. Good features that can effectively represent the properties and characteristics of the objects under investigation could greatly improve the regression model and reduce the subsequent computational burden. An application of kernel PCA in feature extraction for the regression task of modeling the dependence between human signal detection performance and brain event-related potentials (ERPs) was presented in Rosipal et al. (2000). This identified relation could provide physiological approaches to evaluating human performance and could prevent the occurrence of potential errors.

The data were collected from eight subjects experienced in the operation of display systems who performed a signal detection task. This task requires each individual to classify the presented symbol at each trial as a target or nontarget and to indicate his subjective confidence on a three-point scale. The total number of trials for each subject is 1,000. The performance of the individual was evaluated in terms of a measurement variable, $PF1$, which is based on the subject's accuracy, confidence, and speed, written as,

$$PF1 = 0.33 \times \text{Accuracy} + 0.53 \times \text{Confidence} - 0.51 \times \text{Reaction Time}. \quad (6.33)$$

ERPs were recorded from midline frontal, central, and parietal electrodes, referred to average mastoids, filtered digitally to a bandpass of 0.1 to 25 Hz, and decimated to a sampling rate of 50 Hz.

Fig. 6.6 shows the comparison of the performance of the kernel principal component regression (KPCR) and multilayer support vector regression (MLSVR), using the features obtained from kernel PCA, with that of support vector regression (SVR), pre-processed with linear PCA, on three subjects. The performance was assessed in terms of the percentage of the correct prediction of $PF1$ within 10% tolerance. It can be seen that the methods operating on the features extracted from kernel PCA achieve better results than linear PCA for most subjects (8 out of 9; only results for subjects A, B, and C shown here).

Other applications of kernel PCA include handwritten digits denoising (Schölkopf et al., 1999) and nonlinear active shape modeling (Romdhani et al., 1999; Twining and Taylor, 2001).

6.5.2 Prototype Identification within Handwritten Digit Patterns

Handwritten digit recognition is an important and challenging problem in optical character recognition (LeCun et al., 1995) as it has wide and immediate applications in many fields, such as postal ZIP code reading from mail,

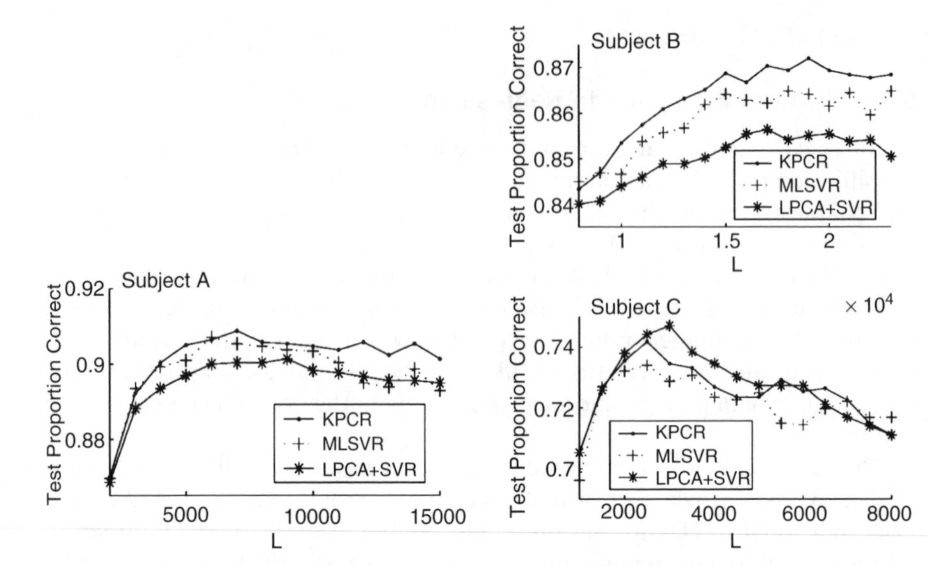

Fig. 6.6. Comparison of the results achieved on subjects, A, B, and C. (From Artificial Neural Networks in Medicine and Biology, Proceedings of the ANNIMAB-1 Conference, 2000, pp. 321–326, Kernel PCA feature extraction of event-related potentials for human signal detection performance, R. Rosipal, M. Girolami, and L. Trejo, Figure 1, Copyright © 2000. With kind permission of Springer Science and Business Media.)

but also faces high variability of handwriting styles. Chiang and Hao (2003) illustrated an application of their MSVC algorithm for identifying the prototypes from handwritten digit patterns, particularly for the digits 2 and 7 (Fig. 6.7a). The data are obtained from the *BR* digit set of the *SUNY CDROM-1* (Hull, 1994) and the *ITRI* database (Chiang and Gader, 1997). Three hundred samples were used for each digit category, with 60-dimensional features.

Figures 6.7b and 6.7c depict the generated prototype digits and the outlier digits with the corresponding membership values. The outlier digits are regarded as the bounded support vectors with small membership values to both digit groups. Membership values larger than 0.5 indicate that the locations of the patterns are inside the corresponding hypersphere in the feature space. We can see from the experimental results that the values of the membership function provide a good way to identify the prototypes while at the same time detecting noise.

6.6. SUMMARY

The process of constructing the sum-of-squared clustering algorithm and kernel PCA presents a typical example to reformulate more powerful nonlinear versions for many existing linear algorithms, provided that the scalar

(a)

membership values for class "2": 0.8031 0.7981 0.7853 0.7782 0.7753

membership values for class "7": 0.1984 0.2371 0.1620 0.1718 0.1742

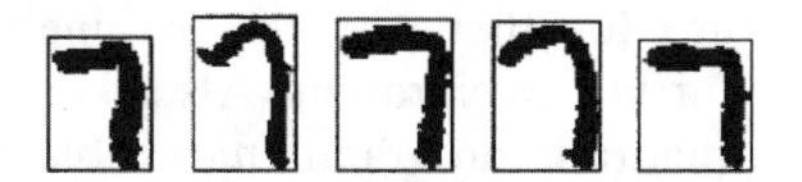

membership values for class "2": 0.2785 0.2387 0.2340 0.2325 0.2248

membership values for class "7": 0.8102 0.7953 0.7920 0.7913 0.7904

(b)

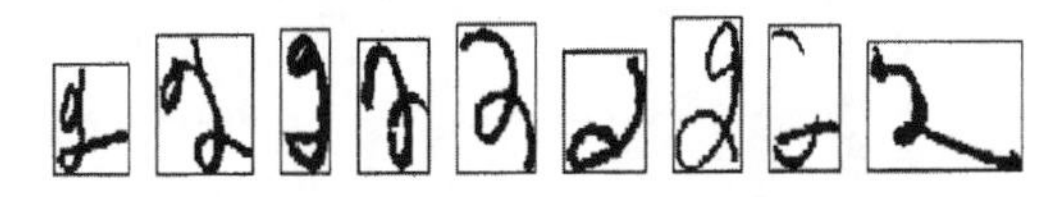

membership values for class "2": 0.3123 0.3742 0.3881 0.3711 0.4203 0.2479 0.2123 0.2611 0.3574

membership values for class "7": 0.1182 0.1606 0.1518 0.1992 0.2159 0.1207 0.1123 0.1103 0.1453

membership values for class "2": 0.2739 0.2297 0.2966 0.2634 0.1873 0.3571

membership values for class "7": 0.3775 0.3968 0.3069 0.2883 0.4072 0.3625

(c)

Fig. 6.7. Handwritten digit prototype identification with MSVC. The experiment focuses on the prototype identification for digits 2 and 7 (a). Prototypes (b) and outlier (c) digits are obtained from MSCV with the corresponding membership values. (J. Chiang and P. Hao. A new kernel-based fuzzy clustering approach: support vector clustering with cell growing. IEEE Transactions on Fuzzy Systems, vol. 11, pp. 518–527, 2003. Copyright © 2003 IEEE.)

product can be obtained. These kernel-based clustering algorithms have many desired properties:

- They are more likely to obtain a linearly separable hyperplane in the high-dimensional, or even infinite, feature space;
- There is no need to explicitly define the nonlinear map, which is bypassed by using the kernel trick;
- Kernel-based clustering algorithms can form arbitrary clustering boundaries other than hypersphere or hyperellipsoid;
- Kernel-based clustering algorithms, like SVC, have the capability of dealing with noise and outliers;
- For SVC, there is no requirement for prior knowledge of the system's topological structure. For squared-error-based methods, the kernel matrix provides a means to estimate the number of clusters (Girolami, 2002).

Meanwhile, there are also some problems requiring further consideration and investigation. Theoretically, it is important to investigate whether these nonlinear variants can retain some useful and essential properties of the original algorithms and how Mercer kernels contribute to the improvement of the algorithms. The effect of different types of kernel functions, which are rich in the literature, together with the parameter selection, is another interesting topic for further exploration. Also, kernel-based methods usually require solving a quadratic programming problem or calculating the kernel matrix, which may cause a high computational burden for large-scale data sets.

CHAPTER 7

SEQUENTIAL DATA CLUSTERING

7.1. INTRODUCTION

Sequential data consist of a sequence of sets of units with possibly variable length and other interesting characteristics, such as dynamic behaviors and time constraints (Gusfield, 1997; Liao, 2005; Sun and Giles, 2000). These properties make sequential data distinct from most types we have seen in the previous chapters and raise many new questions. For instance, for time series clustering, whether the clustering results are meaningful is a key issue (Keogh et al., 2003; Simon et al., 2006). Often, it is infeasible to represent them as points in the multidimensional feature space and use existing clustering algorithms. Note that the sequential data here can generate clusters that would otherwise not happen, as is discussed in Lin and Chen (2002). For example, consider the data points in Fig. 7.1(a). If they are considered to comprise a sequence, the given partition is no longer reasonable because the data point labeled 10 is not continuous with other data points labeled from 1 to 7. The new partition is depicted in Fig. 7.1(b), where data point 10 is grouped with data points 8–9 and 11–16 to form a continuous sequence, even though it is "closer," in the purely Euclidean sense, to the cluster formed by data points 1–7.

Sequential data could be generated from a large number of task sources, such as DNA sequencing, speech processing, text mining, medical diagnosis, stock market analysis, customer transactions, web data mining, and robot

Clustering, by Rui Xu and Donald C. Wunsch, II

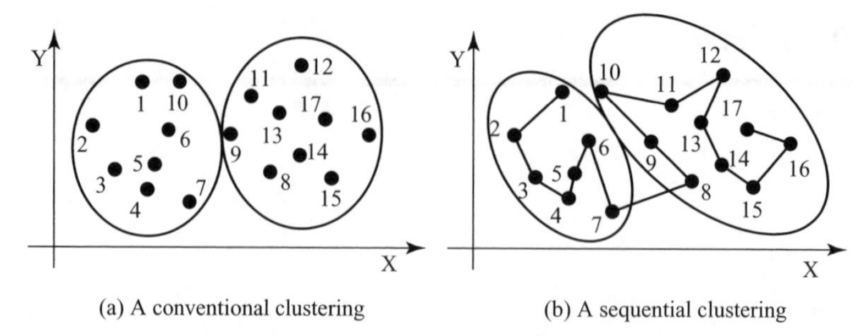

(a) A conventional clustering (b) A sequential clustering

Fig. 7.1. The difference between a conventional clustering and a sequential clustering. (From Lin and Chen, On the optimal clustering of sequential data, Proceedings of the 2nd SIAM International Conference on Data Mining, pp. 141–157, 2002, Copyright ©2002 Society for Industrial and Applied Mathematics. Reprinted with permission.)

sensor analysis, just to name a few (Agrawal and Srikant, 1995; Durbin et al., 1998; D'Urso, 2005; Rabiner, 1989; Sun and Giles, 2000). They can be in the form of continuous data, such as stock time series, or composed of non-numerical symbols, e.g., nucleotides for DNA sequences and amino acids for protein sequences. As emphasized by Sun (2000), "sequential behavior is essential to intelligence, and it is a fundamental part of human activities."

In recent decades, we have seen an enormous growth of sequential data, particularly as a result of the genomic projects and the popularization of the Internet. For example, in genetics, the statistics released on October 15, 2007 (GenBank Release 162.0), shows that there are 81,563,399,765 bases from 77,632,813 sequences in the GenBank database, and release 54.6 of UniProtKB/Swiss-Prot on December 4, 2007, contains 107,100,015 amino acids in 290,484 sequence entries. Moreover, during the 66 days between the close dates for GenBank Releases 162.0 and 161.0, almost 31,738 sequences were added or updated every day, which leads to a total increase of 2,037,840,115 base pairs and 1,486,577 sequence entries. Cluster analysis explores the potential knowledge and information hidden in these huge volumes of sequential data in the context of unsupervised learning, therefore providing an important way to meet current challenges.

In the rest of the chapter, we discuss three major strategies in clustering sequential data. The first scheme, introduced in Section 7.2, is based on the comparison of the distance or similarity among sequences. Section 7.3 focuses on extracting features from the sequences and transforming the sequential data into the familiar samples-features form. In Section 7.4, we present the statistical model-based methods, which are more powerful in disclosing the properties of the sequences. We illustrate the applications of clustering algorithms in genomic sequence analysis in Section 7.5.

G	C	T	T	G	G	A	T	C	C	G
D	M	M	M	D	D	M	D	M	S	M
–	C	T	T	–	–	A	–	C	T	G

M: Match; D: Deletion;
S: Substitution

Fig. 7.2. Illustration of a sequence alignment. A series of edit operations is performed to transform the DNA sequence GCTTGGATCCG into the sequence CTTACTG. The symbol—represents a gap in the second sequence due to the deletion operation. On the contrary, if the sequence CTTACTG is changed to the sequence GCTTGGATCCG, the gaps are replaced with insertion operations.

7.2. SEQUENCE SIMILARITY

Data objects are divided into sets of clusters based on their distance or similarity. If the measure of the distance or similarity between a pair of sequences is available, the proximity-based clustering algorithms, either hierarchical or partitional, could be used directly to group sequences. Since many sequential data are expressed in an alphabetic form, such as DNA or protein sequences, conventional measurement methods discussed in Chapter 2 are inappropriate. If a sequence comparison is regarded as a process of transforming a given sequence into another one with a series of substitution, insertion, and deletion operations, called edit operations, the distance between these two sequences can be measured in terms of the minimum number of required edit operations. The transformation process generates an alignment of the two sequences, as illustrated in Fig. 7.2, where a DNA sequence GCTTGGATCCG is transformed into the sequence CTTACTG with the above edit operations. Accordingly, this defined distance is known as the edit distance or the Levenshtein distance (Gusfield, 1997; Sankoff and Kruskal, 1999). More often, these edit operations are weighted (punished or rewarded) according to some prior domain knowledge, and the distance thereby derived is equivalent to the minimum cost of completing the transformation. In this sense, the distance between two sequences can be reformulated as an optimal alignment problem, which fits well in the framework of dynamic programming (Needleman and Wunsch, 1970; Sankoff and Kruskal, 1999).

In practice, similarity is sometimes used instead of edit distance in the sequence alignment. The difference between the two cases is theoretically trivial, and we only need to change our object from minimizing the cost to maximizing the similarity score. We use similarity in the following discussion due to its wide usage in biological and genomic sequence analysis.

Given two sequences, $\mathbf{x} = (x_1 \ldots x_i \ldots x_N)$ with the length of N and $\mathbf{y} = (y_1 \ldots y_j \ldots y_M)$ with the length of M, the basic dynamic programming-based sequence alignment algorithm, also known as the Needleman–Wunsch algorithm, can be described by the following recursive equation (Durbin et al., 1998; Needleman and Wunsch, 1970),

$$S(i, j) = \max \begin{cases} S(i-1, j-1) + e(x_i, y_j) \\ S(i-1, j) + e(x_i, \phi) \\ S(i, j-1) + e(\phi, y_j) \end{cases}, \tag{7.1}$$

where $S(i, j)$ is the best alignment (similarity) score between sequence segment $(x_1, \ldots x_i)$, $(1 \le i \le N)$ of $\mathbf{x}$ and $(y_1, \ldots y_j)$, $(1 \le j \le M)$ of $\mathbf{y}$, and $e(x_i, y_j)$, $e(x_i, \phi)$, and $e(\phi, y_j)$ represent the scores for aligning x_i to y_j, aligning x_i to a gap (denoted as ϕ), or aligning y_j to a gap, respectively. The computational results for each position at i and j are recorded in a dynamic programming matrix with a pointer that stores current optimal operations and provides an effective path in backtracking the alignment.

An example for aligning DNA sequence GCTTGGATCCG and CTTACTG using the Needleman–Wunsch algorithm is illustrated in Fig. 7.3, which generates the global optimal alignment shown in Fig. 7.2. Here, the scoring scheme is assumed as:

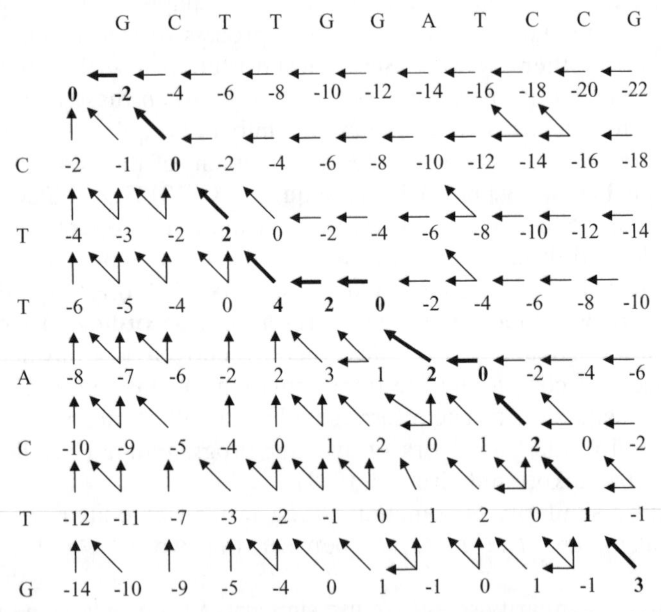

Fig. 7.3. Sequence alignment with the Needleman–Wunsch algorithm. Each value in a cell of the global dynamic programming matrix represents the corresponding best similarity score. The arrows are backtracking pointers. The optimal global alignment path is highlighted in bold, with a total score of 3.

- $e(x_i, y_j) = 2$, if x_i is the same as y_j (match);
- $e(x_i, y_j) = -1$, if x_i is different from y_j (mismatch);
- $e(x_i, \phi) = e(\phi, y_j) = -2$, if a gap is added (gap penalty).

We start with the alignment by creating an $(N + 1) \times (M + 1)$ (here $N = 7$ and $M = 11$) dynamic programming matrix and set the value of the element in the top-left corner, denoted as (0,0). The remaining cells in the first row are filled with values from −2 to −22 because they just correspond to the addition of gaps. Similarly, the remaining cells in the first columns have values from −2 to −14. For element (1,1) in row 2 and column 2, we can calculate its value using Eq. 7.1, as,

$$S(1,1) = \max\begin{cases} S(0,0) + (-1) \\ S(0,1) + (-2) \\ S(1,0) + (-2) \end{cases} = \max\begin{cases} 0 + (-1) \\ (-2) + (-2) \\ (-2) + (-2) \end{cases} = -1, \tag{7.2}$$

and place the resulting score, −1, in the matrix with a pointer pointing back to cell (0,0), which leads to the current maximum similarity score. This process repeats filling cells until reaching the bottom-right corner (7,11), with the maximum alignment score of 3, as shown in Fig. 7.3.

In order to achieve the alignment, a backtracking process is then performed, which works in a reverse order and follows the directions of the pointers. At each step, we move back from the current cell (i, j) to cell $(i - 1, j - 1)$, $(i, j - 1)$, or $(i - 1, j)$, corresponding to the addition of x_i and y_j, gap and y_j, or x_i and gap to the current alignment. In this example, we add $x_7 = \mathrm{G}$ and $y_{11} = \mathrm{G}$ to the alignment at the first step because the pointer points to cell (6,10). Similarly, we add $x_6 = \mathrm{T}$ and $y_{10} = \mathrm{C}$ and $x_5 = \mathrm{C}$ and $y_9 = \mathrm{C}$ to the alignment for the following two steps. Then, as the pointer at cell (4,8) points to cell (4,7), which indicates a deletion of y_8, we add a gap and $y_8 = \mathrm{T}$ to the alignment. The procedure is complete when we finally reach cell (0,0). The resulting optimal global alignment between these two sequences is shown in Fig. 7.2.

For this example, there is only one optimal alignment for the given sequences. However, for some other cases, there may exist more than one optimal alignment that leads to maximum similarity. This fact becomes clearer with the observation of some cells (e.g., cell (7,10)) in Fig. 7.3, which have two or three pointers indicating the equal importance of edit operations in terms of the similarity score. In this case, we may obtain a set of possible optimal alignments with equal similarity scores.

The Needleman-Wunsch algorithm considers the comparison of the whole length of two sequences and therefore performs a global optimal alignment. However, it is also important to find local similarity among sequences in many circumstances, or in other words, to find their subsequences that have the best alignment. The Smith-Waterman algorithm achieves that by allowing a new alignment to begin during the recursive computation and an existing alignment to stop anywhere in the dynamic programming matrix (Durbin et al., 1998; Smith and Waterman, 1980). These changes are reflected in the formula below,

$$S(i, j) = \max \begin{cases} S(i-1, j-1) + e(x_i, y_j), \\ S(i-1, j) + e(x_i, \phi), \\ S(i, j-1) + e(\phi, y_j), \\ 0. \end{cases}, \tag{7.3}$$

together with the backtracking procedure, which does not necessarily start at the bottom-right corner of the matrix. On the contrary, it begins from the maximum similarity score in the matrix and continues until a zero is reached. By not allowing negative scores in the matrix, the Smith-Waterman algorithm terminates the bad alignments and starts a new one. In Fig. 7.4, the Smith-Waterman algorithm finds a subsequence CTT in both sequence GCTTGGATCCG and CTTACTG, which generates a perfect match with a maximum similarity score of 6. Furthermore, the alignments with the second or third highest score may also provide some meaningful insights about the similarity of the sequences of interest.

More detailed discussions of alignments with more complicated models can be found in Durbin et al. (1998) and Gusfield (1997). It is interesting to point

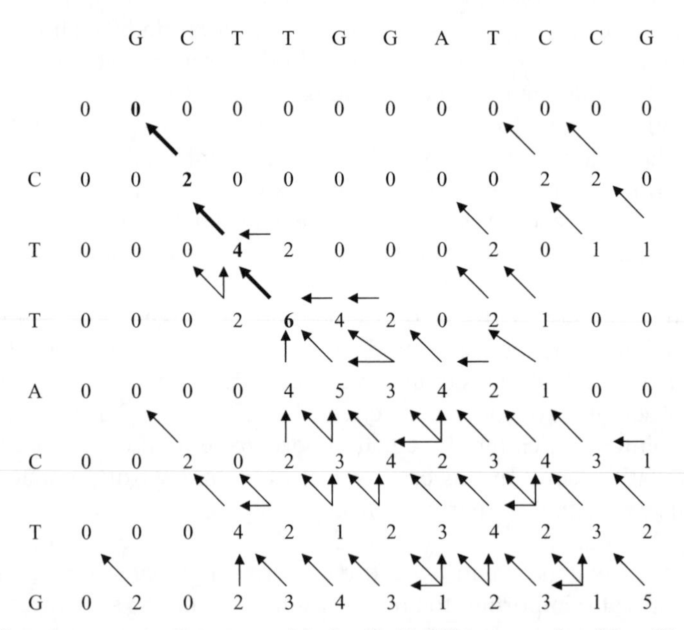

Fig. 7.4. Local sequence alignment with the Smith-Waterman algorithm. The optimal local alignment path is highlighted in bold, with a total score of 6.

out that dynamic programming algorithms, like the Needleman–Wunsch algorithm, are guaranteed to generate the optimal alignment or sets of alignments based on the similarity score (Durbin et al., 1998). However, the computational burden becomes a major obstacle as the lengths of sequences increase. For example, for both the global and local alignment algorithms, the computational complexity is $O(NM)$, which usually is said to be quadratic because N and M are usually comparable. Thus, in order to scale to compare sequences with the order of millions in length, shortcuts to dynamic programming have been developed, which trade speed with sensitivity (Durbin et al., 1998; Gusfield, 1997). Although some best scoring alignments will be missed as a result of this tradeoff, these heuristics still provide effective ways of investigating the sequences' similarity and therefore are widely used in real practice. We will provide more discussion of some heuristic alignment algorithms in Section 7.5 in the context of biological sequences analysis. Other possible examples include applications in speech recognition (Sankoff and Kruskal, 1999) and navigation pattern clustering on websites (Hay et al., 2001).

7.3. INDIRECT SEQUENCE CLUSTERING

The design cycle of cluster analysis involves an important step of feature selection and extraction. These features provide a natural representation of the objects under investigation and their measures are further used as the inputs to the clustering systems. Thus, if we treat the sequential data as raw and unprocessed, we could employ an indirect strategy to extract potential features from them. All the sequences are then mapped onto the points in the transformed feature space, where classical vector space-based clustering algorithms can be used to generate clusters.

Guralnik and Karypis (2001) considered features as the sequential patterns within a pre-specified length range, with a minimum support constraint, which satisfy the three important characteristics summarized as follows:

1. Effectiveness in capturing the sequential relations between the different sets of elements contained in the sequences;
2. Wide presence in the sequences;
3. Completeness of the derived feature space.

Here, a sequential pattern is defined as a list of sets of elements with the percentage of sequences containing it, called support, above some threshold. By indicating the potential dependency between two sequential patterns, Guralnik and Karypis (2001) further suggested both global and the local approaches to prune the initial feature sets, or select the independent features, in order to achieve a better representation of the sequences in the new feature space. Specifically, in the context of global feature selection, the independence of two sequential patterns requires that there is no overlap between them and that there is a trivial intersection of their respective supporting sets. The corresponding features are then selected using a greedy algorithm based on this definition. On the other hand, in local selection, two features are said to be independent if they are supported by non-overlapping segments of the underlying sequences. In this way, a maximal set of independent features is selected for each sequence individually. The resulting clusters are formed by using the K-means algorithm on the generated features.

Similarly, Morzy et al. (1999) used sequential patterns as the basic elements in agglomerative hierarchical clustering and defined a co-occurrence measure as the standard of fusion of smaller clusters.

In the case of sequences with continuous data, Vlachos et al. (2003) calculated the Haar wavelet decomposition (Chan and Fu, 1999) for all sequential data at the first stage of their I-K Means algorithm, which makes it possible to analyze data at different resolutions. K-means is then applied to each level of resolution in the direction of finer levels. At each resolution, the centroids obtained from the previous level are used to initialize the centroids for the current level of clustering. This strategy effectively overcomes the drawback of the original K-means that the clustering results are largely dependent on the initialization. The algorithm also achieves a certain degree of improvement in the performance time because of the dimensionality reduction. Another application of fMRI time series clustering, in terms of the feature space constructed with the cross-correlation function, was proposed by Goutte et al. (1999). Both K-means and hierarchical clustering were used in that study.

7.4. MODEL-BASED SEQUENCE CLUSTERING

The strategies discussed before either directly work on the raw data with a definition of the distance or similarity measure or indirectly extract features from the raw data and use the existing clustering algorithms for feature vector-based data. In this section, we focus on the direct construction of probabilistic models to capture the dynamics of each group of sequences, which can be applied to both numerical and categorical sequences. Now, each cluster of sequences is said to be generated from a certain component density of the mixture density, as a generalization of the mixture density-based clustering in the feature space discussed in Chapter 4. Herein, the component models could be any probabilistic models (Gaffney and Smyth, 1999; Policker and Geva, 2000; Smyth, 1997; Xiong and Yeung, 2004), not necessarily the commonly used Gaussian or t-distribution in non-sequence data clustering. Among all the possible models, the hidden Markov model (HMM) (Oates et al., 2000; Owsley et al., 1997; Smyth, 1997) perhaps is the most important one, which first gained its popularity in the application of speech recognition (Rabiner, 1989).

7.4.1 Hidden Markov Model

A discrete HMM describes an unobservable stochastic process consisting of a set of states, each of which is related to another stochastic process that emits observable symbols. Specifically, an HMM is completely specified by the following:

- A discrete set $\Omega = \{\omega_1, \ldots, \omega_N\}$ with N unobservable states;
- A discrete set $\mathbf{O} = \{o_1, \ldots, o_M\}$ with M observation symbols;
- A state transition probability distribution $\mathrm{A} = \{\alpha_{ij}\}$, where α_{ij} represents the transition probability from state ω_i at time t, denoted as $\omega_i(t)$, to state ω_j at time $t + 1$,

$$\alpha_{ij} = P(\omega_j(t+1)|\omega_i(t)), \quad 1 \le i, j \le N, \tag{7.4}$$

  and satisfies the normalization condition,

$$\sum_{j=1}^{N} \alpha_{ij} = 1, \quad \text{for all } i; \tag{7.5}$$

- A symbol emission probability distribution $\mathrm{B} = \{\beta_{il}\}$, where β_{il} represents the emission probability of the symbol o_l at time t, denoted as $o_l(t)$, in the state ω_i,

$$\beta_{il} = P(o_l(t)|\omega_i(t)), \tag{7.6}$$

and satisfies the normalization condition,

$$\sum_{l=1}^{M} \beta_{il} = 1, \quad \text{for all } i; \tag{7.7}$$

- An initial state distribution $\pi = \{\pi_i\}$, where

$$\pi_i = P(\omega_i(1)), \quad 1 \le i \le N. \tag{7.8}$$

In the following discussion, we use the notation

$$\boldsymbol{\lambda} = \{\mathrm{A}, \mathrm{B}, \pi\} \tag{7.9}$$

to represent the complete parameters of an HMM.

Figure 7.5 shows a 4-state HMM, in which each state could be reached from every other state in the model, i.e., all the transition probabilities are nonzero. For this reason, the HMM is also called ergodic, although in many cases, some transitions between the states do not occur, such as the left-

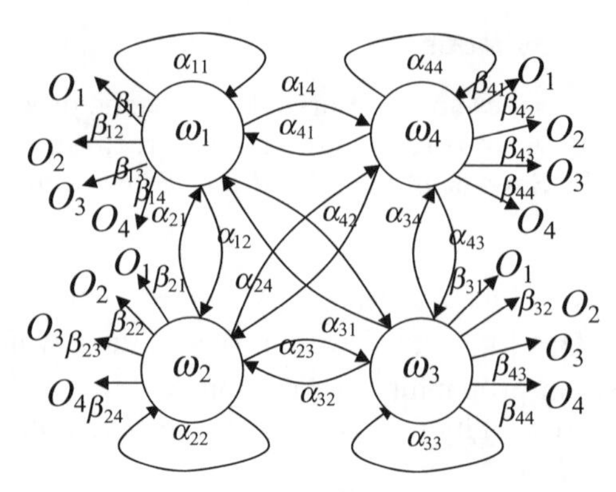

Fig. 7.5. A four-state hidden Markov model. Each hidden state is associated with four visible observations.

right model commonly seen in modeling speech and genomic sequences (Durbin et al., 1998; Rabiner, 1989). An extreme case exists when the probability for self-transition is 1, which makes the process remain at a certain state, known as an absorbing state. It also can be seen from the figure that each hidden state could produce four visible symbols based on the emission probabilities.

After an initial state ω_i is selected according to the initial distribution π, a symbol o_l is emitted with the emission probability β_{il}. The next state ω_j is decided by the state transition probability α_{ij}, and it also produces another symbol o_m based on β_{jm}. The process repeats in this manner until the final state is reached. As a result of this procedure, a sequence of symbol observations is generated, but not the actual states, which explains why the word "hidden" is used to name this type of model.

The relation between an HMM and a recurrent back-propagation network was elucidated in Hwang et al. (1989) from the perspectives of both the retrieving phase and the learning phase. Furthermore, a universal algorithmic framework was formulated to describe both the computational and the structural properties of the HMM and the neural network. Also, the hybrid of neural networks and HMMs, together with the recent advance in HMMs, is reviewed in Bengio (1999).

HMMs are well-founded theoretically (Rabiner, 1989). Particularly, they introduce three central problems that are summarized as follows:

1. Likelihood evaluation. Given an HMM $\boldsymbol{\lambda}$ and a sequence of observations $\mathbf{o}$, compute the probability that this sequence is produced by the model;
2. State interpretation (decoding). Given an HMM $\boldsymbol{\lambda}$ and a sequence of observations $\mathbf{o}$, find a corresponding state sequence that is most likely to generate these observations;

3. Parameter estimation (HMM training). Given a set of training sequences, determine the optimal model parameters that best describe these sequences.

7.4.2 Likelihood Evaluation

Supposing we have an HMM $\boldsymbol{\lambda}$ and a sequence of observations $\mathbf{o} = \{o(1), o(2), \ldots, o(T)\}$, a straightforward way to calculate the conditional probability of $\mathbf{o}$ given λ is to sum the joint probability $P(\mathbf{o}, \boldsymbol{\omega}|\boldsymbol{\lambda})$ over all possible state sequences $\boldsymbol{\omega}^f = \{\omega^f(1), \omega^f(2), \ldots, \omega^f(T)\}$,

$$\begin{aligned} P(\mathbf{o}|\boldsymbol{\lambda}) &= \sum_{f=1}^{N^T} P(\mathbf{o}, \boldsymbol{\omega}^f|\boldsymbol{\lambda}) \\ &= \sum_{f=1}^{N^T} P(\mathbf{o}|\boldsymbol{\omega}^f, \boldsymbol{\lambda}) P(\boldsymbol{\omega}^f|\boldsymbol{\lambda}) \end{aligned}, \tag{7.10}$$

where the probability of the observation sequence given the state sequence is written as

$$\begin{aligned} P(\mathbf{o}|\boldsymbol{\omega}^f, \boldsymbol{\lambda}) &= \prod_{t=1}^{T} P(o(t)|\omega^f(t), \boldsymbol{\lambda}) \\ &= \beta_{\omega^f(1)o(1)} \beta_{\omega^f(2)o(2)} \cdots \beta_{\omega^f(T)o(T)} \end{aligned}, \tag{7.11}$$

and the probability of a state sequence is

$$\begin{aligned} P(\boldsymbol{\omega}^f|\lambda) &= \prod_{t=1}^{T} P(\omega^f(t)|\omega^f(t-1)) \\ &= \pi_{\omega^f(1)} \alpha_{\omega^f(1)\omega^f(2)} \alpha_{\omega^f(2)\omega^f(3)} \cdots \alpha_{\omega^f(T-1)\omega^f(T)} \end{aligned}. \tag{7.12}$$

By combining Eqs. 7.11 and 7.12 into Eq. 7.10, we now have

$$\begin{aligned} P(\mathbf{o}|\boldsymbol{\lambda}) &= \sum_{f=1}^{N^T} \prod_{t=1}^{T} P(o(t)|\omega^f(t), \boldsymbol{\lambda}) P(\omega^f(t)|\omega^f(t-1)) \\ &= \sum_{f=1}^{N^T} \pi_{\omega^f(1)} \beta_{\omega^f(1)o(1)} \alpha_{\omega^f(1)\omega^f(2)} \beta_{\omega^f(2)o(2)} \cdots \alpha_{\omega^f(T-1)\omega^f(T)} \beta_{\omega^f(T)o(T)} \end{aligned}. \tag{7.13}$$

The computational complexity for this exhaustive enumeration method is $O(N^T T)$, which is computationally infeasible in practice. For example, even for a sequence with 10 states and 50 observations, we need on the order 10^{52} computations. In order to compute the likelihood more efficiently, a two-part Forward-Backward algorithm is used. Both algorithms can provide solutions to the problem with the computational complexity $O(N^2 T)$. The above example only requires on the order of 5,000 computations, which reduces the calculation by more than 48 orders of magnitude.

These algorithms create a forward variable (in the Forward algorithm) or backward variable (in the Backward algorithm) that make it possible to calculate the likelihood in a recursive way. The forward variable $\delta_i(t)$ is defined as the probability that the HMM is in the state ω_i at time t and produces a segment of sequences $\{o(1), o(2), \ldots, o(t)\}$,

$$\delta_i(t) = P(o(1)o(2),\ldots o(t), \omega_i(t)|\boldsymbol{\lambda}) = \begin{cases} \pi_i \beta_{io(1)}, & t=1 \\ \left(\sum_{j=1}^{N} \delta_j(t-1)\alpha_{ji}\right)\beta_{io(t)}, & t=2,\ldots,T \end{cases}. \quad (7.14)$$

Beginning with the initialization of $\delta_i(1)$ for all states, the Forward algorithm recursively calculates the forward variable at each time t using Eq. 7.14 until the end of the observation is reached. This operation is depicted in Fig. 7.6, where the forward variable $\delta_2(\tau)$ for the state ω_2 at time τ is obtained by multiplying the emission probability $\beta_{2o(\tau)}$ by the summed product of the forward variable $\delta_j(\tau-1)$ at time $\tau-1$ and the corresponding transition probability α_{j2}. Finally, we can calculate $P(\mathbf{o}|\boldsymbol{\lambda})$ as the sum of the terminal forward variables,

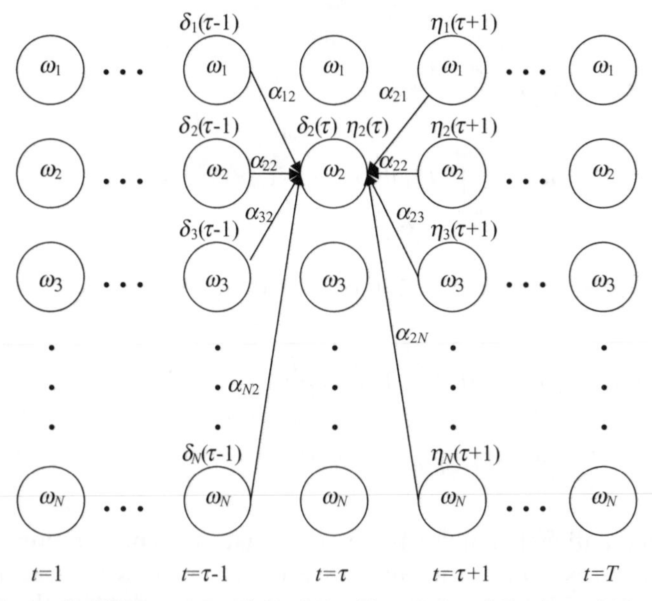

Fig. 7.6. The probability computation of the Forward and Backward algorithm. The HMM is unfolded in time from $t = 1$ to T. The forward variable $\delta_2(\tau)$ and the backward variable $\eta_2(\tau)$ for state ω_2 at time $t = \tau$ are calculated with Eqs. 7.14 and 7.16, respectively.

$$P(\mathbf{o}|\boldsymbol{\lambda}) = \sum_{i=1}^{N} \delta_i(T). \tag{7.15}$$

In contrast, the Backward algorithm operates in a reverse way from the end to the beginning of the observation sequence. Here, the backward variable $\eta_i(t)$ represents the probability that the HMM given in state ω_i at time t produces a segment of sequences $\{o(t+1), o(t+2), \ldots, o(T)\}$,

$$\eta_i(t) = P(o(t+1)o(t+2), \ldots o(T)|\omega_i(t), \boldsymbol{\lambda})$$
$$= \begin{cases} 1, & t = T \\ \sum_{j=1}^{N} \eta_j(t+1)\alpha_{ij}\beta_{jo(t+1)}, & t = T-1, \ldots, 1 \end{cases}. \tag{7.16}$$

The computation of the backward variable is also illustrated in Fig. 7.6.

7.4.3 State Interpretation

In practice, states usually have certain physical meanings and it is important to infer the most likely state sequence given the observance sequence and the model. If we directly calculate the probability that the HMM is in state ω_i at time t, given $\mathbf{o}$ and $\boldsymbol{\lambda}$, using the forward and backward variables,

$$P(\omega_i(t)|\mathbf{o}, \boldsymbol{\lambda}) = \frac{\delta_i(t)\eta_i(t)}{P(\mathbf{o}|\boldsymbol{\lambda})}, \tag{7.17}$$

the individual most likely state at each time t can be obtained via

$$\omega(t) = \arg\max_i (P(\omega_i(t)|\mathbf{o}, \boldsymbol{\lambda})). \tag{7.18}$$

However, since this method considers each state individually instead of the full sequence, it is possible that the resulting path obtained from the connection of all these optimal states is actually not allowed in the model.

A more reasonable method is to keep track of the current best state ω_j at time $t-1$ that leads to the highest joint probability of the observance sequence $\{o(1), o(2), \ldots, o(t)\}$ and state sequence $\{\omega(1), \omega(2), \ldots, \omega(t) = \omega_i\}$. This probability can be calculated recursively, written as

$$\varsigma_i(t) = \left(\max_j \varsigma_j(t-1)\alpha_{jt}\right)\beta_{io}(t), \tag{7.19}$$

and the corresponding best state is recorded in the variable $\zeta_i(t)$. Now, we are ready to present the Viterbi algorithm (Viterbi, 1967), which falls into a category of dynamic programming algorithms (Bellman, 1957):

1. Initialization ($t = 1$):

$$\varsigma_i(1) = \pi_i \beta_{io(1)}, \quad 1 \le i \le N, \tag{7.20}$$

$$\zeta_i(1) = 0, \quad 1 \le i \le N; \tag{7.21}$$

2. Recursion ($t = 2, \ldots T$):

$$\varsigma_i(t) = \left(\max_j \varsigma_j(t-1)\alpha_{jt}\right)\beta_{io(t)}, \quad 1 \le i \le N, \tag{7.22}$$

$$\zeta_i(t) = \arg\max_j \varsigma_j(t-1)\alpha_{ji}, \quad 1 \le i \le N; \tag{7.23}$$

3. Termination:

$$\varsigma^* = \max_j \varsigma_j(T), \tag{7.24}$$

$$\omega^*(T) = \arg\max_j \varsigma_j(T); \tag{7.25}$$

4. Backtracking ($t = T - 1, T - 2, \ldots, 1$):

$$\omega^*(t) = \zeta_{\omega^*(t+1)}(t+1). \tag{7.26}$$

7.4.4 Parameter Estimation

In clustering with HMMs, the parameters $\boldsymbol{\lambda}$ for a certain model are unknown and must be learned from the training observance sequences. This is constructed as a model learning or parameter estimation problem with the goal of maximizing the probability $P(\mathbf{o}|\boldsymbol{\lambda})$. However, there exists no known method that could provide the optimal solutions. As we recall the expectation-maximization (EM) algorithm used for parameter estimation in mixture clustering, it is clear that the EM algorithm can also be applied here to estimate the HMM parameters, which is known as the Baum-Welch algorithm in this context (Rabiner, 1989).

Specifically, the Baum-Welch algorithm considers the estimation of the state transition probability from ω_i to ω_j as the ratio between the expected number of transitions from ω_i to ω_j and the expected total number of transitions from ω_i. Similarly, the estimation of the emission probability of o_l in the state ω_i is obtained by calculating the ratio between the expected number of times o_l is emitted in state ω_i and the expected total number of times in ω_i. The estimation of the initial probability at state ω_i is just the expected frequency in ω_i at time $t = 1$. These required frequencies can be calculated via the probability of transition from $\omega_i(t)$ to $\omega_j(t + 1)$, given $\boldsymbol{\lambda}$ and $\mathbf{o}$, written as,

$$v_{ij}(t) = P(\omega_i(t), \omega_j(t+1) | \mathbf{o}, \boldsymbol{\lambda}) = \frac{\delta_i(t)\alpha_{ij}\beta_{jo(t+1)}\eta_j(t+1)}{P(\mathbf{o}|\boldsymbol{\lambda})}. \quad (7.27)$$

By using $v_{ij}(t)$, we can write the estimation of HMM parameters as,

$$\hat{\pi}_i = \sum_{j=1}^{N} v_{ij}(1), \quad (7.28)$$

$$\alpha_{ij} = \frac{\sum_{t=1}^{T-1} v_{ij}(t)}{\sum_{t=1}^{T-1}\sum_{j=1}^{N} v_{ij}(t)}, \quad (7.29)$$

$$\hat{\beta}_{il} = \frac{\sum_{t=1, o(t)=o_l}^{T} \sum_{j=1}^{N} v_{ij}(t)}{\sum_{t=1}^{T}\sum_{j=1}^{N} v_{ij}(t)}. \quad (7.30)$$

Accordingly, the basic procedure of the Baum-Welch algorithm is summarized in Fig. 7.7.

7.4.5 HMMs with Continuous Observation Densities

The discussion above is based on the assumption that the emission probability is discrete, i.e., there are only a finite number of symbols that could be generated for every state. However, in practice, it is not rare to have observation sequences with continuous probability densities. In this case, the emission distribution takes the form of finite mixtures (e.g., the Gaussian mixtures commonly used) represented as,

$$\beta_{i\mathbf{o}} = \sum_{j=1}^{M} w_{ij}\Psi(\mathbf{o}; \boldsymbol{\mu}_{ij}, \boldsymbol{\Sigma}_{ij}), \quad (7.31)$$

where w_{ij} is the mixture coefficient and satisfies the constraint

$$w_{ij} \geq 0, 1 \leq i \leq N, 1 \leq j \leq M, \quad (7.32)$$

and

$$\sum_{j=1}^{M} w_{ij} = 1, \quad 1 \leq i \leq N, \quad (7.33)$$

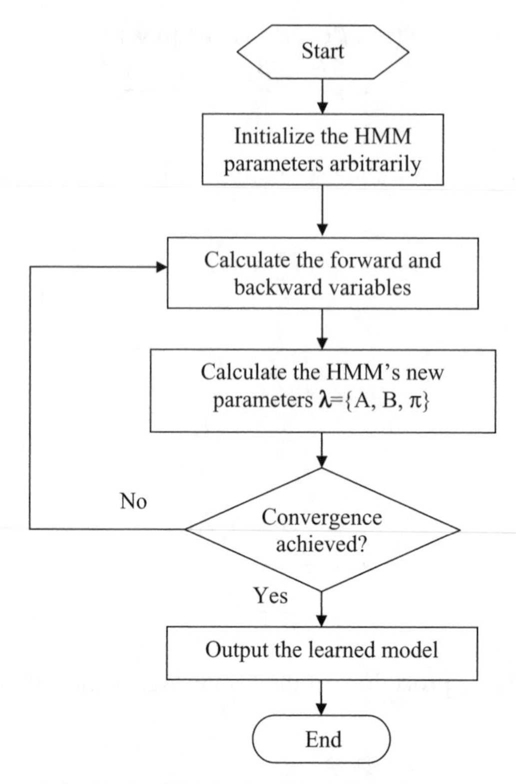

Fig. 7.7. Flowchart of the Baum-Welch algorithm. The algorithm iteratively estimates the model parameters until the changes of the estimated parameters are below some threshold, a satisfying likelihood is obtained, or just some prespecified maximum number of iterations is reached.

and $\Psi(\mathbf{o}; \boldsymbol{\mu}, \boldsymbol{\Sigma})$ is a density function with mean vector $\boldsymbol{\mu}$ and covariance matrix $\boldsymbol{\Sigma}$.

Following the similar strategy used in discrete HMMs, we could estimate the parameters of the mixture density with the following formulas,

$$\hat{w}_{ij} = \frac{\sum_{t=1}^{T} \xi_{ij}(t)}{\sum_{t=1}^{T} \sum_{j=1}^{M} \xi_{ij}(t)}, \tag{7.34}$$

$$\hat{\boldsymbol{\mu}}_{ij} = \frac{\sum_{t=1}^{T} \xi_{ij}(t) \cdot \mathbf{o}(t)}{\sum_{t=1}^{T} \xi_{ij}(t)}, \tag{7.35}$$

$$\widehat{\Sigma}_{ij} = \frac{\sum_{t=1}^{T} \xi_{ij}(t) \cdot (\mathbf{o}(t) - \hat{\boldsymbol{\mu}}_{ij})(\mathbf{o}(t) - \hat{\boldsymbol{\mu}}_{ij})^T}{\sum_{t=1}^{T} \xi_{ij}(t)}, \tag{7.36}$$

where the superscript T in Eq. 7.36 represents vector transpose and $\xi_{ij}(t)$ is the probability of state ω_i at time t with $\mathbf{o}(t)$ produced by the jth mixture component, defined as,

$$\xi_{ij}(t) = \frac{\delta_i(t)\eta_i(t)}{\sum_{l=1}^{N} \delta_l(t)\eta_l(t)} \times \frac{w_{ij}\Psi(\mathbf{o}(t), \boldsymbol{\mu}_{ij}, \boldsymbol{\Sigma}_{ij})}{\sum_{m=1}^{M} w_{im}\Psi(\mathbf{o}(t), \boldsymbol{\mu}_{im}, \boldsymbol{\Sigma}_{im})}. \tag{7.37}$$

7.4.6 HMM-Based Clustering

In the framework of finite mixture-based clustering, given a set of L sequences $\{\mathbf{o}_1, \mathbf{o}_2, \ldots \mathbf{o}_L\}$ from K clusters $\{C_1, C_2, \ldots C_K\}$, the mixture probability density can be expressed as

$$p(\mathbf{o}|\boldsymbol{\theta}) = \sum_{i=1}^{K} p(\mathbf{o}|C_i, \boldsymbol{\theta}_i)P(C_i), \tag{7.38}$$

where $p(\mathbf{o}|C_i, \boldsymbol{\theta}_i)$ is the component density with parameters $\boldsymbol{\theta}_i$, and the prior probability $P(C_i)$ for the cluster C_i satisfied the constraint $\sum_{i=1}^{K} P(C_i) = 1$. The difference between this definition and the one in Eq. 4.28 of Chapter 4 is that data take the form of sequences instead of vectors in the feature space. Correspondingly, the component model could be any probabilistic model that is suitable for describing this change. A discussion of the situation in which data contain both sequential components and static features was given in Smyth (1999). In the following discussion, we will focus on the component models in the form of HMMs. In this context, the parameters $\boldsymbol{\theta}$ include initial state probabilities, state transition probabilities, and observation emission probabilities, i.e., $\boldsymbol{\lambda} = \{\pi, \mathrm{A}, \mathrm{B}\}$.

Smyth (1997) suggested that the mixture model of HMMs could be described in terms of a single composite HMM with the transition matrix

$$\mathrm{A} = \begin{pmatrix} \mathrm{A}_1 & & & 0 \\ & \mathrm{A}_2 & & \\ & & \mathrm{O} & \\ 0 & & & \mathrm{A}_K \end{pmatrix}, \tag{7.39}$$

where A_i is the transition distribution for a certain component model associated with the cluster C_i. The initial distribution of this HMM is determined based on the prior probability $P(C_i)$ for each cluster. Therefore, in order to produce a sequence, block A_i along the diagonal is first selected with the probability $P(C_i)$, and the data are generated using the distribution A_i and B_i. Based on the composite HMM, the basic procedure for the HMM clustering algorithm is summarized as follows (Smyth, 1997):

1. Model each sequence $\mathbf{o}_i$, $1 \leq i \leq L$, with an N-state HMM;
2. Calculate the log-likelihood $\log(P(\mathbf{o}_i|\boldsymbol{\lambda}_j))$ of each sequence $\mathbf{o}_i$ with respect to a given model λ_j, $1 \leq j \leq L$;
3. Group all sequences into K clusters using the distance measure based on the log-likelihood;
4. Model each cluster with an HMM, and initialize the overall composite HMM using the derived K HMMs;
5. Train the composite HMM with the Baum-Welch algorithm.

Intrinsically, the first four steps provide an initialization to the composite HMM so that the EM-based Baum-Welch algorithm can achieve effective performance. In essence, the HMM clustering algorithm is a hybrid of both distance-based and model-based methods. Smyth (1997) used the symmetrized distance,

$$D(\boldsymbol{\lambda}_i, \boldsymbol{\lambda}_j) = \frac{P(\mathbf{o}_i|\boldsymbol{\lambda}_j) + P(\mathbf{o}_j|\boldsymbol{\lambda}_i)}{2}, \tag{7.40}$$

to measure the dissimilarity between a pair of HMMs and hierarchical clustering to divide the sequences into K clusters in step 3. A Monte-Carlo cross validation method was also introduced to estimate the possible number of clusters K. An extension of the HMM clustering algorithm produced by combining the effects of context variables, on which the HMMs are conditioned, was proposed in Hoey (2002) and applied to facial display sequence clustering.

Oates et al. (2000) addressed the initial problem by pre-grouping the sequences with agglomerative hierarchical clustering, which operates on the proximity matrix determined by the dynamic time warping (DTW) technique (Sankoff and Kruskal, 1999). Given a pair of sequences, $\mathbf{o}^1 = \{o^1(1) \ldots o^1(i) \ldots o^1(T_1)\}$ with the length of T_1 and $\mathbf{o}^2 = \{o^2(1) \ldots o^2(j) \ldots o^2(T_2)\}$ with the length of T_2, DTW aligns them while attempting to achieve the minimal difference. The warping path with the optimal distance can be obtained by dynamic programming, with the cumulative distance $D_{cum}(i,j)$ calculated using the following recursive equation,

$$D_{cum}(i, j) = d(o^1(i), o^2(j)) + \min(D_{cum}(i-1, j-1), D_{cum}(i, j-1), D_{cum}(i-1, j)), \tag{7.41}$$

where $d(o^1(i), o^2(j))$ represents the current distance between a pair of points $o^1(i)$ and $o^2(j)$. In visualization, the area formed between one original sequence and the new sequence, generated by warping the time dimension of another original sequence, reflects the similarity of the two sequences.

Li and Biswas (1999) suggested a scheme with four-level nested searches, from the outermost to the innermost level, for:

1. the number of clusters in a partition, based on the partition mutual information (PMI) measure;
2. the structure for a certain partition with a fixed size, based on K-means or depth-first binary divisive clustering;
3. the HMM structure for each cluster, selected with the marginal likelihood criterion calculated with the Bayesian information criterion (Schwarz, 1978) and Cheeseman-Stutz approximation (Cheeseman and Stutz, 1996); and
4. the parameters for each HMM according to a segmental K-means procedure.

Particularly, the PMI measure functions as an object criterion to determine the partitional size K, which is defined as,

$$PMI = \frac{\sum_{j=1}^{K} \sum_{i=1}^{N_j} MI_i}{K}, \tag{7.42}$$

where N_j is the number of data objects in the cluster C_j, and MI_i is the average mutual information between the observation sequence $\mathbf{o}_i$ and all K HMM models, written as,

$$\begin{aligned} MI_i &= \log P(\boldsymbol{\lambda}_i|\mathbf{o}_i) \\ &= \log P(\mathbf{o}_i|\boldsymbol{\lambda}_i)P(\boldsymbol{\lambda}_i) - \log \sum_{j=1}^{K} P(\mathbf{o}_i|\boldsymbol{\lambda}_i)P(\boldsymbol{\lambda}_i). \end{aligned} \tag{7.43}$$

7.4.7 Other Models for Sequence Clustering

The component models in Eq. 7.38 could also take other forms, such as mixtures of Markov chains (Cadez et al, 2000a; Ramoni et al., 2002; Sebastiani et al, 2000; Smyth, 1999), polynomial models (Bagnall et al., 2003; DeSarbo and Cron, 1988; Gaffney and Smyth, 1999), and autoregressive moving average (ARMA) models (Bagnall and Janacek, 2004; Maharaj, 2000; Xiong and Yeung, 2004).

7.4.7.1 *Mixtures of ARMA Models* Given a sequence $\mathbf{o} = \{o(1), \ldots o(i), \ldots o(T)\}$ with length T, the ARMA model with autoregressive order p and moving average order q, denoted as ARMA(p,q), is written as,

$$o(i) = \phi_0 + \sum_{j=1}^{p} \phi_j o(i-j) + \sum_{j=1}^{q} \psi_j \varepsilon(i-j) + \varepsilon(i), \quad \text{for } i = 1, \ldots T, \tag{7.44}$$

where ϕ_j and ψ_j are the coefficients for the model, and $\varepsilon(i)$ are independent and identically distributed Gaussian white noise terms with mean zero and variance σ^2. Note that in theory, finite order ARMA models are equivalent to infinite order AR models, denoted as AR(∞) (Box et al., 1994). More often, truncated AR(∞) models with finite order are used in practice.

Maharaj (2000) fitted each time series $\mathbf{o}_i$ with an AR(p) model with order p. The structure and parameters of the AR(p) model, denoted as $\boldsymbol{\theta}_i$, are estimated through the generalized least squares estimators and Akaike's information criterion (Akaike, 1974) or Bayesian information criterion (Schwarz, 1978). The time series are grouped using a procedure similar to agglomerative hierarchical clustering, which is based on the p-values of a chi-square distributed test statistic. This test statistic examines the null hypothesis that there is no difference between the generating processes of a pair of time series $\mathbf{o}_i$ and $\mathbf{o}_j$, i.e., $\boldsymbol{\theta}_i = \boldsymbol{\theta}_j$. The series are clustered together only when their associated p-values are greater than some pre-specified significance level. Like the HMM clustering algorithm, this method is a combination of model-based and distance-based strategies.

Bagnall and Janacek (2004) investigated the effects of clipping, which discretizes continuous sequential data into binary sequences by using 1 or 0 to represent points above or below the median or mean, respectively, on the time series clustering. The clipped data are fitted with ARMA models and clustered with K-means and K-medoids algorithms (Chapter 4) with the Euclidean distance on the fitted parameters as the distance measure. Based on their experimental results, they concluded that the clustering accuracy on the clipped sequence does not decrease significantly from the original sequence, as long as the sequence is long enough. Also, when outliers exist in the data, the clustering performance could achieve significant improvement on the clipped data. Another advantage when using the clipping procedure lies in the reduction of both time and space complexity. Similar conclusions are also reached when mixtures of regression models are used (Bagnall et al., 2003).

In contrast to the approaches discussed above, Xiong and Yeung (2004) used the EM algorithm to estimate parameters for the ARMA mixtures. Supposing we have a set of L sequences $\{\mathbf{o}_1, \mathbf{o}_2, \ldots \mathbf{o}_L\}$, the goal of the EM algorithm is to estimate the mixture model parameter $\boldsymbol{\theta}$ that maximizes the log-likelihood,

$$l(\boldsymbol{\theta}; \mathbf{o}) = \sum_{j=1}^{L} \ln p(\mathbf{o}_j|\boldsymbol{\theta}), \tag{7.45}$$

where the logarithm of the conditional likelihood function for a component model is

$$\ln p(\mathbf{o}|\boldsymbol{\theta}_i) = -\frac{T}{2}\ln(2\pi\sigma^2) - \frac{1}{2\sigma^2}\sum_{t=1}^{T}\varepsilon^2(t). \tag{7.46}$$

where T is the length of the sequence. Furthermore, the complete data log-likelihood function is written as,

$$\begin{aligned} Q(\boldsymbol{\theta}, \hat{\boldsymbol{\theta}}^t) = & \sum_{i=1}^{L}\sum_{j=1}^{K} \ln p(C_j|\mathbf{o}_i, \hat{\boldsymbol{\theta}}^t) \ln p(\mathbf{o}_i|C_j, \boldsymbol{\theta}_j) + \\ & \sum_{i=1}^{L}\sum_{j=1}^{K} p(C_j|\mathbf{o}_i, \hat{\boldsymbol{\theta}}^t) \ln P(C_j). \end{aligned} \tag{7.47}$$

For each iteration, the E-step calculates the posterior probabilities and likelihood using the current parameter estimation, and M-step seeks the best parameters that maximize the Q function. The number of clusters in the mixture model is estimated using the Bayesian information criterion (Schwarz, 1978).

7.4.7.2 Mixtures of Markov Chains In the case of the mixture components in Eq. 7.38 taking the form of Markov chains, the states are no longer hidden, and they correspond to a set of observable values with state transition probabilities α_{ij} from state ω_i at time t to state ω_j at time $t + 1$. The initial state is determined by the initial state probabilities π_i. Again, we are interested in learning the model parameters α_{ij} and π_i from a set of sequences, which can be achieved with the EM algorithm. In this context, the probability of a particular sequence $\mathbf{o}$, given a component model $\boldsymbol{\lambda}_i$ in the form of a Markov chain, is defined as,

$$P(\mathbf{o}|C_i, \boldsymbol{\lambda}_i) = \pi_{o(1)}\prod_{t=1}^{T-1}\alpha_{o(t)o(t+1)}. \tag{7.48}$$

Based on this, the EM procedure could be constructed, as illustrated by Cadez et al. (2000a), who further generalized a universal probabilistic framework to model sequence clustering. An implementation of mixtures of Markov chains in clustering web user behaviors, called WebCANVAS, is also given by Cadez et al. (2000b).

Ramoni et al. (2002) proposed a hybrid of model-based and distance-based approaches, called Bayesian clustering by dynamics (BCD), to cluster discrete time series. BCD first transforms each time series into a Markov chain, which summarizes the time series dynamics through the generated state transition probability distribution. These state transition probabilities α_{ij} from state ω_i to state ω_j are learned with a Bayesian estimation method considering the observed frequencies n_{ij} in the sequences and the prior knowledge represented with the hyper-parameters κ_{ij}, given as,

$$\alpha_{ij} = \frac{\kappa_{ij} + n_{ij}}{\sum_{j=1}^{N} \kappa_{ij} + \sum_{j=1}^{N} n_{ij}}, \tag{7.49}$$

where N is the number of possible states. The Markov chains are then clustered in an agglomerative hierarchical way, where the merge of a pair of Markov chains is based on the symmetrized Kullback-Leibler distance. Suppose A_1 and A_2 are transition probability matrices for two Markov chains; the average symmetrized Kullback-Leibler distance between A_1 and A_2 is defined as,

$$D(A_1, A_2) = \sum_{i=1}^{N} \frac{d(\alpha_i^1, \alpha_i^2) + d(\alpha_i^2, \alpha_i^1)}{2N}. \tag{7.50}$$

where $d(\alpha_i^1, \alpha_i^2)$ is the Kullback-Leibler distance of the probability distributions α_{ij}^1 and α_{ij}^2 in A_1 and A_2, written as,

$$d(\alpha_i^1, \alpha_i^2) = \sum_{j=1}^{N} \alpha_{ij}^1 \log \frac{\alpha_{ij}^1}{\alpha_{ij}^2}. \tag{7.51}$$

The merge continues until no better model could be obtained in terms of $p(\mathbf{o}|\boldsymbol{\lambda})$.

7.4.7.3 Mixtures of Polynomial Models Now, we consider placing the regression models into the mixture clustering in Eq. 7.38. In this context, each measurement of a sequence $\mathbf{o} = (o(1) \ldots o(i) \ldots o(T))$ with length T is a function of an independent variable or set of variables $\mathbf{x}$. This changes Eq. 7.38 to

$$p(\mathbf{o}|\mathbf{x}, \boldsymbol{\theta}) = \sum_{i=1}^{K} p(\mathbf{o}|\mathbf{x}, C_i, \boldsymbol{\theta}_i) P(C_i). \tag{7.52}$$

Furthermore, we can write the q-order regression equation for a particular observance as,

$$o(i) = \sum_{j=0}^{q} x(i)^j b_{lj} + \varepsilon_l(i), \quad i = 1, \ldots, T, \ l = 1, \ldots, K, \tag{7.53}$$

where b_{lj} are the regression coefficients for the lth cluster, and $\varepsilon_l(i)$ are the zero mean Gaussian error terms with σ_l^2 variance, or in the matrix form for $\mathbf{o}$,

$$\mathbf{o}^T = \mathbf{X}\mathbf{B}_l + \varepsilon_l, \tag{7.54}$$

where the first column of $\mathbf{X}$ contains all ones. It is interesting to see that this construction is exactly equivalent to using a Gaussian distribution with mean $\mathbf{X}\mathbf{B}_l$ and covariance $\sigma_l^2\mathbf{I}$ as the component density $p(\mathbf{o}|\mathbf{x}, C_i, \boldsymbol{\theta}_i)$ (Gaffney and Smyth, 1999). The EM algorithm and weighted least squares can be used to estimate the parameters $\boldsymbol{\theta}_i$ (Gaffney and Smyth, 1999).

An extension combining linear random effects models with regression mixtures, called random effects regression mixtures, was proposed by Gaffney and Smyth (2003) in order to model heterogeneous behaviors. A maximum a posteriori-based EM algorithm was developed to perform parameter inference. Lenk and DeSarbo (2000) also discussed the mixtures of generalized linear models with random effects based on full Bayesian inference, which is approximated through Markov chain Monte Carlo algorithms.

Bar-Joseph et al. (2002) used mixtures of cubic splines to cluster gene expression time series data. Under this model, the measurement $o_j(t)$ at time t for a sequence $\mathbf{o}_j$ belonging to category C_i is represented as,

$$o_j(t) = s(t)(\mu_i + \gamma_j) + \varepsilon_j, \tag{7.55}$$

where s(t) is the $p \times 1$ vector of spline basis functions evaluated at time t, p is the number of spline control points, μ_i is the average value of the spline coefficients for sequence $\mathbf{o}_j$ in cluster C_i, γ_j is a Gaussian vector with mean zero and cluster spline control points covariance matrix Σ_i, and ε_j is Gaussian noise with mean zero and variance σ^2. Again, all the parameters are estimated with an EM algorithm for maximum likelihood of the training data. A continuous alignment process is also used to provide a comparison of the clustering results. Gaffney and Smyth (2004) suggested a unified probabilistic framework for joint clustering and alignment of time series. This goal is achieved by integrating alignment models into the mixture models, such as regression mixtures and spline mixtures.

7.5. APPLICATIONS—GENOMIC AND BIOLOGICAL SEQUENCE CLUSTERING

7.5.1 Introduction

Genome sequencing projects have achieved great advances in recent years. The first draft of the human genome sequence was completed in February

2001, several years earlier than expected (Consortium, I.H.G.S., 2001; Venter et al., 2001). The genomic sequence data for other organisms, e.g., *Drosophila melanogaster* and *Escherichia coli*, are also abundant (Baldi and Brunak, 2001). With the huge volumes of sequences in hand, investigating the functions of genes and proteins and identifying their roles in the genetic process become increasingly important. A large number of computational methods have already been used to accelerate the exploration of the life sciences. For instance, the similarity between the newly sequenced genes or proteins and the annotated ones usually offers a cue to identify their functions. Searching corresponding databases for a new DNA or protein sequence has already become routine in genetic research. Not just restricted to sequence comparison and search, cluster analysis provides a more effective means of discovering complicated relations and revealing hidden structures among DNA and protein sequences, and it is particularly useful for helping biologists investigate and understand the activities of uncharacterized genes and proteins and further, the systematic architecture of the whole genetic network. We summarize the following clustering applications for DNA and protein sequences:

1. Function recognition of uncharacterized genes or proteins (Guralnik and Karypis, 2001; Heger and Holm, 2000; Kasif, 1999);
2. Structure identification of large-scale protein databases (Heger and Holm, 2000; Liu and Rost, 2002; Lo Conte et al., 2000);
3. Redundancy decrease of large-scale DNA or protein databases (Holm and Sander, 1998; Li et al., 2001, 2002);
4. Domain identification (Enright and Ouzounis, 2000; Guan and Du, 1998; Marcotte et al., 1999; Yona et al., 1999);
5. Expressed Sequence Tag (EST) clustering (Burke et al., 1999; Kim et al., 2005; Miller et al., 1999; Wang et al., 2004).

7.5.2 Genetics Basics

We begin our discussion of genomic and biological sequence clustering with a brief introduction of some genetic basics. (See Griffiths et al. (2000) for more detail.)

Genetic information that determines structures, functions, and properties of all living cells is stored in deoxyribonucleic acid (DNA). A DNA molecule is a double helix consisting of two strands, each of which is a linear sequence composed of four different nucleotides—adenine, guanine, thymine, and cytosine, abbreviated as the letters A, G, T, and C, respectively. Each letter in a DNA sequence is also called a base. Two single-stranded DNA molecules bind to each other based on the following complementary base paring rules, or Watson-Crick base pairing rules: base A always pairs with base T, while base C always pairs with base G. Proteins are encoded from coding regions of DNA

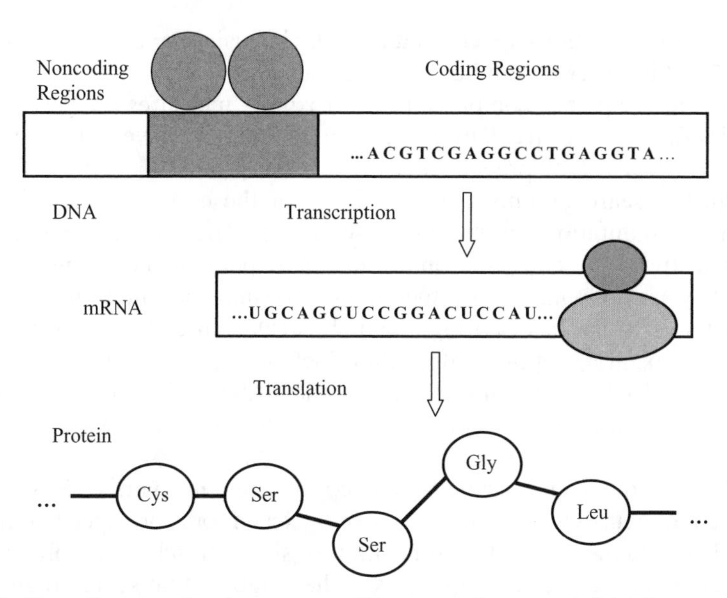

Fig. 7.8. Encoding proteins from genes. Genes are first transcribed into messenger ribonucleic acid (mRNA) molecules, which are then translated into proteins, using the genetic code, where each string of triplets, referred to as codons, encodes a corresponding type of amino acid. And proteins are simply chains of amino acids. Note that neither mRNA nor DNA are proteins themselves.

sequences, known as genes, through a two-stage process of transcription and translation, as shown in Fig. 7.8, and are regarded as the ultimate productions that determine the structures, functions, properties, and regulatory mechanisms of a cell. The primary structure of a protein is also a linear and alphabetic chain with the difference that each unit represents an amino acid, which has twenty types in total. Investigation of the relations between DNA and proteins, as well as their own functions and properties, is one of the important research directions in both genetics and bioinformatics.

7.5.3 Applications

7.5.3.1 Biological and Genomic Sequence Comparison As described in Section 7.2 on sequence similarity comparison, classical dynamic programming algorithms for global and local sequence alignment are too intensive in computational complexity, although they are guaranteed to find the optimal similarity score. This problem becomes more serious because current DNA or protein databases consist of a large volume of nucleic acids and amino acids; for example, bacteria genomes are from 0.5 to 10 Mbp (million base pairs), fungi genomes range from 10 to 50 Mbp, while the human genome is around 3,310 Mbp (Baldi and Brunak, 2001). Thus, conventional dynamic programming algorithms are computationally infeasible, and faster algorithms are required in order to handle such large sequential data.

In practice, sequence comparisons or proximity measures are achieved via heuristics that attempt to identify the regions that may have potentially high matches, with a list of pre-specified, high-scoring words, at an early stage. Thus, further searches only need to focus on these small regions without expensive computation for the entire sequences. These heuristic algorithms trade sensitivity for computationally effective performance time. In other words, they are no longer guaranteed to give the maximal similarity score, and they may miss some best scoring alignments (Durbin et al., 1998). The most well-known examples of these algorithms include BLAST (Basic Local Alignment Search Tool) (Altschul et al., 1990) and FASTA (Pearson and Lipman, 1988), together with many of their variants (Altschul et al., 1997; Kim et al., 2005).

BLAST searches all so-called high scoring segment pairs (HSPs) of two compared sequences whose similarity scores exceed some pre-specified threshold T_1. This search starts with the location of short stretches or words among the query sequence and the sequences in the database that score greater than or equal to some threshold T_2. Typically, the length of the words is 3 for proteins and 11 for nucleic acids (Durbin et al., 1998). The algorithm then uses these words as seeds to extend them in both directions until there is no further improvement for the score of the extensions. As we can see, the threshold T_2 controls the tradeoff between computational efficiency and sensitivity. A higher value of T_2 leads to a faster search while increasing the probability of missing some potential meaningful alignment. Intrinsically, BLAST is based on the assumption that a statistically significant alignment is more likely to contain a high-scoring pair of aligned words (Altschul et al., 1997). The original BLAST does not consider gaps during the alignment, a capability that became available in the later version (Altschul et al., 1997). Table 7.1 illustrates the comparison of the performance of the Smith-Waterman algorithm as well as three versions of BLAST, to the original BLAST on 11 query sequences in SWISS-PROT in terms of the number of sequences that align with E-value ≤ 0.01 and the average ratio of running times. See Altschul et al. (1997) for details on the parameter selections and the scoring models used. It can be seen that all versions of BLAST can perform at least 36 times faster than the Smith-Waterman algorithm. Particularly, gapped BLAST is >100 times faster than the Smith-Waterman algorithm and misses only 8 out of the 1,739 alignments in the 11 queries (Altschul et al., 1997). The iterative version of BLAST, called PSI-BLAST (Position-Specific Iterated BLAST), is more sensitive and outputs more alignments.

FASTA stands for "FAST-All," because it works for both protein (FAST-P) and nucleotide (FAST-N) alignment. FASTA implements a hash table to store all words of length T (typically, 1 or 2 for protein and 4 or 6 for DNA (Durbin et al., 1998)) in the query sequence. Sequences in the database are

TABLE 7.1. The number of SWISS-PROT sequences yielding alignments with E-value ≤ 0.01 and relative running times for Smith-Waterman and various versions of BLAST.

Protein Family	*Query*	*Smith-Waterman*	*Original BLAST*	*Gapped BLAST*	*PSI-BLAST*
Serine protease	P00762	275	273	275	286
Serine protease inhibitor	P01008	108	105	108	111
Ras	P01111	255	249	252	375
Globin	P02232	28	26	28	623
Hemagglutinin	P03435	128	114	128	130
Interferon α	P05013	53	53	53	53
Alcohol dehydrogenase	P07327	138	128	137	160
Histocompatibility antigen	P10318	262	241	261	338
Cytochrome P450	P10635	211	197	211	224
Glutathione transferase	P14942	83	79	81	142
H+-transporting ATP synthase	P20705	198	191	197	207
Normalized running time		36	1.0	0.34	0.87

(From S. Altschul, T. Madden, A. Schäffer, J. Zhang, Z. Zhang, W. Miller, and D. Lipman. Gapped BLAST and PSI-BLAST: A new generation of protein database search programs. Nucleic Acids Research, 1997, vol. 25, no. 17, pp. 3389–3402, by permission of Oxford University Press.)

then scanned to mark all the matches of words, and the 10 best diagonal regions with the highest densities of word matches are selected for a rescore in the next step, where these diagonals are further extended similar to the strategy in BLAST. The third step examines whether the obtained ungapped regions could be joined by considering gap penalties. Finally, standard dynamic programming is applied to a band around the best region to yield the optimal alignments.

Recognizing the benefit of the separation of word matching and sequence alignment to the reduction of computational burden, Miller et al. (1999) described the following three algorithms for sequence comparison: RAPID (Rapid Analysis of Pre-Indexed Datastructures) for word search and PHAT (Probabilistic Hough Alignment Tool) and SPLAT (Smart Probabilistic Local Alignment Tool) for alignment. The implementation of the scheme for large database vs. database comparison exhibits approximately one order of magnitude improvement in computational time compared with BLAST, while maintaining good sensitivity. Kent and Zahler (2000) designed a three-pass algorithm, called WABA (Wobble Aware Bulk Aligner), for aligning large-scale genomic DNA sequences of different species. The specific consideration of WABA is to deal with a high degree of divergence in the third, or "wobble," position of a codon. The seeds used to start an alignment take the form of XXoXXoXX, where the X's indicate a must-match and the o's do not, instead of requiring a perfect match of 6 consecutive bases. A seven-state pairwise hidden Markov model (Durbin et al., 1998) was used for more effective alignments, where the states correspond to the long inserts in the target sequence, the long inserts in the query sequence, highly conserved regions, lightly conserved regions, and three coding regions. The application of WABA in aligning 8 million bases of *Caenorhabditis Briggsae* genomic DNA against the entire 97 million bases of the *Caenorhabditis elegans* genome was also presented. See also Delcher et al. (1999), Morgenstern et al. (1998), Sæbø et al. (2005), Schwartz et al. (2000), and Zhang et al. (2000) for more algorithms for sequence comparison and alignment. Miller (2001) summarized the current research status of genomic sequence comparison and suggested the following valuable directions for further research efforts:

1. Improved software for aligning two genomic sequences, with a rigorous statistical basis;
2. An industrial-strength gene prediction system, combining genomic sequence comparisons, intrinsic sequence properties, and results from searching databases of protein sequences and ESTs in an effective way;
3. Reliable and automatic software for aligning more than two sequences;
4. Better approaches for visualizing the aligning results;
5. Benchmark datasets and improved protocols for evaluating and comparing the performance of sequence alignment software.

7.5.3.2 Biological and Genomic Sequence Clustering Many clustering algorithms have been applied to organize DNA or protein sequence data. Some directly operate on a proximity matrix with a similarity or distance measure for each pair of sequences, some transform the sequential data into feature vectors via feature extraction approaches, and the others are constructed based on some particular probabilistic models.

Somervuo and Kohonen (2000) illustrated an application of self-organizing feature maps (SOFMs) (Chapter 5) in clustering protein sequences of the SWISS-PROT database, release 37, which contains 77,977 protein sequences. Each node of SOFM is represented by a prototype sequence, which is the generalized median of the neighborhood, based on the sequence similarity calculated by FASTA. The resulting 30-by-20 hexagonal SOFM grid is shown in Fig. 7.9, which provides a visualized representation of the relations within the entire sequence database. We can see that the large clusters of protein sequences could be obtained within the light regions.

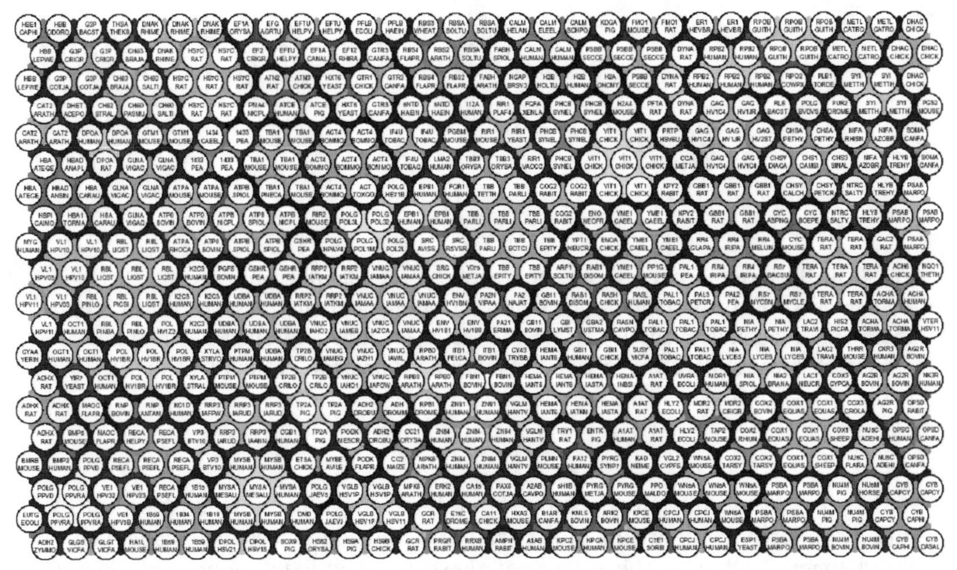

Fig. 7.9. A 30-by-20-unit hexagonal SOFM grid. The SOFM was constructed using all the 77,977 protein sequences of the SWISS-PROT release 37. Each node contains a prototype sequence and a list of data sequences. The labels on the map nodes are the SWISS-PROT identifiers (Bairoch and Apweiler, 1999) of the prototypes sequences. The upper label in each map node is the mnemonic of the protein name and the lower label is the mnemonic of the species name. The similarities of the neighboring prototype sequences on the map are indicated by shades of gray. The light shades indicate a high degree of similarity, and the dark shades a low degree of similarity, respectively. Light areas on the map reveal large clusters of similar sequences. (From Figure 2, P. Somervuo and T. Kohonen. Clustering and visualization of large protein sequence databases by means of an extension of the self-organizing map. Proceedings of the 3rd International Conference on Discovery Science, Lecture Notes in Artificial Intelligence 1967, pp. 76–85, 2000. Copyright © 2000. With kind permission of Springer Science and Business Media.)

Sasson et al. (2002) clustered the protein sequences in SWISS-PROT in a hierarchical structure using an agglomerative hierarchical clustering technique based on the similarity measure of gapped BLAST. (A recent release of the clustering system, called ProtoNet 4.0, is described in Kaplan et al. (2005). The inter-cluster similarity measures for merging two groups are based on the averages of the *E*-score from BLAST. They compare and contrast averaging based on the arithmetic mean, the square mean, the geometric mean, and the harmonic mean.) The advantages as well as the risks of transitivity of homology were also discussed by Sasson et al. (2002). In this context, two sequences that do not have high sequence similarity by virtue of direct comparison may be homologous (having a common ancestor) if there exists an intermediate sequence similar to both of them, as illustrated in Fig. 7.10. This property

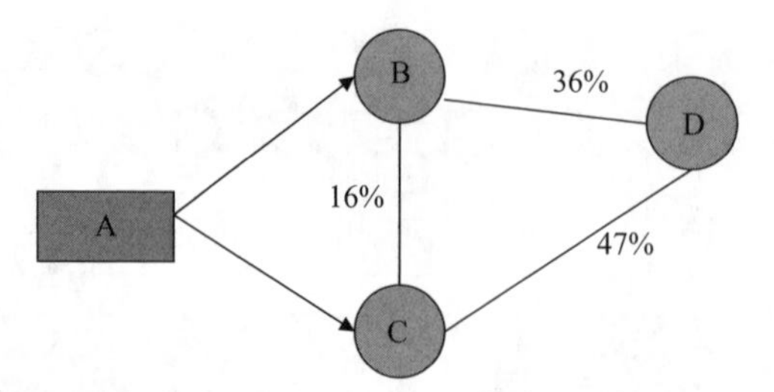

Fig. 7.10. Illustration of transitivity of homology. The fact that proteins B and C have a common ancestor A could not be inferred from their similarity score, which is only 16%. However, this relation can be indirectly identified through an intermediate protein D, because both B and C are significantly similar to protein D with sequence similarities of 36% and 47%, respectively.

makes it possible to detect remote homologues that cannot be observed by direct similarity comparison. However, unrelated sequences may also be clustered together due to the effects of these intermediate sequences (Sasson et al., 2002). Bolten et al. (2001) addressed this problem by the construction of a directed graph in which each protein sequence corresponds to a vertex, and an edge between two nodes is weighted based on the Smith-Waterman alignment score of the corresponding two sequences and the self-alignment score of each sequence. Clusters were generated through the search of strongly connected components (SCCs), defined as a maximal subset of vertices such that, for each pair of vertices u and v in the subset, there exist two directed paths from u to v and v to u, respectively. Two more graph theory-based clustering algorithms in protein family detection and domain information extraction were given in Abascal and Valencia (2002) and Guan and Du (1998), respectively. The former uses a minimum normalized cut algorithm, and the latter is based on the application of the minimum spanning tree (MST).

In contrast with the aforementioned similarity-based methods, Guralnik and Karypis (2001) transformed protein or DNA sequences into a feature space based on the detected sub-patterns or motifs, treated as sequence features, and further performed clustering with the K-means algorithm. This method avoids all-against-all expensive sequence comparison and is suitable for analyzing large-scale databases. An example of this strategy in clustering 43,569 sequences from 20 different protein families is illustrated in Table 7.2, where each row represents a particular cluster (20 in total, as set in advance). 22,672 motifs of length 3–6 were found by the pattern discovery algorithm, and 21,223 of them were kept as independent features. Among all the 20 functional classes, 13 of them can be distinguished, and most clusters, except 18 and 20, include sequences from, at most, 2 major protein families.

Krogh et al. (1994) released a system called SAM (Sequence Alignment and Modeling), applied hidden Markov models (HMMs) in genomic sequence modeling and clustering of protein families. Figure 7.11 depicts a typical

TABLE 7.2 Clustering of 43,569 protein sequences with a feature-based algorithm.

Number of Sequences	Cluster Similarity	Cluster Entropy	Functional Classes																			
			F1	F2	F3	F4	F5	F6	F7	F8	F9	F10	F11	F12	F13	F14	F15	F16	F17	F18	F19	F20
2,466	0.33	0.01	2,463	2	0	0	0	0	0	0	0	0	0	0	0	0	0	0	0	0	0	1
2,169	0.15	1.39	11	1,578	0	0	15	5	193	2	0	1	3	1	267	0	4	2	0	1	80	6
3,581	0.67	0	0	0	3,581	0	0	0	0	0	0	0	0	0	0	0	0	0	0	0	0	0
582	0.67	0.02	0	0	581	0	0	0	0	0	0	0	0	0	1	0	0	0	0	0	0	0
1,291	0.52	0	0	0	0	1,291	0	0	0	0	0	0	0	0	0	0	0	0	0	0	0	0
1,522	0.53	0	0	0	0	0	1,522	0	0	0	0	0	0	0	0	0	0	0	0	0	0	0
1,573	0.37	0.75	0	0	1	0	0	1,240	0	0	0	0	0	0	332	0	0	0	0	0	0	0
1,159	0.35	0.02	0	0	1	0	0	0	1,157	0	0	0	0	0	0	0	0	0	0	0	0	1
1,773	0.43	0	0	0	0	0	0	0	0	1,773	0	0	0	0	0	0	0	0	0	0	0	0
1,712	0.39	0.02	0	0	0	0	4	0	0	0	1,708	0	0	0	0	0	0	0	0	0	0	0
1,219	0.2	0.5	0	0	0	0	0	0	1	72	0	1,123	1	1	0	0	0	1	10	1	8	1
718	0.49	0.06	0	0	0	0	0	0	0	0	0	0	714	0	0	1	1	0	1	0	1	0
1,005	0.25	0.67	0	0	0	0	0	0	0	3	0	91	882	0	0	0	2	26	0	0	0	1
1,708	0.22	1.07	0	0	0	0	0	19	4	1	0	2	0	1,182	479	1	0	0	0	0	20	0
802	0.54	0.04	0	0	0	0	0	0	1	0	0	0	0	0	799	0	0	2	0	0	0	0
1,050	0.29	1.7	0	0	0	0	320	0	2	0	40	0	0	2	506	0	0	0	0	178	0	2
1,129	0.22	1.54	177	0	0	0	0	0	20	0	243	1	0	2	1	676	0	1	0	0	1	7
2,916	0.13	2.26	8	4	0	0	11	1	46	184	56	7	229	21	30	2	1,595	283	8	400	2	6
534	0.68	0.02	0	0	0	0	0	0	0	0	1	0	0	0	0	0	0	533	0	0	0	0
14,660	0.12	3.65	84	721	21	125	702	66	68	66	159	3,161	650	1,055	211	958	665	676	1,905	956	1,080	1,331

(From V. Guralnik and G. Karypis. A scalable algorithm for clustering sequential data. In Proceedings of the 1st IEEE International Conference on Data Mining—ICDM 2001, pp. 179–186, 2001.)

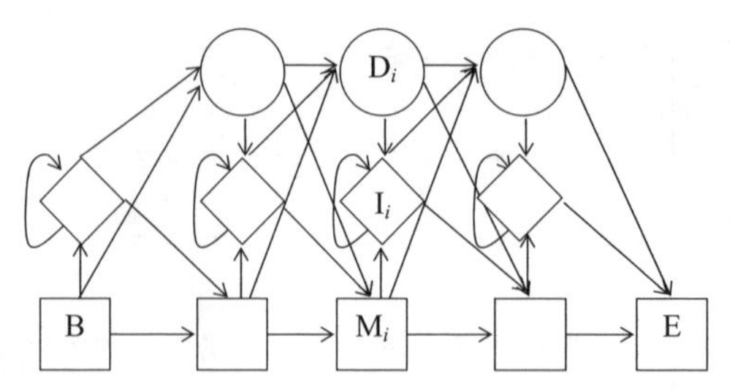

Fig. 7.11. An HMM architecture (Krogh et al., 1994). There are three different states, match (M), insert (I), and delete (D), represented as rectangles, diamonds, and circles, respectively. A begin (B) and end (E) state are also introduced to represent the start and end of the process. This process goes through a series of states according to the transition probability and emits either a 4-letter nucleotide or a 20-letter amino acid alphabet based on the emission probability.

HMM, in which match states (abbreviated with the letter M), insert states (I), and delete states (D) are represented as rectangles, diamonds, and circles, respectively (Durbin et al., 1998; Hughey et al, 2003; Krogh et al., 1994). These states correspond to substitution, insertion, and deletion in edit operations. Letters, either from the 4-letter nucleotide alphabet or 20-letter amino acid alphabet, are produced from match and insert states according to emission probability distributions. Delete states do not produce any symbols and are used to skip the match states. For convenience, a begin state and an end state are added to the model, denoted by the letters B and E. They do not generate any amino acid, either. So, the initial state distribution is considered as the transition probabilities from the begin state to the first state. Given an HMM with completely specified parameters, a DNA or protein sequence can be generated using the following procedure:

1. Starting at the begin state B, transit to a match, insert, or delete state based on the state transition probabilities;
2. If the state is
    a. match or insert state, generate an amino acid or nucleotide in the sequence according to the emit probabilities;
    b. delete state, generate no amino acid or nucleotide.
3. Transit to a new state with the state transition probabilities. If it is not the end state, return to step 2. Otherwise, terminate the procedure and output the genomic sequence.

In order to group a set of sequences into K clusters, or families (subfamilies), K HMMs are required, each a component of a mixture model. The parameters are obtained through the EM algorithm. Krogh et al. (1994) clustered subfamilies of 625 globins with average length of 145 amino acids and 3 non-globins, where the number of component HMMs was set at 10 with initial lengths randomly selected between 120 and 170. HMM clustering identified three major globin subfamilies as alpha, beta, and myoglobin. Further discussion and applications can be found in Durbin et al. (1998), Hughey and Krogh (1996), and Karplus et al. (1998).

7.6. SUMMARY

Sequential data has quite different properties from data described as vectors in the feature space. In contrast to the feature vectors with fixed length, the length of the sequences in a particular data set usually varies. These properties make direct applications of feature vector-based clustering infeasible. Many clustering algorithms have been proposed to group sequential data, which either convert the sequential data into the form that could be represented in the feature space (e.g., feature extraction or selection from the raw data), or reinterpret clustering algorithms in order to deal with the new characteristics of the sequential data (definition of new distance or similarity measures and introduction of probabilistic models in finite mixture model clustering).

The conversion from sequential data to static vectors requires the extraction of new features from the raw data, which makes the process of feature extraction essential to the performance effectiveness of this type of algorithm. The ideal features maintain the potential structure of the sequential data and greatly reduce the computational complexities, in contrast to the all-against-all similarity comparison, which makes it possible to apply it to large-scale sequence databases analysis. However, the process of feature extraction or selection inevitably causes the loss of some information in the original sequences, and poor feature selection may distort the similarity between the sequences, leading to meaningless clusters. Therefore, caution is advised.

The distance based methods directly calculate the proximity matrix of the sequences, which makes it easy to partition data using the conventional clustering algorithms, such as hierarchical clustering. However, since the characteristics of the sequences could be unclear, it is not trivial to determine an appropriate distance or similarity measure that could effectively compare the sequences. Moreover, sequential data sets in practice usually contain large volumes of data with long sequences. The requirement of all-against-all comparison of these sequences inevitably causes a bottleneck in the computational time and space.

Model-based clustering models clusters directly from raw data without additional processes that may cause information loss. The uncertainty in the cluster membership is naturally included in the mixture model. These methods provide more intuitive ways to capture the dynamics of data and more flexible means to deal with variable length sequences. For models like HMMs, there also exists a direct relationship between the states and the real situations, which makes the model easily interpreted. Furthermore, this paradigm is usually well-grounded mathematically and the model parameters could be derived in a well-defined manner. However, as the major challenge in cluster validation, determining the number of model components remains a complicated process (Oates et al., 2000; Smyth, 1997). Also, the model should have sufficient complexity to interpret the data.

CHAPTER 8

LARGE-SCALE DATA CLUSTERING

8.1. INTRODUCTION

In Chapter 7, we saw the demanding requirements for clustering sequential data, which are just examples of more applications that require the capability of scalability for clustering algorithms. Scientific data from many fields, such as astronomy, finance, genomics, geology, and physics, are measured and stored in hundreds of gigabytes or even terabytes. With the further advances of database, Internet, and other technologies, which increase the complexity of data, scalability becomes increasingly important in clustering. The data complexity has two aspects: data volume and dimensionality. Table 8.1 summarizes the computational complexity of some typical clustering algorithms, together with several approaches specifically designed to deal with large-scale data sets.

Table 8.1 shows that classical hierarchical clustering algorithms, including single linkage, complete linkage, average linkage, centroid linkage, and median linkage, are not appropriate for large-scales due to the quadratic computational complexities in both execution time and storage space. In contrast, the K-means algorithm shows a desirable time complexity of $O(NKdT)$ (recall that d is the number of features and T is the number of iterations) and space complexity of $O(N+K)$. Since N is usually much larger than both K and d, both time and space complexity become near linear to the number of samples in the data sets. Therefore, the K-means algorithm scales well to large-scale data sets. However, it should be noted that there are also other disadvantages

Clustering, by Rui Xu and Donald C. Wunsch, II

TABLE 8.1. Computational complexity of clustering algorithms. Algorithms like BIRCH and WaveCluster can scale linearly with the input size and handle very large data sets.

Cluster Algorithm	Complexity	Suitable for High Dimensional Data
K-means	$O(NKd)$ (time) $O(N + K)$ (space)	No
Fuzzy c-means	Near $O(N)$	No
Hierarchical clustering*	$O(N^2)$ (time and space)	No
PAM	$O(K(N - K)^2)$	No
CLARA⁺	$O(K(40 + K)^2 + K(N - K))$ (time)	No
CLARANS	Quadratic in total performance	No
BIRCH	$O(N)$ (time)	No
DBSCAN	$O(N \log N)$ (time)	No
CURE	$O(N_{sample}^2 \log N_{sample})(\text{time})O(N_{sample})(\text{space})$	Yes
WaveCluster	$O(N)$ (time)	No
DENCLUE	$O(N \log N)$ (time)	Yes
FC	$O(N)$ (time)	Yes
STING	O(Number of cells at the bottom layer)	No
CLIQUE	Linear with the number of objects, Quadratic with the number of dimensions	Yes
OptiGrid	Between $O(Nd)$ and $O(Nd \log N)$	Yes
ORCLUS	$O(K_0^3 + K_0Nd + K_0^2d^3)(\text{space})O(K_0d^2)(\text{space})$ #	Yes

*Include single-linkage, complete-linkage, average-linkage, etc.

⁺Based on the heuristic for drawing a sample from the entire data set (Kaufman and Rousseeuw, 1990)

#K_0 is the number of initial seeds.

and limitations inherent with the K-means algorithm, as discussed in Chapter 4.

The performance of K-means can be further improved. We discussed in Chapter 5 that Adaptive Resonance Theory (Carpenter and Grossberg, 1987a) was a type of varying-K-means clustering, which was pointed out as early as (Moore, 1989). Another example of improving K-means is the usage of a *kd*-tree data structure to reduce the requirement for the nearest centroid search (Pelleg and Moore, 1999). A speedup factor of up to 170 was reported on the Sloan Digital Sky Survey data set. Another application of a *kd*-tree in K-means data storage was given by Kanungo et al. (2000). Parallel techniques for accelerating K-means were also developed (Stoffel and Belkoniene, 1999). The implementation of K-means in the relational database management system was given in Ordonez and Omiecinski (2004) and Ordonez (2006). Furthermore, Bradley et al. (1998) proposed a scalable clustering framework, in the context of K-means, which considers seven important requirements in dealing with very large-scale databases that cannot be fully loaded into the core memory:

1. Take at most one scan of the database, with early termination preferred when appropriate.
2. Be always able to provide an online "best" solution.
3. Be suspendable, stoppable, and resumable. Incremental progress can be saved to resume a stopped job.
4. Be able to incrementally incorporate additional data with existing models.
5. Be able to work within a user-allocated limited memory buffer.
6. Use a variety of scan modes, such as sequential, index, and sampling scans.
7. Be able to operate on forward-only cursor over a view of the database.

An application of the framework to EM mixture models was also illustrated (Bradley et al., 1998, 2000b).

Another observation from Table 8.1 is that many algorithms have been specifically developed for large-scale data sets cluster analysis, especially in the context of data mining (Bradley et al., 1998, 2000b; Ester, 1996; Hinneburg and Keim, 1998; Ng and Han, 2002; Sheikholeslami et al., 1998). Some algorithms scale linearly with the input size and can handle very large data sets. The capability of these algorithms in dealing with high dimensional data is also summarized in the table. However, the performances of many of these algorithms degenerate with the increase of input dimensionality. Some algorithms, like FC and DENCLUE, have shown some successful applications in such cases, but these are still far from completely effective. Discussions on the data visualization, dimensionality reduction, and high dimensional data clustering are elaborated in Chapter 9. In the following sections of the chapter, we divide and discuss the algorithms in large-scale data clustering in the following categories:

- Random sampling
- Data condensation
- Density-based approaches
- Grid-based approaches
- Divide and conquer
- Incremental learning

Many proposed algorithms combine more than one method to be scalable to large-scale data cluster analysis and thus belong to at least two categories above. For example, the algorithm DENCLUE relies on both density-based and grid-based notions of clustering. The algorithm FC processes data points in an incremental way, and it also represents cluster information with a series of grids.

8.2. RANDOM SAMPLING METHODS

Clustering algorithms that use a random sampling approach in large-scale data clustering are applied to a random sample of the original data set instead of the entire data set. The key point of the random sampling approach is that an appropriate-sized sample can maintain the important geometrical properties of potential clusters, while greatly reducing the requirement for both computational time and storage space. The lower bound of the minimum sample size can be estimated in terms of Chernoff bounds, given the low probability that clusters are missing in the sample set (Guha et al., 1998).

Let N be the size of the entire data set and N_C represent the number of points in a cluster C. Cluster C is considered to be included in the sample, or to have a low probability of missing data, if there are at least ηN_C data points in the sample set, where η is a parameter in the range of [0, 1]. The sample size S_C, which is required to assure that the sample contains fewer than ηN_C data points from cluster C with a probability less than δ, $0 \leq \delta \leq 1$, can be obtained by using Chernoff bounds (Guha et al., 1998):

$$S_C \geq \eta N + \frac{N}{N_C}\log(1/\delta) + \frac{N}{N_C}\sqrt{(\log(1/\delta))^2 + 2\eta N_C \log(1/\delta)}. \tag{8.1}$$

Thus, when C^* is the cluster with the smallest size, the corresponding S_{C^*} can make Eq. 8.1 hold true for all $N_C \geq N_{C^*}$. Therefore, S_{C^*} represents the minimum sample size that guarantees the inclusion of any cluster C in the sample, with a high probability $1 - \delta$.

The algorithms CURE (Clustering Using REpresentatives) (Guha et al., 1998) and CLARA (Clustering LARge Applications) (Kaufman and Rousseeuw, 1990) use the random sampling scheme. CURE chooses a set of well-scattered and center-shrunk points to represent each cluster and is discussed in Chapter 3. CLARA combines random sampling with the clustering algorithm PAM (discussed in Chapter 4), which represents each cluster with a medoid, as defined in Chapter 4. Given a sample drawn from the entire data set, PAM is applied to find the corresponding medoids and the remaining data points are assigned to the nearest medoids based on a dissimilarity measure. Multiple samples (usually 5) are drawn, and the one with the lowest average distance is used as output. CLARA generates more samples by initially including the previously-found medoids, which represent a set of points that have the smallest average distance to the entire data set obtained so far. In Kaufman and Rousseeuw (1990), the sample size is suggested as $40 + 2K$.

Both PAM and CLARA are generalized in a graph theory framework, where each node in a graph G corresponds to a set of K medoids, or a clustering partition (Ng and Han, 2002). The algorithm CLARANS (Clustering Large Applications based on RANdomized search) is also developed under such a framework and sees the clustering as a search process in the graph (Ng and Han, 2002). CLARANS starts with an arbitrary node as the current node and examines a sample of its neighbors, defined as the node consisting of only one different data object, to seek a better solution. In other words, any neighbor with a lower cost becomes the current node. This differentiates CLARANS from PAM, which checks all the neighbors, and CLARA, which works on the sub-graphs of the original graph. In its implementation, the maximum number of neighbors is designed as a user-specified parameter and is usually determined empirically. The basic procedure of CLARANS is summarized in Fig. 8.1. Though CLARANS achieves better performance than algorithms PAM and CLARA, the total computational time is still quadratic, which makes CLARANS ineffective in very large data sets.

In (uniform) random sampling, each data point is selected to the sample with the same probability. The sampling process can be performed in a biased way so that each point has a different probability to be included in the sample based on the specific clustering requirements (Kollios et al., 2003). For example, over-sampling the sparse regions reduces the likelihood of missing the small clusters in the data. The advantages of biased sampling are further explained through a theorem by Kollinos et al. (2003), which identifies that the sample size required by biased sampling is smaller than, or at lest equal to, that required by uniform random sampling under the same condition. More specifically, suppose the data points $\mathbf{x}_i$, $i \in (1, \ldots, N)$ are sampled into a cluster C using the following rule:

$$P(\mathbf{x}_i \text{ is included in the sample}) = \begin{cases} p & \text{if } \mathbf{x}_i \in C \\ 1-p & \text{otherwise} \end{cases}, \quad 0 \le p \le 1. \tag{8.2}$$

The following inequality holds true for the sample size S_R based on the rule in Eq. 8.2 and the sample size S_C based on uniform random sampling, both of which guarantee that cluster C is included in the sample with a probability more than $1 - \delta, 0 \le \delta \le 1$

$$S_R \le S_C \quad \text{if} \quad p \ge \frac{N_C}{N}. \tag{8.3}$$

Because the probability distribution is usually unknown, a kernel function is used to provide density estimation. One such kernel function is the

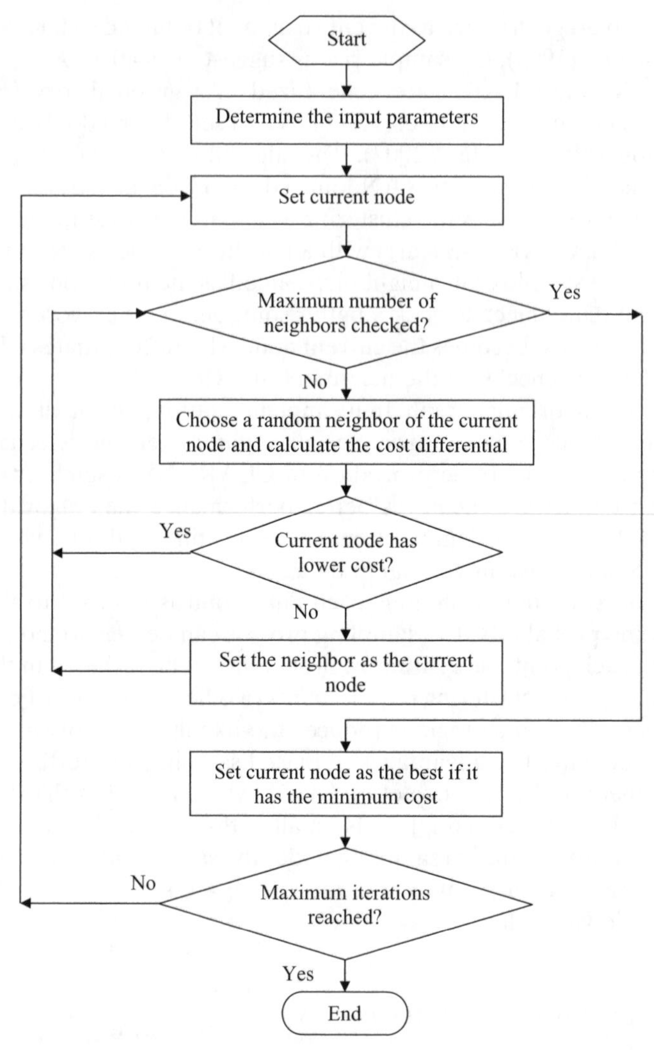

Fig. 8.1. Flowchart of CLARANS. CLARANS performs a search process in the constructed graph, where each node corresponds to a medoid. A sample of neighbors of the current node is examined at each iteration, and the neighbor with a lower cost is set as the current node.

d-dimensional 0-centered Epanechnikov kernel function (Cressie, 1993), defined as,

$$K(x_1,\dots,x_d)=\left(\frac{3}{4}\right)^d \frac{1}{B_1B_2\cdots B_d}\prod_{1\le i\le d}\left(1-\left(\frac{x_i}{B_i}\right)^2\right), \tag{8.4}$$

where B_i is the bandwidth of the kernel function.

8.3. CONDENSATION-BASED METHODS

Condensation-based approaches perform clustering by using the calculated summary statistics of the original data rather than the entire data set. In this way, the requirement for the storage of and the frequent operations on the large volume of data is greatly reduced, and large-scale data sets can be clustered with reasonable time and space efficiency. The algorithm BIRCH (Balanced Iterative Reducing and Clustering using Hierarchies) (Zhang et al., 1996), discussed in Chapter 3, is such an example, which has an important impact on many other condensation-based methods.

The basic unit of BIRCH in representing a cluster is called a clustering feature (CF), which is a 3-tuple, $\mathbf{CF} = (N_c, \mathbf{LS}, SS)$, containing the number of data points in the cluster N_c, the linear sum of the data points $\mathbf{LS}$, and the square sum of the data points SS. This data summary maintains sufficient information for further clustering operations, such as the merge of two clusters, and largely reduces the computational burden. The clustering features or data summaries are stored in a height-balanced tree, called a CF tree, which is built dynamically with the data points incrementally inserted.

The preclustering phase of BIRCH is also used in an algorithm aiming to cluster both continuous and categorical features (Chiu et al., 2001). In this case, the cluster feature of a cluster C_j is described in terms of four elements, i.e., $\mathbf{CF}_j = (N_j, \mathbf{S}_{Aj}, S_{Aj}^2, \mathbf{N}_{Bj})$, where N_j is the number of data points in the cluster, $\mathbf{S}_{Aj}$ is the sum of continuous features, S_{Aj}^2 is the square sum of continuous features, and $\mathbf{N}_{Bj}$ is a vector given as $\mathbf{N}_{Bj} = (\mathbf{N}_{Bj1}, \ldots, \mathbf{N}_{BjK_B})$, where $\mathbf{N}_{Bji} = (\mathbf{N}_{ji1}, \ldots, \mathbf{N}_{jiL_l-1})$ and N_{jil} is the number of data points in C_j whose i^{th} categorical feature has the value of lth category, $l = 1, \ldots, L_k - 1$ and $k = 1, \ldots K_B$. The CF for a new cluster, generated by merging two other clusters, is calculated by adding corresponding entries in both clusters. The distance measure between a pair of clusters is defined as the decrease in log-likelihood function L resulting from the merge of the two clusters,

$$D(C_i, C_i) = L_{C_i} + L_{C_j} - L_{C_i \cup C_j}. \tag{8.5}$$

The BIRCH* framework is a generalization of the concept of CF and CF-tree of BIRCH into the general metric spaces, where only the distance measure is available for data points, which is further realized as two algorithms named BUBBLE and BUBBLE-FM (Ganti et al., 1999). Under this framework, the CF for a cluster C contains the information on the row sum, which is a sum of the squared distance of a data point to other points in the cluster, the clustroid $\mathbf{x}_o$ of the cluster, which is the data point corresponding to the smallest row sum, and the radius r of the cluster, which is defined as

$$r = \sqrt{\frac{\sum_{i=1}^{N_C} D^2(\mathbf{x}_i, \mathbf{x}_o)}{N_C}}, \tag{8.6}$$

where N_C is the number of data points in the cluster and D is a distance function. For large clusters, only a subset of data points, with their corresponding row sums, is maintained in order to assure practical efficiency. Similar to BIRCH, all these cluster features are stored and organized in a CF-tree. Compared with BUBBLE, BUBBLE-FM combines the algorithm FastMap (Faloutsos and Lin, 1995) in order to reduce the requirement for distance computation.

Bradley et al. (1998) considered performing data compression, which represents data points with their sufficient statistics, during different phases in the scalable clustering framework. Data points were classified into three categories based on their importance to the clustering model: retained set (RS), discard set (DS), and compression set (CS). RS contains data points that must be assessed all the time and therefore is always kept in the memory. DS includes data points that are unlikely to move to a different cluster and is determined in the primary phase of the algorithm. (Two methods were proposed to compress data in this phase. The first method calculates the Mahalanobis distance of a point to the cluster mean and compresses all data points within a certain radius. The second method emphasizes the creation of a worst case scenario by disturbing the cluster means within calculated confidence intervals.) CS consists of data points that are further compressed in the secondary phase to generate sub-clusters represented with sufficient statistics. Like CF of BIRCH, the sufficient statistics are 3-tuples, constituting the number of data points in the cluster and the sum and the square sum of the data points. In BIRCH we can consider that all data points are put into the DS. Farnstrom et al. (2000) simplified the clustering framework above and described a simple single pass K-means algorithm. Each time, all data points in the buffer are discarded after the corresponding sufficient statistics are updated. Analysis shows that, compared with the algorithm of Bradley et al. (1998), this method effectively reduces the computational complexity in both time and space while maintaining the quality of clusters.

Two recent algorithms that use data summaries to represent sub-clusters are based on Gaussian mixture models (Jin et al., 2005). The basic algorithm EMADS (EM Algorithm for Data Summaries) works directly on the summary statistics and provides an approximation of the aggregate behavior for sub-clusters under the Gaussian mixture model. The combination of EMADS with BIRCH and the grid-based data summarization procedure, which partitions a data space with a multidimensional grid structure and treats each cell as a sub-cluster, leads to the bEMADS and gEMADS algorithm, respectively.

8.4. DENSITY-BASED METHODS

Density-based approaches rely on the density of data points for clustering and have the advantage of generating clusters with arbitrary shapes and good scalability. The density of points within a cluster is considerably higher than the density of points outside of the cluster. Specifically, the algorithm DBSCAN (Density Based Spatial Clustering of Applications with Noise) (Ester et al., 1996) implements the concept of density-reachability and density-connectivity to define clusters. Letting $N_{Eps}(\mathbf{x}) = \{\mathbf{y} \in \mathbf{X} \mid D(\mathbf{x},\mathbf{y}) \le Eps\}$, where $D(\cdot)$ is a distance function and *Eps* is the given radius, be the *Eps*-neighborhood of a data point $\mathbf{x}$, the point $\mathbf{x}$ is said to be density-reachable from a point $\mathbf{y}$ if there exists a finite sequence of points $\mathbf{x}_1 = \mathbf{x}, \ldots, \mathbf{x}_s = \mathbf{y}$ such that for each $\mathbf{x}_{i+1}$,

1. $\mathbf{x}_{i+1} \in N_{Eps}(\mathbf{x}_i)$, and
2. $|NEps(\mathbf{x}_i)| \ge Minpts$, where *Minpts* is a user-specified parameter that identifies the minimum number of data points that must be included in the neighborhood of the point $\mathbf{x}_i$ in a cluster.

According to condition 2, each point that lies inside of a cluster should contain enough points in its neighborhood, i.e., the threshold *Minpts*. Such points are known as core points. If two core points are within each other's neighborhood, they belong to the same cluster. It also can be seen that density-reachability is symmetric for core points. In comparison, there exist two other types of data points, border points and noise points. Border points, those on the border of a cluster, do not contain enough points in their neighborhood to be the core points, but they belong to the neighborhood of some core points. The points that are neither core points nor border points are regarded as noise or outliers.

Because two border points belonging to the same cluster are possibly not density reachable from each other, density-connectivity is used to describe the relations of the border points, which requires that both points should be density reachable from a common core point. Obviously, two core points are also density-connected. The notations of both density-reachability and density-connectivity are illustrated in Fig. 8.2 on a two-dimensional feature space.

We now can define a cluster C using density-reachability and density-connectivity in terms of the following two conditions:

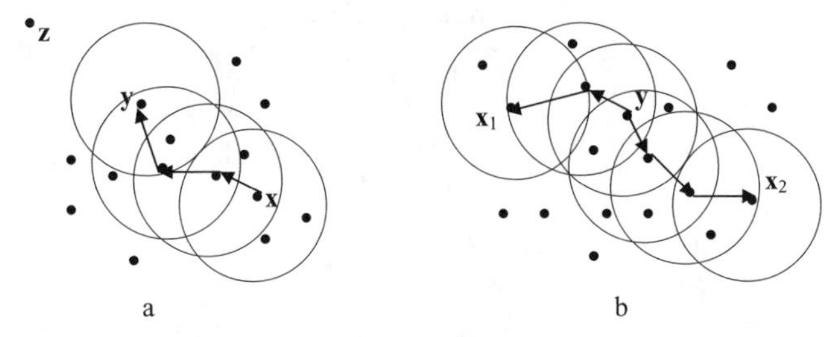

Fig. 8.2. Notions of density-reachability and density-connectivity (Ester et al., 1996). The parameter *Minpts* is set as 4. (a) $\mathbf{x}$ is a core point, $\mathbf{y}$ is a border point, and $\mathbf{z}$ is an outlier. y is density-reachable from $\mathbf{x}$, but $\mathbf{x}$ is not density-reachable from $\mathbf{y}$. (b) $\mathbf{x}_1$ and $\mathbf{x}_2$ are density-connected to each other in terms of $\mathbf{y}$.

1. For any pair of points $\mathbf{x}_i, \mathbf{x}_j \in \mathbf{X}$, if $\mathbf{x}_i \in C$ and $\mathbf{x}_j$ is density reachable from $\mathbf{x}_i$, then $\mathbf{x}_j \in C$.
2. For any pair of points $\mathbf{x}_i, \mathbf{x}_j \in C$, $\mathbf{x}_i$ is density connected to $\mathbf{x}_j$.

In its realization, DBSCAN creates a new cluster from a core point by absorbing all points in its neighborhood. In order to calculate the neighborhood, DBSCAN uses an R^*-tree structure for effective queries, which has a time complexity of $O(\log N)$ per search.

DBSCAN requires users to identify two important parameters, *Eps* and *Minpts*, which is not trivial. Different, more complicated regions of data may ask for different parameters. Rather than remaining limited to one parameter setting, the algorithm OPTICS (Ordering Points To Identify the Clustering Structure) (Ankerst et al., 1999) focuses on the construction of an augmented ordering of data representing its density-based clustering structure and processes a set of neighborhood radius parameters $0 \le Eps_i \le Eps$. For each data point, OPTICS stores its corresponding information for further clustering as core-distance and reachability-distance. The core-distance of a point $\mathbf{x}$ refers to the distance *Eps* between $\mathbf{x}$ and its *Minpts* nearest neighbor with the condition that $\mathbf{x}$ is a core point with respect to *Eps*. Otherwise, the core-distance is undefined. The reachability-distance of a point $\mathbf{x}$ with respect to another point $\mathbf{y}$ is the smallest distance such that $\mathbf{x}$ is in the *Eps*-neighborhood of $\mathbf{y}$ if $\mathbf{y}$ is a core point, or undefined if $\mathbf{y}$ is not a core point. The concepts of core-distance and reachability-distance are also depicted in Fig. 8.3. The run time of OPTICS is slightly slower than that of DBSCAN: 1.6 times the run time of DBSCAN based on experimental results (Ankerst et al., 1999).

The algorithm DBCLASD (Distribution Based Clustering of LArge Spatial Databases) (Xu et al., 1998) reduces dependence on user-specified parameters by assuming that data points inside a cluster follow a uniform distribution, and attempting to build a characteristic probability distribution of the distance to

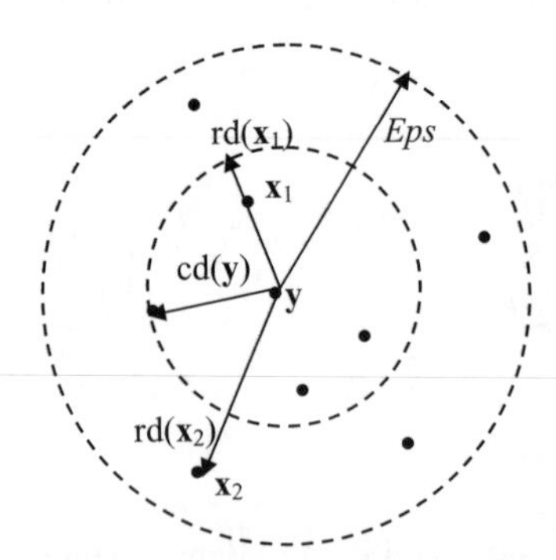

Fig. 8.3. Notions of core-distance and reachability-distance (Ankerst et al., 1999). The core-distance of $\mathbf{y}$, denoted as cd($\mathbf{y}$), the reachability-distance of $\mathbf{x}_1$ and $\mathbf{x}_2$ with respect to $\mathbf{y}$, denoted as rd($\mathbf{x}_1$) and rd($\mathbf{x}_2$), respectively, when *Minpts* is set as 5.

the cluster's nearest neighbors. This is used to decide whether a neighboring point belongs to the cluster. So, a cluster C is defined as a non-empty subset of a set of data points $\mathbf{X}$ and satisfies the following conditions:

1. Its nearest neighbor distance set, which contains all values of the nearest neighbor distance of any point belonging to C, has the expected distribution with a required confidence level;
2. C is a maximal connected set.

From condition 2, it can be seen that any further extension of the cluster by its neighboring points will violate condition 1. Regarding connectivity, a grid-based method is used to determine cluster-approximating polygons, and each pair of points in the cluster is connected via a path of occupied grid cells, which contains at least one point of the cluster.

DBCLASD processes one data point at a time, which is known as incremental or online learning and will be discussed in Section 8.7. As long as a set of candidates is generated through a region query (circle query in this case), DBCLASD examines each candidate by augmenting the current cluster and testing whether the nearest neighbor distance set of the augmented cluster still fits the expected distance distribution based on χ^2–test. Although the run time of DBCLASD is about two to three times that of DBSCAN, which has certain limits in very large-scale data clustering, the nonparametric property still makes DBCLASD a promising algorithm in many clustering applications.

DBSCAN is also generalized in terms of the re-definition of neighborhood and the cardinality of neighborhood to deal with extended data objects such as polygons (Sander et al., 1998). Any symmetric and reflexive binary predicate can be used to define a neighborhood, and non-spatial attributes can provide measures of points in the neighborhood rather than a direct and simple count. The generalized algorithm is called GDBSCAN and also has a time complexity of $O(N\log N)$ with the R^*-tree for indexation.

Unlike the aforementioned density-based methods, the algorithm DENCLUE (DENsity-based CLUstEring) (Hinneburg and Keim, 1998) models the overall density function over the corresponding data space based on the influence of each point on its neighborhood. This influence can be modeled mathematically using an influence function. Specifically, for a point $\mathbf{y}$, its influence function is defined as

$$f^{\mathbf{y}}(\mathbf{x}) = f(\mathbf{x}, \mathbf{y}). \tag{8.7}$$

Typical influence functions include a square wave influence function,

$$f(\mathbf{x}, \mathbf{y}) = \begin{cases} 0, & \text{if } D(\mathbf{x}, \mathbf{y}) > \sigma \\ 1, & \text{otherwise} \end{cases}, \tag{8.8}$$

where σ is a threshold parameter, and a Gaussian influence function,

$$f(\mathbf{x},\mathbf{y})=\exp\left(-\frac{D(\mathbf{x},\mathbf{y})^2}{2\sigma^2}\right). \quad (8.9)$$

where σ is the kernel width parameter.

The density function in a data set $\mathbf{X}$ with N data points, denoted as $f^{\mathbf{X}}(\mathbf{x})$, is then defined as the sum of the influence function of all points,

$$f^{\mathbf{X}}(\mathbf{x})=\sum_{i=1}^{N} f(\mathbf{x},\mathbf{x}_i). \quad (8.10)$$

The overall density function has local peaks, or local maxima, called density attractors, which could be obtained by hill-climbing methods if the influence function is continuous and differentiable. A point $\mathbf{y}$ is said to be density attracted to a density attractor $\mathbf{x}^*$ if the following condition holds:

$$D(\mathbf{x}^k,\mathbf{x}^*)\le\varepsilon \quad \text{with} \quad \mathbf{x}^0=\mathbf{x}, \mathbf{x}^i=\mathbf{x}^{i-1}+\delta\frac{\nabla f^{\mathbf{X}}(\mathbf{x}^{i-1})}{\|\nabla f^{\mathbf{X}}(\mathbf{x}^{i-1})\|}, \quad (8.11)$$

where

$$\nabla f^{\mathbf{X}}(\mathbf{x})=\sum_{i=1}^{N}(\mathbf{x}_i-\mathbf{x})f(\mathbf{x},\mathbf{x}^i). \quad (8.12)$$

A center-defined cluster is then identified through a density attractor, with its overall density above a density threshold ξ, and a set of points density attracted to the density attractor. If the overall density is below ξ, the points density attracted to the local maximum are regarded as noise. Further, if a pair of density attractors can be connected by a path and each point on the path has a density above ξ, these clusters merge and form a cluster with arbitrary shape.

It is interesting to point out that DENCLUE provides a generalization of both K-means and DBSCAN algorithms. By using the Gaussian influence function and choosing an appropriate value of σ, DENCLUE can provide a globally optimal clustering with K center-defined clusters corresponding to the K clusters of K-means, which generalizes K-means because K-means can only generate local optimal partitioning. On the other hand, by using a square wave influence function and setting $\sigma = Eps$ and $\xi = Minpts$, DENCLUE can form clusters with arbitrary shapes that are equivalent to those generated by DBSCAN. Moreover, if σ is increased from a very small initial value, a hierarchical clustering structure can also be obtained.

Before performing clustering, DENCLUE includes a pre-processing step to create a map of hyperrectangle cubes with edge length 2σ. Because of this property, DENCLUE is also considered a grid-based method, which will be discussed in the following section. The cubes are then numbered according to their relative positions to a certain origin to map d-dimensional cubes containing data points to one-dimensional keys. Information of an occupied cube, such as the key, the number of points in the cube, the linear sum of points in the cube, the pointers to these points, and the connections to neighboring cubes, are stored for further clustering. Note that during the clustering step, DENCLUE only processes the highly-occupied cubes and cubes connected to them.

8.5. GRID-BASED METHODS

Grid-based approaches divide a data space into a set of cells or cubes by a grid. This space partitioning is then used as a basis for determining the final data partitioning.

The algorithm STING (STatistical INformation Grid) (Wang et al., 1997) uses a hierarchical structure within the division of the data space. Cells are constructed at different levels in the hierarchy corresponding to different resolutions. The hierarchy starts with one cell at the root level and each cell at a higher level has l children (four by default). Information in each cell is stored in terms of a feature independent parameter, i.e., the number of points in the cell, and feature dependent parameters, i.e., mean, standard deviation, minimum, maximum, and distribution type. Parameters at higher-level cells can be obtained from parameters at lower-level cells. For example, supposing that N^i and M^i are the number of points and mean value in current level cells, respectively, they can be calculated from the corresponding parameters of lower-level cells N^{i-1} and M^{i-1} as,

$$N^i = \sum_j N_j^{i-1}, \tag{8.13}$$

$$M^i = \frac{\sum_j M_j^{i-1} N_j^{i-1}}{N^i}. \tag{8.14}$$

Clustering is performed using a top-down method, starting with the root level or some intermediate layer. Cells that are relevant to certain conditions are determined based on their data summaries, and only those cells that are children of the relevant cells are further examined. After the bottom level is reached, a breadth-first search can be used to find the clusters that have densities greater than a prespecified threshold. Thus, STING combines both data condensation and density-based clustering strategies. The clusters formed by STING can approximate the result from DBSCAN when the granularity of the bottom level approaches zero (Wang et al., 1997). STING has a run time complexity of $O(L)$, where L is the number of cells at the lowest level. Because L is usually much smaller than the number of points in the data set, STING achieves faster performance in simulation studies than other algorithms, such as BIRCH. STING is also extended as STING+ (Wang et al., 1999) to deal with dynamically evolving spatial data while maintaining the similar hierarchical structure. STING+ supports user-defined triggers, which are decomposed into sub-triggers associated with cells in the hierarchy. STING+ considers four categories of triggers based on the absolute or relative condition on certain regions or features.

WaveCluster (Sheikholeslami et al., 1998) considers clustering data in the feature space from a signal processing perspective. Cluster boundaries, which display rapid changes in the distribution of data points, correspond to the high-frequency parts of the signal, while the interiors of clusters, which have high densities, correspond to the low frequency parts of the signal with high amplitude. Signal processing techniques, such as wavelet transform, can be used to identify the different frequency subbands of the signal and therefore generate the clusters. Wavelet transform demonstrates many desirable properties in cluster identification, particularly with the benefits of effective filters, outlier detection, and multi-resolution analysis. For example, the hat-shaped filters make the clusters more distinguishable by emphasizing dense regions while suppressing less dense areas in the boundaries. Low-pass filters have the advantage of automatically eliminating noise and outliers. Multi-resolution representation of a signal with wavelet transform allows the identification of clusters at different scales, i.e., coarse, medium, and fine.

The basic steps of WaveCluster are described in Fig. 8.4. Data points are first assigned to a set of cells of a grid dividing the original feature space uniformly. The size of the grid will vary corresponding to different scales of transform. A discrete wavelet transform, such as Harr and Cohen-Daubechies-Feauveau transforms, is then used on these cells to map the data into the new feature space, where the clusters, represented as the connected components in the space, are detected. Different resolutions of wavelet transform lead to different sets of clusters. Because the clusters are formed in the transformed space, a lookup table is created to associate cells in the original feature space with cells in the new feature space. The assignment of points in the original space to the corresponding clusters becomes trivial.

The algorithm FC (Fractal Clustering) (Barbará and Chen, 2000) combines the concepts of both incremental clustering and fractal dimension. Fractal dimensions are used to characterize fractals, which have the property of self-similarity (Mandelbrot, 1983). The basic procedure of FC is to incrementally add data points to the clusters, specified through an initial nearest neighbor-based process and represented as cells in a grid, with the condition that the fractal dimension of clusters must remain relatively stable. Therefore, the data

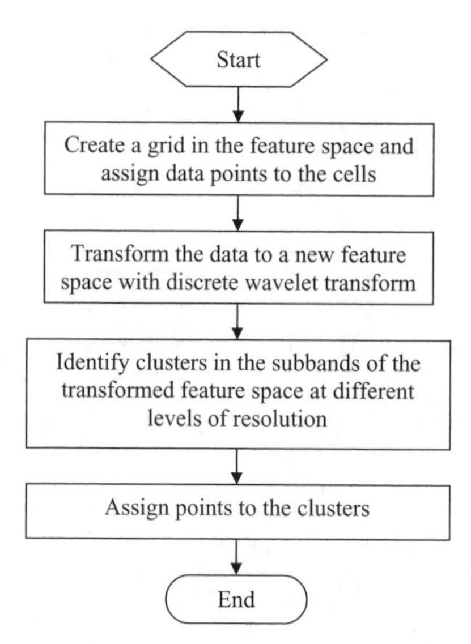

Fig. 8.4. Flowchart of WaveCluster. Clusters are identified in the transformed feature space. The assignment of data points to the corresponding clusters is achieved via a lookup table, which associate cells in the original feature space with cells in the transformed feature space.

points that are included into a cluster will not greatly change the fractal dimension of the cluster. The data points that have the smallest fractal impact exceeding a threshold is declared as noise. In order to calculate the fractal dimension, the Hausdorff dimension is used, which can be obtained as the negative value of the slope of a log-log scale plot, known as a box-counting plot, depicting the number of cells occupied by points in the data set versus a grid size. Specifically, clusters are described through a series of layers, or grid representations. The sizes of grids in each layer decrease with regard to the previous layer. For example, we can divide the cardinality of each dimension in the layer with the largest grids by 2, the next layer by 4, and so on. The information on the number of points in each grid is saved, and only occupied grids are kept in the memory. FC has $O(N)$ time complexity; however, FC is largely dependent on the clusters generated in the initialization step, and the algorithm also relies on the presentation order of data points, which is a common problem for incremental learning.

8.6. DIVIDE AND CONQUER

When the size of a data set is too large to be stored in the main memory, it is possible to divide the data into different subsets that can fit the main memory and to use the selected cluster algorithm separately to these subsets. The final

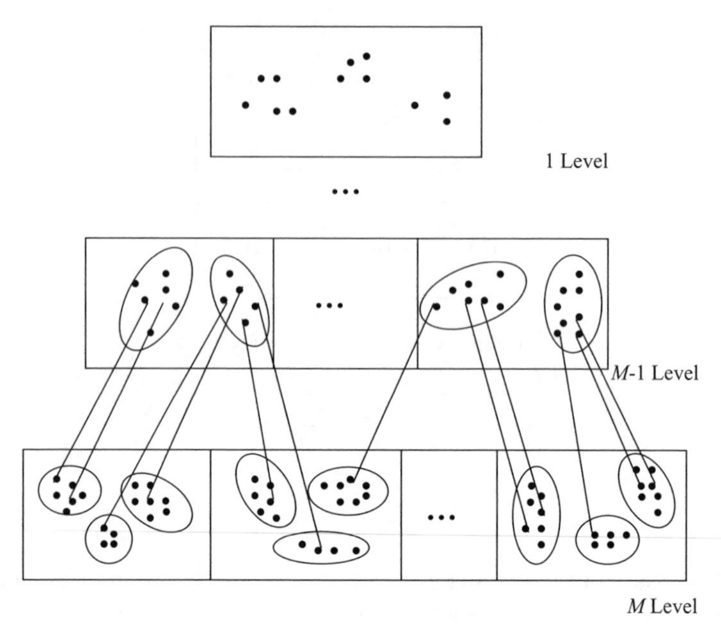

Fig. 8.5. *M*-level divide-and-conquer approach in large-scale data clustering. Commonly, *M* is set as 2, which leads to a two-level divide-and-conquer clustering algorithm.

clustering result is obtained by merging the previously formed clusters. This approach is known as divide and conquer (Guha et al., 2003; Jain et al., 1999), and an *M*-level divide-and-conquer algorithm is depicted in Fig. 8.5.

Specifically, given a data set with *N* points stored in a secondary memory, the divide-and-conquer algorithm first divides the entire data set into *r* subsets with approximately similar sizes. Each of the subsets is then loaded into the main memory and is divided into a certain number of clusters with a clustering algorithm. Representative points of these clusters, such as the centers of the clusters, are then picked for further clustering. These representatives may be weighted based on some rule, e.g., the centers of the clusters could be weighted by the number of points belonging to them (Guha et al., 2003). The algorithm repeatedly clusters the representatives obtained from the clusters in the previous level until the highest level is reached. The data points are then put into corresponding clusters formed at the highest level based on the representatives at different levels.

Stahl (1986) illustrated a two-level divide-and-conquer clustering algorithm applied to a data set with 2,000 data points. The leader algorithm (Duda et al., 2001) is first used to form a large number of clusters from the original data. The obtained representatives of these clusters are then clustered with a hierarchical clustering algorithm.

8.7. INCREMENTAL CLUSTERING

In contrast to batch clustering, which requires loading the entire data set into the main memory, an incremental or online clustering approach does not require the storage of all these data points, instead handling the data set one point at a time. If the current data point displays enough closeness to an existent cluster according to some predefined criteria, it is assigned to the cluster. Otherwise, a new cluster is created to represent the point. Because only the representation of each cluster must be stored in the memory, an incremental clustering strategy saves a great deal of space. A typical example that is based on incremental learning is the adaptive resonance theory (ART) family (Carpenter and Grossberg, 1987a, b, 1988), discussed in Chapter 5. Several other clustering algorithms in this chapter, such as DBCLASD and FC, also process the input data points incrementally.

As previously mentioned, one of the major problems for incremental clustering algorithms is that they are order dependent, which means that different presentation orders of the input points cause different partitions of the data set (Carpenter and Grossberg, 1987a; Moore, 1989). Obviously, this is not an appealing property because of the problem caused in cluster validation. DBCLASD (Xu et al., 1998) uses two methods to decrease the reliance on data ordering. The first heuristic retains the unsuccessful candidates rather than discarding them and then tries them again to the clusters. The second heuristic allows data points to change their cluster membership.

8.8. APPLICATIONS

8.8.1 Traveling Salesman Problem

The Traveling Salesman Problem (TSP) is one of the most studied examples in an important class of problems known as NP-complete problems. Given a complete undirected graph $G = (V,E)$, where V is a set of vertices and E is a set of edges each relating two vertices with an associated non-negative integer cost, the most general form of the TSP is equivalent to finding any Hamiltonian cycle, which is a tour over G that begins and ends at the same vertex and visits other vertices exactly once. The more common form of the problem is the optimization problem of trying to find the shortest Hamiltonian cycle, and in particular, the most common is the Euclidean version, where the vertices and edges all lie in the plane. Mulder and Wunsch (2003) applied a divide-and-conquer clustering technique, with ART networks, to scale the problem to 1 million cities, and later, to 25 million cities (Wunsch and Mulder, 2004). The divide-and-conquer paradigm provides the flexibility to hierarchically break large problems into arbitrarily small clusters depending on the desired trade-off between accuracy and speed. In addition, the sub-problems provide an excellent opportunity to take advantage of parallel systems for further optimization.

Specifically, the proposed algorithm combines both ART and the Lin-Kernighan (LK) local optimization algorithm (Lin and Kernighan, 1973) to divide-and-conquer instances of the TSP. The Lin-Kernighan algorithm takes a randomly selected tour and optimizes it by stages until a local minimum is found. This result is saved, and a new random tour is selected to begin the process again. Given enough time, the optimal tour can be found for any instance of the problem, although in practice this is limited to small instances for a global optimum due to time constraints. A popular variant and the most effective known algorithm for large TSP is Chained LK. With Chained variants of LK, instead of beginning with a random tour at each new iteration, the algorithm perturbs the previous tour by some amount and optimizes from there.

As the first stage of the divide-and-conquer algorithm, an ART network is used to sort the cities into clusters, dividing the problem into smaller sub-problems. The vigilance parameter is used to set a maximum distance from the current pattern, and a vigilance parameter between 0 and 1 is considered as a percentage of the global space to determine the vigilance distance. Values are chosen based on the desired number and size of individual clusters. Each individual cluster is then passed to a version of the LK algorithm. Since the size of the tours is controlled and kept under one thousand cities, the LK is allowed to search an infinite depth as long as the total improvement for a given swap series remains positive. After a first pass through the LK algorithm, a simple intersection removal algorithm is applied. This algorithm is based on the idea that any tour containing an intersection between two edges is demonstrably sub-optimal and capable of making double-bridge swaps that the LK algorithm is unable to discover. This double-bridge property is the same as that used in the Chained LK algorithms. The last step involves the combination of a number of sub-tours back into one complete tour, which is achieved by adding the tours in order of increasing distance from the origin.

One major factor involved with merging the tours is running time. Since this is a potentially global operation, care must be exercised in the nature of the algorithm. For example, attempting to find an optimal linking between the two tours could be at least an $O(N_g^2)$ algorithm, which is unacceptable because the N involved would be the total number of cities, not just the cities in a tour. To avoid such a global operation, the centroid of the cluster to be added is identified. This is just the average of the X and Y coordinates of each city in the cluster and is easily calculated in $O(N_c)$. The k nearest cities to that centroid in the combined tour are then found, which requires $O(N_g)$ time. Next, each of the k cities from the main tour is considered to determine the cost of inserting the cluster tour in place of the following edge. This involves comparing k cities to N_c cities to determine the lowest cost matching, yielding a running time of $O(k*N_c)$, where $k << N_g$. Finally, the cluster tour is inserted into the merged tour at the best location discovered.

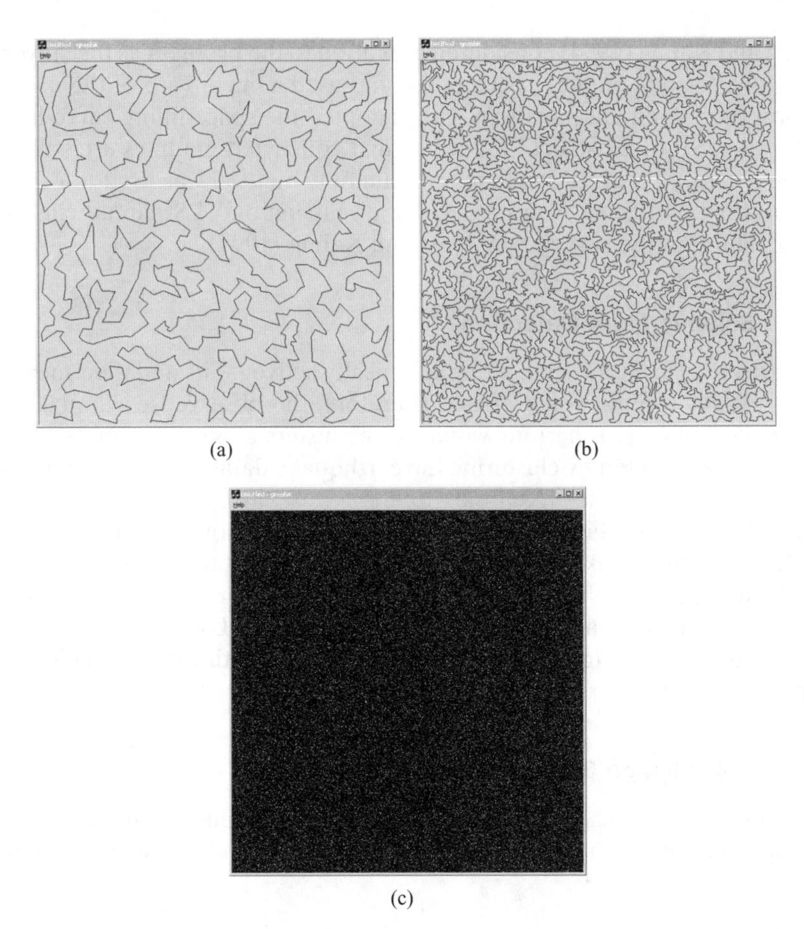

Fig. 8.6. Clustering divide-and-conquer TSP resulting tours for (a) 1k, (b) 10k, (c) 1M cities. The clustered LK algorithm achieves a significant speedup and shows good scalability.

Tours with good quality for city levels up to 1,000,000 were obtained within 25 minutes on a 2 GHz AMD Athlon MP processor with 512 M of DDR RAM. For the 25 million city problem, the algorithm takes 13,500 seconds, while Chained LK can not solve the problem at all within the memory constraints of the machine. The visualizing results for 1,000, 10,000, and 1,000,000 cities are shown in Fig. 8.6, respectively.

It is worthwhile to emphasize the relation between the TSP and VLSI (Very Large-Scale Integrated) circuit clustering, which partitions a sophisticated system into smaller and simpler sub-circuits to facilitate the circuit design. The object of the partitions is to minimize the number of connections among the components. One strategy for solving the problem is based on geometric representations, either linear or multi-dimensional (Alpert and Kahng, 1995).

Alpert and Kahng (1994) considered a solution to the problem as the "inverse" of the divide-and-conquer TSP method and used a linear tour of the modules to form the sub-circuit partitions. They adopted the space-filling curve heuristic for the TSP to construct the tour so that connected modules are still close in the generated tour. A dynamic programming method was used to generate the resulting partitions. A more detailed discussion on VLSI circuit clustering can be found in the survey by Alpert and Kahng (1994).

8.8.2 Seismic Fault Detection

Earthquake epicenters occur along seismically active faults and their measurements always contain some errors. Therefore, the observed earthquake epicenters during certain periods would form clusters along such faults. Seismic faults can be detected by clustering the earthquake databases that record locations of the earthquakes (Xu et al., 1998).

The left part of Fig. 8.7 illustrates the locations of the earthquakes in California's earthquake database. An earthquake catalog recording over a 40,000 km^2 region of the central coast ranges in California from 1962 to 1981 was used for further analysis. Clusters obtained with DBCLASD are shown in the right part of Fig. 8.7. Four clusters, depicted with different colors, corresponding to the two main seismic faults, were detected by DBCLASD.

8.8.3 Color Image Segmentation

Segmentation, or the partition of an image into regions that have certain homogenous image characteristics, is important in the identification of signifi-

Fig. 8.7. Seismic fault detection based on California's earthquake database. (From X. Xu, M. Ester, H. Kriegel, and J. Sander. A distribution-based clustering algorithm for mining in large spatial databases. In Proceedings of 14th International Conference on Data Engineering, pp. 324–331, 1998. Copyright © 1998 IEEE.)

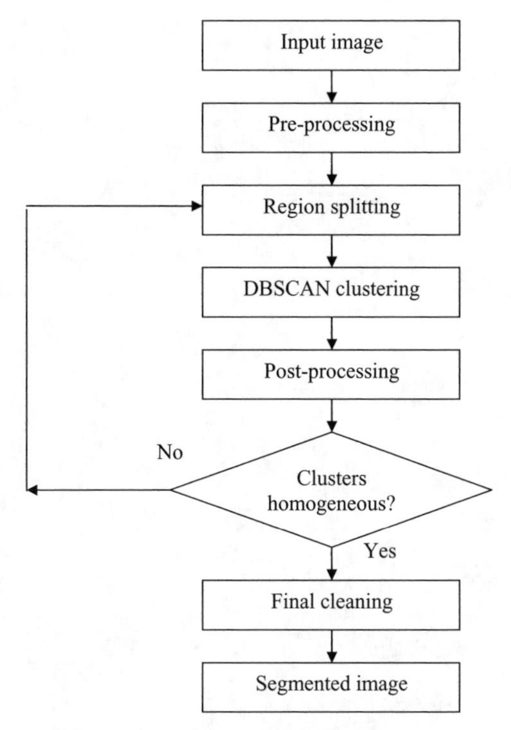

Fig. 8.8. Flowchart of the segmentation procedure. (From M. Celebi, Y. Aslandogan, and P. Bergstresser. Mining biomedical images with density-based clustering. In Proceedings of the International Conference on Information Technology: Coding and Computing, vol. 1, pp. 163–168, 2005. Copyright © 2005 IEEE.)

cant features. The basic steps for segmenting and identifying homogenous color regions in images based on the clustering algorithm DBSCAN as summarized in Fig. 8.8 (Celebi et al., 2005). After the pre-processing step, which eliminates the effects of some extraneous factors, a region splitting process is performed according to a homogeneity criterion. The function of DBSCAN is to merge these small regions to generate the clusters. The regions that are too small to be perceived by humans are also merged into the closest clusters in the final cleaning step.

The segmentation results for a set of 135 biomedical images (18 of them were used for parameter tuning) with 256×256 dimensions are shown in Fig. 8.9. The object was to identify the border between the healthy skin and the lesion, and furthermore, to find the sub-regions inside the lesion with varying coloring. An expert examination showed that the lesion borders in 80% of the images were successfully detected by the DBSCAN-based segmentation system. Further analysis on the comparisons of color sub-region identification by the proposed system and humans also suggested a significant agreement (Celebi et al., 2005).

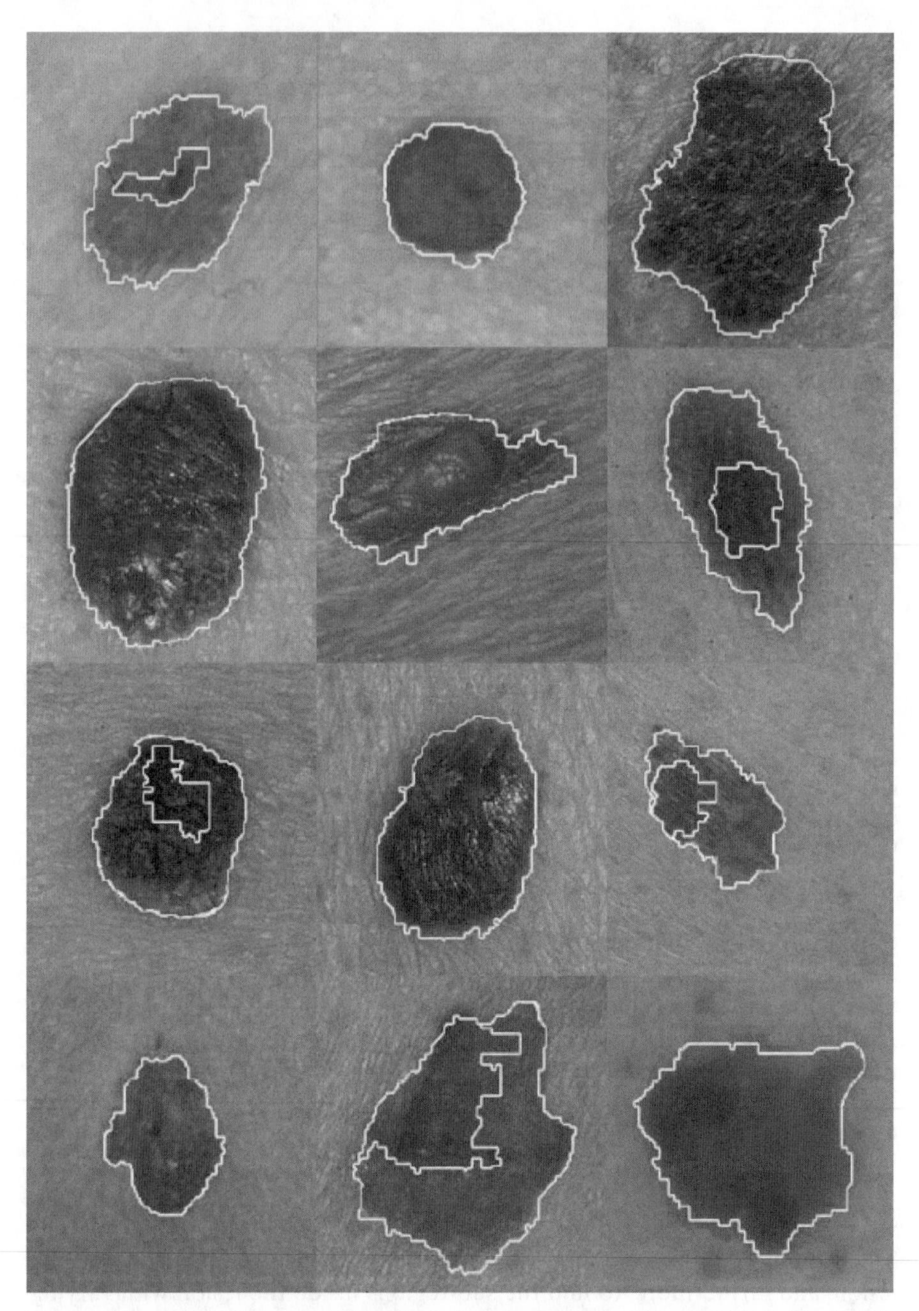

Fig. 8.9. Examples of segmentation results for a set of clinical images. (From M. Celebi, Y. Aslandogan, and P. Bergstresser. Mining biomedical images with density-based clustering. In Proceedings of the International Conference on Information Technology: Coding and Computing, vol. 1, pp. 163–168, 2005. Copyright © 2005 IEEE.)

8.9. SUMMARY

The enormous proliferation of large-scale data sets brings new challenges and requirements to cluster analysis. Scalability is often critical. We have discussed clustering algorithms specifically designed for massive data sets. These algorithms combine incremental learning with approaches, such as random sampling, data condensation, or divide and conquer, and have achieved great improvements in clustering large-scale data. In addition to the aforementioned approaches, parallel algorithms, which can more effectively use computational resources and can greatly improve overall performance in the context of both time and space complexity, also play a significant role in clustering large-scale data sets (Dahlhaus, 2000; Olson, 1995; Stoffel and Belkoniene, 1999). As mentioned in the chapter, the performance of most of these algorithms degrades as the data to be processed exhibits higher dimensionalities. In the following chapter, we focus on the algorithms that address this problem.

9.4 SUMMARY

[illegible]

CHAPTER 9

DATA VISUALIZATION AND HIGH-DIMENSIONAL DATA CLUSTERING

9.1. INTRODUCTION

High dimensionality is a major contributor to data complexity. Technology makes it possible to automatically and systematically obtain a large amount of measurements. However, they often do not precisely identify the relevance of the measured features to the specific phenomena of interest. Data observations with thousands of features, or more are now common, such as genomic data, financial data, web document data, sensor data, and satellite image data. As a result of such high dimensionalities, the assumption of $N > d$, on which many approaches are based, no longer hold true. In many cases, the applications require clustering algorithms to process data with more features than the number of observances. Cluster analysis on gene expression data analysis is one such typical example, with data sets usually containing thousands of genes (features) but less than 100 samples. The lag of experimental condition (e.g., sample collection and experimental cost) and the rapid advancement of micro-array and sequencing technologies cause this situation. Most of the algorithms summarized in Table 8.1 in Chapter 8 are not sufficient for analyzing high-dimensional data, although they provide effective means to deal with large-scale data.

The term "curse of dimensionality," which was first used by Bellman (1961) to indicate the exponential growth of complexity in the case of multivariate function estimation under a high dimensionality situation, is generally used to describe the problems accompanying high dimensional spaces (Beyer et al.,

Clustering, by Rui Xu and Donald C. Wunsch, II

1999; Haykin, 1999). Assume we have a Cartesian grid of spacing 1/10 on the unit cube in 5 dimensions; we then have 10^5 points. The number of points increases to 10^{10} if the cube is in 10 dimensions, and 10^{20} if in 20 dimensions. Therefore, the exhaustive search of such a discrete space for function estimation would become infeasible with the requirement for evaluating the function tens of millions of times. As indicated by Friedman (1995), "A function defined in high-dimensional space is likely to be much more complex than a function defined in a lower-dimensional space, and those complications are harder to discern."

High dimensionality also causes a problem in the separation of data points. Beyer et al. (1999) showed that the distance between the nearest point and a query point is no different from that of other points when the dimensionality of the space is high enough (10–15 dimensions). Therefore, algorithms that are based on the distance measure may no longer be effective in a high-dimensional space. Fortunately, in practice, although data are represented with a large set of variables, many of them are only included in the data as a result of a subjective measurement choice and contribute nothing to the description of the real structure of the data. On the other hand, the minimum number of free variables that provide sufficient information in representing data is referred to as intrinsic dimensionality, d_0, which is much lower than the original dimension d (Camastra, 2003; Cherkassky and Mulier, 1998). In this way, data are embedded in a low-dimensionality and compact subspace, and the standard and commonly used algorithms for low-dimensional data analysis can then be applied.

Therefore, the goal of dimension reduction is to find a mapping F that maps the input data from the space $\mathcal{R}^d$ to a lower-dimension feature space $\mathcal{R}^{d'}$, denoted as

$$F(\mathbf{x}): \Re^d \rightarrow \Re^{d'}. \tag{9.1}$$

This process is usually associated with the optimization of a risk function, which determines the information to be kept or discarded during the mapping. Dimension reduction is important in cluster analysis, which not only makes the high-dimensional data addressable and reduces the computational cost, but can provide users with a clearer picture and visual examination of the data of interest (Fodor, 2002). However, because dimensionality reduction methods inevitably cause some loss of information or may damage the interpretability of the results, even distorting the real clusters, extra caution is advised.

In the remaining sections of this chapter, we discuss dimension reduction and data visualization technologies as either linear or nonlinear. The liner transforms include technologies like principal component analysis, independent component analysis, and projection pursuit, introduced in Section 9.2. Section 9.3 concentrates on the nonlinear approaches, such as nonlinear principal component analysis, multidimensional scaling, ISOMAP, locally linear embedding, diffusion maps, etc. Self-organizing feature maps (SOFM) can also provide good visualization for high-dimensional input patterns (Kohonen, 1990). Discussion, together with applications of SOFM, was given in Chapter 5. Projected and space clustering algorithms are discussed in Section 9.4. Section 9.5 illustrates several examples of the corresponding dimension reduction algorithms.

9.2. LINEAR PROJECTION ALGORITHMS

9.2.1 Principal Component Analysis

One natural strategy for dimensionality reduction is to extract important components from the original data, which can effectively represent the potential structure of the data. Principal component analysis (PCA), or Karhunen-Loéve transformation, is one of the best known approaches, which is concerned with constructing a linear combination of a set of vectors that can best describe the variance of data (Duda et al., 2001; Jollife, 1986).

Given a set of N d-dimensional input patterns $\{\mathbf{x}_1, \ldots, \mathbf{x}_i, \ldots, \mathbf{x}_N\}$, each of which can be written as a linear combination of a set of d orthonormal vectors,

$$\mathbf{x}_i = \sum_{j=1}^{d} c_{ij} \mathbf{v}_j, \tag{9.2}$$

PCA aims to approximate the data by a linear l-dimension subspace based on the squared error criterion,

$$J = \sum_{i=1}^{N} \|\mathbf{x}_i - \mathbf{x}_i'\|^2, \tag{9.3}$$

where $\mathbf{x}_i'$ is an approximation of $\mathbf{x}_i$ when only a subset of l vectors $\mathbf{v}_j$ is kept, represented as,

$$\mathbf{x}_i' = \sum_{j=1}^{l} c_{ij} \mathbf{v}_j + \sum_{j=l+1}^{d} e_j \mathbf{v}_j, \tag{9.4}$$

where e_j are constants used to replace the original coefficients c_{ij} ($l + 1 \leq j \leq d$).

Now we recall Eq. 9.2

$$\mathbf{x}_i = \sum_{j=1}^{d} c_{ij} \mathbf{v}_j, \tag{9.5}$$

and by multiplying $\mathbf{v}_j^T$ on both sides, obtain

$$c_{ij} = \mathbf{v}_j^T \mathbf{x}_i. \tag{9.6}$$

Now we replace Eq. 9.4 into Eq. 9.3 and considering the orthonormality relation of $\mathbf{v}_j$, the error criterion function is now written as

$$J = \sum_{i=1}^{N} \sum_{j=l+1}^{d} (c_{ij} - e_j)^2. \tag{9.7}$$

By setting $\frac{\partial J}{\partial e_j} = 0$, we obtain

$$e_j = \frac{1}{N} \sum_{i=1}^{N} c_{ij} = \mathbf{v}_j^T \mathbf{m}, \tag{9.8}$$

where $\mathbf{m} = \frac{1}{N} \sum_{i=1}^{N} \mathbf{x}_i$ is the mean vector.

We then substitute Eqs. 9.6 and 9.8 into Eq. 9.7 to obtain

$$\begin{aligned} J &= \sum_{i=1}^{N} \sum_{j=l+1}^{d} \left(\mathbf{v}_j^T (\mathbf{x}_i - \mathbf{m})\right)^2 \\ &= \sum_{j=l+1}^{d} \sum_{i=1}^{N} \mathbf{v}_j^T (\mathbf{x}_i - \mathbf{m})(\mathbf{x}_i - \mathbf{m})^T \mathbf{v}_j, \\ &= \sum_{j=l+1}^{d} \mathbf{v}_j^T \mathbf{S} \mathbf{v}_j \end{aligned} \tag{9.9}$$

where $\mathbf{S}$ is the scatter matrix, defined as,

$$\mathbf{S} = \sum_{i=1}^{N} (\mathbf{x}_i - \mathbf{m})(\mathbf{x}_i - \mathbf{m})^T, \tag{9.10}$$

which is exactly $N - 1$ times the covariance matrix and is also discussed in Chapter 4.

Using the method of Lagrange optimization, the minimization of J is obtained when the vectors $\mathbf{v}_j$ are the eigenvectors of the scatter matrix,

$$\mathbf{S}\mathbf{v}_j = \lambda_j \mathbf{v}_j, \tag{9.11}$$

where λ are the corresponding eigenvalues. The resulting minimum error criterion function is

$$J_{\min} = \sum_{j=l+1}^{d} \lambda_j. \tag{9.12}$$

Eq. 9.12 indicates that the minimum error can be obtained by retaining the l largest eigenvalues and the corresponding eigenvectors of **S**, or discarding the $p - l$ smallest eigenvalues and eigenvectors. The retained eigenvectors are called the principal components.

PCA calculates the eigenvectors that are constituted of the matrix of the linear transform while minimizing the sum of squares of the error of approximating the input patterns. In this sense, PCA can be realized in terms of a three-layer neural network, called an auto-associative multilayer perceptron (Baldi and Hornik, 1989; Lerner et al., 1999; Oja, 1992), as depicted in Fig. 9.1. Linear activation functions are used for all the neurons in the network, rather than the commonly used sigmoid transfer function. The input patterns are presented to both the input and output layers of the network in order to provide a map for the input patterns onto themselves. The network's training is based on the minimization of the sum of squared error, which can be achieved using gradient descent-based methods (Haykin, 1999). The trained network captures the l principal components in the linear hidden layers. Another category of PCA networks is the one-layer network using the Hebbian type learning rules (Oja, 1982, 1992). A framework for generalizing such algorithms is proposed in Weingessel and Hornik (2000). Also, the intrinsic connections between PCA and K-means (Chapter 4) are discussed in Mirkin (2005, 2007).

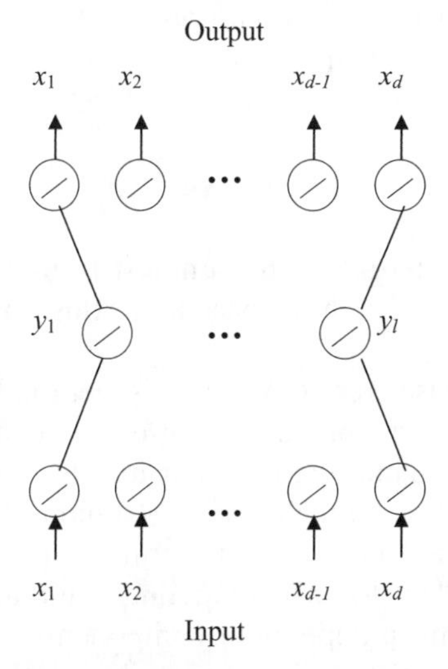

Fig. 9.1. Three-layer, auto-associative neural network for PCA. Inputs are mapped onto themselves, and the network is trained to minimize a sum of squared error.

9.2.2 Independent Component Analysis

PCA is appropriate for Gaussian distributions because it relies on second-order relationships stored in the scatter (covariance) matrix. Other linear transforms, such as independent component analysis (ICA) and projection pursuit (discussed in Section 9.2.3), consider higher-order statistical information and therefore are used for non-Gaussian distributions (Cherkassky and Mulier, 1998; Hyvärinen, 1999; Jain et al., 2000). The basic goal of ICA is to find the components that are most statistically independent from each other (Hyvärinen, 1999; Jutten and Herault, 1991). In the context of blind source separation, ICA aims to separate the independent source signals from the mixed observation signals.

Given a d-dimensional random vector $\mathbf{x}$, Hyvärinen (1999) summarized three different formulations of ICA:

1. General model. ICA seeks a linear transform

$$\mathbf{s} = \mathbf{W}\mathbf{x}, \tag{9.13}$$

so that s_i in the vector $s = (s_1, \ldots, s_d)^T$ are as independent as possible while maximizing some independence measure function $f(s_1, \ldots, s_d)$.

2. Noisy ICA model. ICA estimates the model

$$\mathbf{x} = \mathbf{A}\mathbf{s} + \boldsymbol{\varepsilon}, \tag{9.14}$$

where the components s_i in the vector $\mathrm{s} = (s_1, \ldots, s_l)^T$ are statistically independent from each other, $\mathbf{A}$ is a nonsingular $d \times l$ mixing matrix, and ε is a d-dimensional random noise vector.

3. Noise-free ICA model. ICA estimates the model without considering noise

$$\mathbf{x} = \mathbf{A}\mathbf{s}. \tag{9.15}$$

The noise-free ICA model takes the simplest form among the three definitions, and most ICA research is based on this formulation (Hyvärinen, 1999).

To ensure that the noise-free ICA model is identifiable, all the independent components s_i, with the possible exception of one component, must be non-Gaussian. On the other hand, the constant matrix $\mathbf{A}$ must be of full column rank, and d must be no less than l, i.e., $d \geq l$ (Comon, 1994). The non-Gaussian condition can be justified because for normal random variables, statistical independence is equivalent to uncorrelatedness, and any decorrelating representation would generate independent components, which cause ICA to be ill-posed from a mathematical point of view. A situation in which the components s_i are nonnegative was discussed by Plumbley (2003).

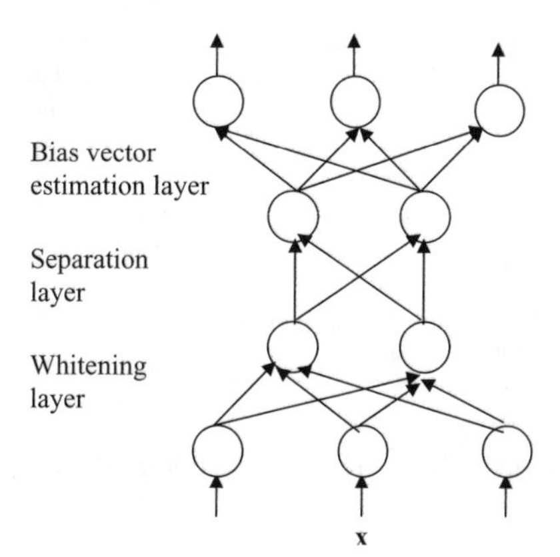

Fig. 9.2. Realization of ICA via multilayer perceptrons. The network is composed of whitening, separation, and basis vector estimation layers.

The estimation of the ICA model consists of two major steps: the construction of an objective function, or contrast function, and the development and selection of an optimization algorithm for maximizing or minimizing the objective function. Hyvärinen (1999) reviewed a large number of such objective functions and the corresponding optimization algorithms. For example, mutual information between the components is regarded as the "most satisfying" objective function in the estimation of the ICA model (Hyvärinen, 1999). The minimization of such an objective function can be achieved through a gradient descent method.

ICA can also be realized by virtue of multilayer perceptrons, as illustrated in Fig. 9.2 (Karhunen et al., 1997). The proposed ICA network includes whitening, separation, and basis vector estimation layers, with the corresponding learning algorithm for each layer. The relation between the ICA network and the auto-associative neural network is also discussed.

9.2.3 Projection Pursuit

Projection pursuit is another statistical technique for seeking low-dimensional projection structures for multivariate data (Friedman, 1987; Huber, 1985). Generally, each projection is associated with an index to measure the potential structured information. Optimization algorithms are then used to maximize or minimize these indices with respect to certain parameters in order to find the "interesting" projections. More specifically, it is found that most projections of high-dimensional data take the form of normal distributions. Therefore, projection pursuit regards the normal distribution as the least interesting projection and optimizes the projection indices that measure the degree of nonnormality (Friedman, 1987; Huber, 1985). One such example is differential entropy (Huber, 1985). Given a random variable $\mathbf{x}$ with density function $p(\mathbf{x})$, the differential entropy H is defined as

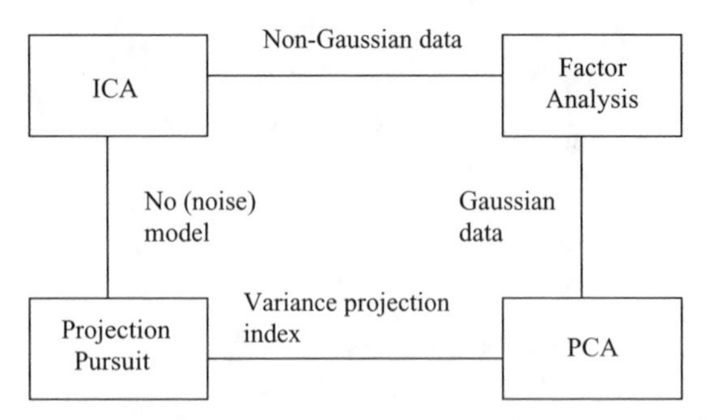

Fig. 9.3. Relations among PCA, ICA, Projection Pursuit, and Factor Analysis (Hyvärinen, 1999).

$$H(\mathbf{x}) = -\int p(\mathbf{x}) \log p(\mathbf{x}) d\mathbf{x}. \tag{9.16}$$

Among all possible distributions with mean μ and variances σ^2, the entropy for the Gaussian distribution is the largest. We are interested in the projection directions that can minimize H.

The relationships among PCA, ICA, Projection Pursuit, and Factor Analysis are summarized in Fig. 9.3 (Hyvärinen, 1999). PCA can be considered as a special example of projection pursuit when the variance is used as the projection index (Cherkassky and Mulier, 1998), while the noise-free ICA model can also be regarded as a special case of projection pursuit by using the criteria for independent component estimation as the index for exploring interesting directions. Both ICA and PCA can be considered as a type of factor analysis (Johnson and Wichern, 2002) under the contradictory assumptions of normality and non-normality, respectively.

9.3. NONLINEAR PROJECTION ALGORITHMS

9.3.1 Nonlinear PCA

PCA projects data into a lower-dimensional linear subspace of the feature space that best accounts for the data in terms of a sum of squares criterion. However, in many cases, it is important to extract a more complicated nonlinear data structure, where PCA may overestimate the true dimensionality. A data set consisting of data points lying on a perimeter of a circle is one such example (Bishop, 1995). PCA suggests a dimension of 2 with two equal eigenvalues, rather than the true dimension of 1. To obtain the nonlinear components, nonlinear PCA was developed that mainly includes two types of approaches: kernel PCA based on Mercer kernels (Schölkopf et al., 1998) and autoassociative neural networks (Bishop, 1995; Kramer, 1991).

As discussed in Chapter 6, kernel PCA first maps the input patterns into a high-dimensional feature space with an appropriate nonlinear function. The

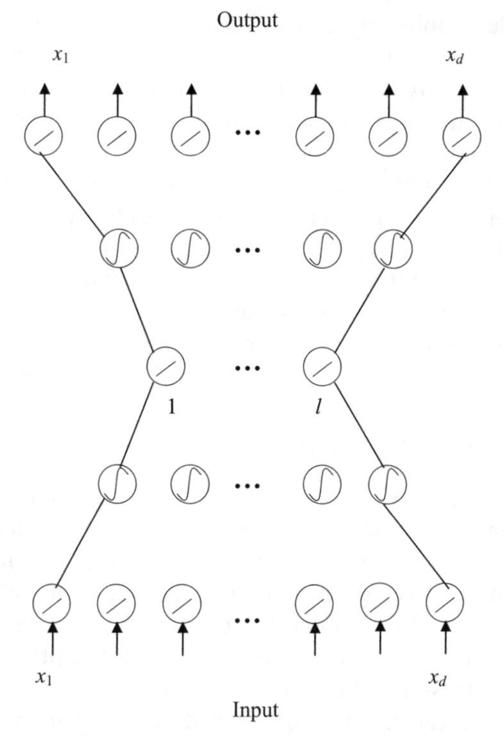

Fig. 9.4. Nonlinear PCA neural networks. Two extra hidden layers with sigmoid activation functions are added into the network in Fig. 9.1. The nonlinear components are disclosed in the linear hidden layer consisting of $l<d$ neurons.

dot product of the nonlinear function can be described in terms of the Mercer kernels, which avoid the requirement to explicitly define the nonlinear function. Steps similar to PCA are then applied to solve the eigenvalue problem with the new covariance matrix in the transformed space,

$$\Sigma^{\Phi} = \frac{1}{N}\sum_{j=1}^{N}\Phi(\mathbf{x}_j)\Phi(\mathbf{x}_j)^T, \tag{9.17}$$

where $\Phi(\cdot)$ performs the nonlinear mapping. We obtain the principal components by calculating the eigenvectors $\mathbf{e}$ and eigenvalues $\lambda > 0$ of Σ^{Φ},

$$\lambda\mathbf{e} = \Sigma^{\Phi}\mathbf{e}. \tag{9.18}$$

On the other hand, nonlinear PCA can be achieved by adding two extra hidden layers with sigmoid functions into the auto-associative network shown in Fig. 9.1 (Bishop, 1995; Duda et al., 2001; Kramer, 1991). The resulting five-layer neural network in Fig. 9.4, has three hidden layers named the mapping, bottleneck, and demapping layer, respectively. Like the linear PCA network, each data pattern is presented to both the input and the output layer as input and target simultaneously. The mapping and demapping layers have the same number of neurons, which is smaller than in the input and output layers, but larger than in the bottleneck layer, therefore generating a bottleneck structure for the network. The nonlinear PCA network is trained with backpropogation using a sum of squared error criterion, and the nonlinear subspace is found in the bottleneck layer that has $l < d$ units. Malthouse (1998) discussed the sub-optimal projections of nonlinear PCA onto curves and surfaces and indicated its limitations in modeling curves and surfaces that intersect themselves and parameterizations that have discontinuous jumps.

9.3.2 Multidimensional Scaling

Multidimensional scaling (MDS) is a nonlinear projection method that fits original multivariate data into a low (usually two or three) dimensional structure while tending to maintain the proximity information as much as possible (Borg and Groenen, 1997; Young and Hamer, 1987). In other words, data points that are close in the original feature space should also be near each other in the projected space. The distortion is measured through some criterion functions or stress functions. For example, Sammon (1969) defined a stress function based on the sum of squared error between the original distance and the projection distance,

$$J_S = \frac{1}{\sum_{i<j} D(\mathbf{x}_i, \mathbf{x}_j)} \sum_{i<j} \frac{\left(D'(\mathbf{x}_i, \mathbf{x}_j) - D(\mathbf{x}_i, \mathbf{x}_j)\right)^2}{D(\mathbf{x}_i, \mathbf{x}_j)}, \tag{9.19}$$

where $D(\cdot)$ is the distance between a pair of data points $\mathbf{x}_i$ and $\mathbf{x}_j$ in the original feature space, and $D'(\cdot)$ is the distance of the projections of $\mathbf{x}_i$ and $\mathbf{x}_j$ in the new space.

As the stress function is constructed, the optimal configuration of the lower-dimensional image points $\mathbf{x}'_i$ of the original data points $\mathbf{x}_i$ can be achieved by minimizing the stress function using a gradient descent algorithm. The image points $\mathbf{x}'_i$ are initially randomly placed into the mapping space and are iteratively moved in the directions that lead to the greatest decrease rate for the stress function. The gradients for the stress function in Eq. 9.19 can be calculated as,

$$\nabla_{\mathbf{x}'_{ki}} J_S = \frac{2}{\sum_{i<j} D(\mathbf{x}_i, \mathbf{x}_j)} \sum_{j \neq k} \frac{D(\mathbf{x}_k, \mathbf{x}_j) - D'(\mathbf{x}_k, \mathbf{x}_j)}{D(\mathbf{x}_k, \mathbf{x}_j)} \frac{\mathbf{x}'_j - \mathbf{x}'_k}{D'(\mathbf{x}_k, \mathbf{x}_j)}. \tag{9.20}$$

9.3.3 ISOMAP

The isometric feature mapping (ISOMAP) algorithm is a local version of MDS, tending to explore more complicated nonlinear structures in the data (Tenenbaum et al., 2000). ISOMAP is interested in the estimation of the geodesic distances between all pairs of data points, which are the shortest paths between the points on a manifold and provide the best representation of the intrinsic geometry of the data. In order to calculate the geodesic distances, ISOMAP first constructs a symmetric neighborhood graph by connecting a pair of points $\mathbf{x}_i$ and $\mathbf{x}_j$ if $\mathbf{x}_i$ is one of K nearest neighbors of $\mathbf{x}_j$, or if $\mathbf{x}_i$ is in the ε-neighborhood of $\mathbf{x}_j$. The graph is also weighted using the Euclidean distance between neighboring points. The geodesic distances are then calculated as the shortest path distances among edges using an algorithm like Floyd's algorithm (Tenenbaum et al., 2000). Let $D(\mathbf{x}_i,\mathbf{x}_j)$ represent the Euclidean distance between points $\mathbf{x}_i$ and $\mathbf{x}_j$. The geodesic distance $D_G(\mathbf{x}_i,\mathbf{x}_j)$ is initially set as,

$$D_G(\mathbf{x}_i, \mathbf{x}_j) = \begin{cases} D(\mathbf{x}_i, \mathbf{x}_j) & \text{if } \mathbf{x}_i \text{ and } \mathbf{x}_j \text{ are connected} \\ \infty & \text{otherwise} \end{cases}, \tag{9.21}$$

and further calculated for each data point $\mathbf{x}_k$, $k = 1, \ldots, N$ in turn as

$$D_G(\mathbf{x}_i, \mathbf{x}_j) = \min(D_G(\mathbf{x}_i, \mathbf{x}_j), D_G(\mathbf{x}_i, \mathbf{x}_k) + D_G(\mathbf{x}_k, \mathbf{x}_j)). \tag{9.22}$$

With the obtained geodesic distance matrix $\mathbf{G} = \{D_G(\mathbf{x}_i, \mathbf{x}_j)\}$, MDS can be applied to construct the embedding of the data in a lower-dimensional space. Let F be the new space with $\mathbf{E} = \{D(\mathbf{x}'_i, \mathbf{x}'_j)\}$ as the Euclidean distance matrix, where $\mathbf{x}'_i$ and $\mathbf{X}'_j$ are corresponding points, the criterion function for seeking an optimal configuration of these points is defined as,

$$J = \|\tau(\mathbf{G}) - \tau(\mathbf{E})\|_{L^2}, \tag{9.23}$$

where $\|\mathbf{\Gamma}\|_{L^2} = \sqrt{\sum_{i,j} \gamma_{ij}^2}$ and τ is an operator converting distances to inner products, defined as,

$$\tau(\mathbf{\Gamma}) = -\mathbf{HSH}/2, \tag{9.24}$$

where

$$\mathbf{S} = \{S_{ij} = \gamma_{ij}^2\}, \tag{9.25}$$

and

$$\mathbf{H} = \{H_{ij} = \delta_{ij} - 1/N\}. \tag{9.26}$$

The minimization of the criterion function is achieved when the lth component of the coordinates $\mathbf{x}_i'$ is set by using the lth eigenvalue λ_l (in decreasing order) and the ith component of the lth eigenvector v_{li} of the matrix $\tau(\mathbf{G})$,

$$x_{il}' = \sqrt{\lambda_l}\, v_{li}. \tag{9.27}$$

9.3.4 Locally Linear Embedding

Locally linear embedding (LLE) emphasizes the local linearity of the manifold (Roweis and Saul, 2000). It assumes that the local relations in the original d-dimensional data space are also preserved in the projected low-dimensional space (L-dimensional). Such relations are represented through a weight matrix $\mathbf{W}$, describing how a point $\mathbf{x}_i$ can be reconstructed from its neighbors, determined by either K nearest neighbors or ε-neighborhood, as done in ISOMAP. $\mathbf{W}$ is calculated by solving a constrained least-squares problem with the minimization of the criterion function,

$$J(\mathbf{W}) = \sum_i \left| \mathbf{x}_i - \sum_j w_{ij}\mathbf{x}_j \right|^2. \tag{9.28}$$

Because each point is only constructed from its neighbors, $\mathbf{W}$ is a sparse matrix. Another constraint for $\mathbf{W}$ is that each row of $\mathbf{W}$ sums to unity.

This reduces the problem to finding L-dimensional vectors $\mathbf{x}_i'$ so that the locally linear embedding criterion function, defined as

$$J(\mathbf{X}') = \sum_i \left| \mathbf{x}_i' - \sum_j w_{ij}\mathbf{x}_j' \right|^2, \tag{9.29}$$

is minimized. The optimization is performed with the weights fixed.

Other local nonlinear dimensionality reduction approaches include Laplacian eigenmaps (Belkin and Niyogi, 2002) and Hessian eigenmaps (Donoho and Grimes, 2003).

An interesting observance indicated by Ham et al. (2003) is that ISOMAP, LLE, and Laplacian eigenmaps all can be described in the framework of kernel PCA with different kernel matrices. Ham et al. (2003) states that "They construct a kernel matrix over the finite domain of the training data that preserves some aspect of the manifold structure from the input space to a feature space. Diagonalization of this kernel matrix then gives rise to an embedding that captures the low-dimensional structure of the manifold." For example, the kernel matrix for ISOMAP can be written as

$$\mathbf{K}_{ISOMAP} = -\frac{1}{2}(\mathbf{I} - \mathbf{e}\mathbf{e}^T)\mathbf{S}(\mathbf{I} - \mathbf{e}\mathbf{e}^T), \tag{9.30}$$

where $\mathbf{S}$ is the squared distance matrix and $\mathbf{e} = N^{-1/2}(1, \ldots , 1)^T$ is the uniform vector of unit length. The eigenvectors corresponding to the largest eigenvalues of $\mathbf{K}_{ISOMAP}$ determine the embedding of the original space into a lower-dimensional space.

9.3.5 Elastic Maps

Similar to SOFM discussed in Chapter 5, elastic maps use an ordered system of nodes that are placed in the multidimensional space to achieve an approximation of the cloud of data points (Gorban et al., 2001). Let $\mathbf{Y} = \{y_i, i = 1, \ldots , P\}$ denote a set of graph nodes and $\mathbf{E} = \{e_i, i = 1, \ldots , Q\}$ denote the corresponding edges connecting a pair of nodes. The connected, unordered graph $G(\mathbf{Y}, \mathbf{E})$ is called an elastic net. The adjacent edges can be combined in pairs to form the collection of elementary ribs $\mathbf{R} = \{r_i, i = 1, \ldots , S\}$, where $r_i = \{e_i, e_j\}$. Every edge e_i starts at the node $e_i(0)$ and ends at the node $e_i(1)$. Similarly, every elementary rib has the starting node $r_i(1)$, central node $r_i(0)$, and end node $r_i(2)$. Several typical examples of elastic nets are illustrated in Fig. 9.5.

Each graph G is considered to be associated with an energy function U that is defined to summarize the energy of every node, edge, and elementary rib,

$$U = U^{(Y)} + U^{(E)} + U^{(R)}, \tag{9.31}$$

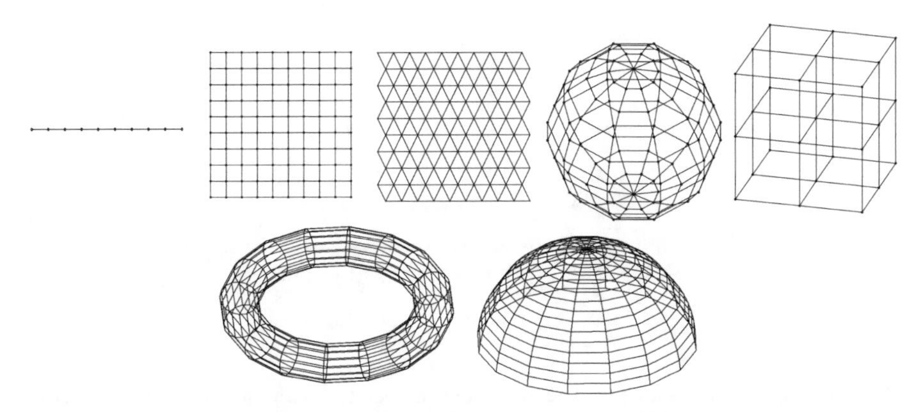

Fig. 9.5. Elastic nets used in practice. The examples include simple polyline, planar rectangular grid, planar hexagonal grid, non-planar graph spherical grid, non-planar cubical grid (upper row: from left to right), torus, and hemisphere (bottom row: from left to right). (From A. Gorban, A. Zinovyev, and D. Wunsch, II, Application of the method of elastic maps in analysis of genetic texts, In Proceedings of International Joint Conference on Neural Networks—IJCNN'03, vol. 3, pp. 1826–1831, Copyright © 2003 IEEE.)

where

$$U^{(Y)} = \frac{1}{N} \sum_{i-1}^{P} \sum_{x_j \in T_i} \|x_j - y_i\|^2, \tag{9.32}$$

$$U^{(E)} = \sum_{i=1}^{Q} \lambda_i \|e_i(1) - e_i(0)\|^2, \tag{9.33}$$

$$U^{(R)} = \sum_{i=1}^{S} \mu_i \|r_i(1) + r_i(2) - 2r_i(0)\|^2. \tag{9.34}$$

Here, T_i is a set of closest nodes to y_i. The coefficients λ_i and μ_i represent the stretching elasticity of every edge e_i and bending elasticity of every rib r_i, respectively. Under this framework, every node is considered to be connected by elastic bonds to the closest points and simultaneously to the adjacent nodes, as shown in Fig. 9.6.

With the energy function U, the optimal configuration of nodes can be achieved by minimizing U through an iterative procedure. Then the net is used as a nonlinear screen to visualize the distribution of data points by projecting them onto the manifold, constructed using the net as a point approximation. For example, a piecewise linear manifold is the simplest approach, although other, more intricate methods could also be used.

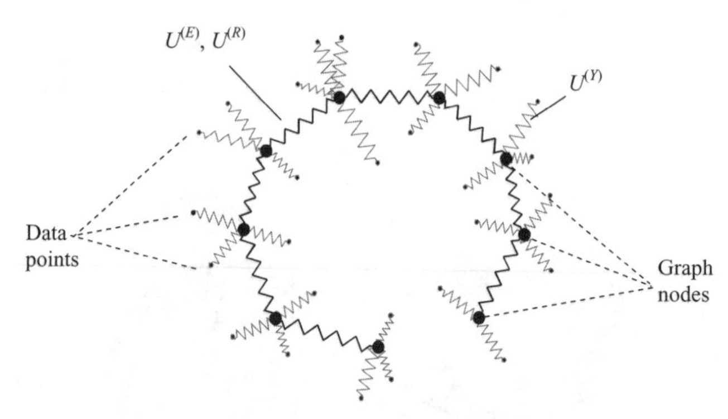

Fig. 9.6. Energy of elastic net. The graph G is associated with an energy function consisting of three parts. $U^{(Y)}$ is the average square of the distance between y_i and points in T_i, $U^{(E)}$ is the analogue of summary energy of elastic stretching, and $U^{(R)}$ is the analogue of summary energy of elastic deformation of the net. (From A. Gorban, A. Zinovyev, and D. Wunsch, II, Application of the method of elastic maps in analysis of genetic texts, In Proceedings of International Joint Conference on Neural Networks—IJCNN'03, vol. 3, pp. 1826–1831, Copyright © 2003 IEEE.)

9.3.6 Diffusion Maps

In the framework of diffusion processes, the eigenfunctions of Markov matrices are interpreted as systems of coordinates on the original data set used to obtain efficient representation of data geometric descriptions (Coifman and Lafon, 2006; Lafon and Lee, 2006). Kernel eigenmap methods, such as LLE, Laplacian eigenmaps, and Hessian eigenmaps, can also be regarded as special cases under the framework of diffusion maps (Coifman and Lafon, 2006).

Given a data set $\mathbf{X} = \{\mathbf{x}_i, i = 1, \ldots, N\}$ on a d-dimensional data space, a finite graph with N nodes corresponding to N data point can be constructed on $\mathbf{X}$. Every two nodes in the graph are connected by an edge weighted through a non-negative and symmetric similarity measure. Typically, a Gaussian kernel, defined as

$$w(\mathbf{x}_i, \mathbf{x}_j) = \exp\left(-\frac{\|\mathbf{x}_i - \mathbf{x}_j\|^2}{2\sigma^2}\right), \tag{9.35}$$

where σ is the kernel width parameter, satisfies such constraints and reflects the degree of similarity between $\mathbf{x}_i$ and $\mathbf{x}_j$.

Let

$$d(\mathbf{x}_i) = \sum_{\mathbf{x}_j \in \mathbf{X}} w(\mathbf{x}_i, \mathbf{x}_j) \tag{9.36}$$

be the degree of $\mathbf{x}_i$. The Markov or affinity matrix $\mathbf{P}$ is then constructed by calculating each entry as

$$p(\mathbf{x}_i, \mathbf{x}_j) = \frac{w(\mathbf{x}_i, \mathbf{x}_j)}{d(\mathbf{x}_i)}. \tag{9.37}$$

Because of the definition of the weight function, $p(\mathbf{x}_i, \mathbf{x}_j)$ can be interpreted as the transition probability from $\mathbf{x}_i$ to $\mathbf{x}_j$ in one time step. This idea can be extended further by considering $p^t(\mathbf{x}_i, \mathbf{x}_j)$ in the tth power $\mathbf{P}^t$ of $\mathbf{P}$ as the probability of transition from $\mathbf{x}_i$ to $\mathbf{x}_j$ in t time steps (Lafon and Lee, 2006). Therefore, the parameter t defines the granularity of the analysis. With the increase of the value of t, local geometric information of data is also integrated. Changing t makes it possible to control the generation of more specific or broader clusters. It thus generates a sequence of partitions of the data and provides a hierarchical clustering algorithm (Xu et al., 2007).

Because of the symmetry property of the weight function, for each $t \geq 1$, we can obtain a sequence of N eigenvalues of $\mathbf{P}$ $1 = \lambda_0 \geq \lambda_1 \geq \ldots \geq \lambda_N$, with the corresponding eigenvectors $\{\boldsymbol{\phi}_j, j = 1, \ldots, N\}$, satisfying,

$$\mathbf{P}^t \phi_j = \lambda^t \phi_j. \tag{9.38}$$

Using the eigenvectors as a new set of coordinates on the data set, the mapping from the original data space to the L-dimensional ($L < d$) Euclidean space $\mathcal{R}^L$ can be defined as

$$\mathbf{\Psi}_t : \mathbf{x}_i \rightarrow \left(\lambda_1^t \phi_1(\mathbf{x}_i), \ldots, \lambda_L^t \phi_L(\mathbf{x}_i)\right)^T. \tag{9.39}$$

Correspondingly, the diffusion distance between a pair of points $\mathbf{x}_i$ and $\mathbf{x}_j$,

$$D_t(\mathbf{x}_i, \mathbf{x}_j) = \|p_t(\mathbf{x}_i, \cdot) - p_t(\mathbf{x}_j, \cdot)\|_{1/\phi_0}, \tag{9.40}$$

where ϕ_0 is the unique stationary distribution

$$\phi_0(\mathbf{x}) = \frac{d(\mathbf{x})}{\sum_{\mathbf{x}_i \in \mathbf{X}} d(\mathbf{x}_i)}, \tag{9.41}$$

is approximated with the Euclidean distance in $\mathcal{R}^L$, written as

$$D_t(\mathbf{x}_i, \mathbf{x}_j) = \|\mathbf{\Psi}_t(\mathbf{x}_i) - \mathbf{\Psi}_t(\mathbf{x}_j)\|. \tag{9.42}$$

It can be seen that the more paths that connect two points in the graph, the smaller the diffusion distance is.

9.3.7 Other Methods

A shrinking-based method proposed by Shi et al. (2005) is also based on the division of a data space into a finite number of varying-size grids. A preprocessing step condenses data by moving it along the density gradient. Clusters are then identified as the connected components of dense units and are further evaluated via the computation of their compactness, which considers both intercluster and intracluster relationships.

Principal curves are defined as self-consistent smooth curves that pass through the "middle" of multidimensional probability distributions (Hastie and Stuetzle, 1989). The property of self-consistency requires that each point of the curve is the average of all data points that project onto it. An iterative algorithm is used to refine the principal curve until the change in the average squared distance between the curve and the data points is less than a threshold. Principal curves can also be regarded as generalizations of principal components obtained from PCA because linear principal curves are in fact principal components.

The generative topographic mapping (GTM) (Bishop et al., 1998) algorithm consists of a constrained mixture of Gaussians. The parameters of the models can be obtained via expectation-maximization algorithm. Different from other data visualization methods that usually define a mapping from the original data space to a lower dimensional projected space, GTM considers a nonlinear transformation from the projected space to the data space. Bayes' theorem can then be used to invert the mapping to provide a posterior distribution in the projected space.

Applications of the wavelet transform in clustering high-dimensional data are illustrated in Murtagh et al. (2000) and Yu et al. (1999). The former method employs the wavelet transform to data that are ordered based on matrix reordering schemes. The latter method, called WaveCluster$^+$, constructs a hash table to represent the data set. The wavelet transform is applied to the hashed feature space to generate a new hash table including only significant cells of grids. Clusters are then identified as the connected components in the new hash table. A clustering and indexing scheme, abbreviated as Clindex, is shown in Li et al. (2002). Clindex tends to achieve an approximately similar search in high-dimensional space with high efficiency.

9.4. PROJECTED AND SUBSPACE CLUSTERING

Methods like PCA, ICA, MDS, and diffusion maps linearly or nonlinearly create new dimensions based on the combination of information in the original dimensions. These new dimensions can prove difficult to interpret, making the results hard to understand. Projected clustering, or subspace clustering, addresses the high-dimensional challenge by restricting the search in subspaces of the original data space. According to Aggarwal and Yu (2002), "A generalized projected cluster is a subset ε of vectors together with a subset C of data points such that the points in C are closely clustered in the subspace defined by the vectors ε." For example, for a four-dimensional data space with independent features A_1, A_2, A_3, and A_4, assume cluster C_1 is present in the subspace formed by A_1 and A_3, while cluster C_2 exists in the subspace formed by A_1, A_2, and A_4. In the former case, features A_2 and A_4 become noise variables and affect the identification of the embedded cluster. In the latter case, feature A_3 is the noise.

9.4.1 CLIQUE, and Variants MAFIA, and ENCLUS

CLIQUE (CLustering In QUEst) (Agrawal et al., 1998) is perhaps one of the earliest projected clustering algorithms. The basic idea of CLIQUE comes from the observation that dense regions can always be found in any $(k - 1)$-dimensional subspace for a cluster that is defined to represent dense regions of a k-dimensional subspace. Therefore, CLIQUE starts with identifying all dense regions in a one-dimensional subspace, which is achieved by partitioning each dimension into ξ intervals with equal length and selecting the units with densities exceeding a threshold. All two-dimensonal dense units are then obtained based on the information from the lower level (one dimensonal), and each is required to be associated with two one-dimensonal dense units. CLIQUE employs this bottom-up scheme to generate k-dimensional dense units from (k – 1)-dimensional units by self-joining (k – 1)-dimensional units that share the first k – 2 dimensions. This algorithm can be sped up by only retaining the dense units in "interesting" subspaces, which are determined by the minimal description length (MDL) principle (Rissanen, 1978, 1996). All the subspaces are sorted in descending order based on their coverage of dense units, and subspaces with small coverage are pruned. The optimal cut point is determined by minimizing the code length for a selection of data encoding.

In CLIQUE, a cluster is defined as a maximal set of connected dense units in k dimensions (Agrawal et al., 1998). In other words, clusters are equivalent to the connected components in a graph whose nodes stand for the dense units. Two nodes are connected with an edge if, and only if, the corresponding dense units share a common face. If $\mathbf{u}_1 = \{u_{11}, \ldots, u_{1k}\}$ and $\mathbf{u}_2 = \{u_{21}, \ldots, u_{2k}\}$ (where $u_{ij} = [l_{ij}, r_{ij}]$ represents a right-open interval) are two k-dimensional units, then they are said to have a common face if there are k – 1 dimensions so that $u_{1i} = u_{2i}$. The connected components can be found by using a depth-first search algorithm. Each cluster is identified by starting with a certain unit and finding all units connected to it. The resulting clusters are described with DNF (disjunctive normal form) expressions for the purpose of simplicity. As shown in Table 8.1, CLIQUE has a computational complexity that is linear with the number of objects and quadratic with the number of dimensions.

MAFIA (Merging of Adaptive Finite Intervals) modifies CLIQUE by allowing the construction of adaptive grids in each dimension (Nagesh et al., 2001). Thus, the intervals in each dimension are determined according to the data distribution in the dimension without the requirement to specify the interval size in advance as CLIQUE does. At first, histograms are calculated for each dimension that has at least 1,000 intervals. Contiguous windows are merged if the difference between their values is below some threshold. The resulting variable-sized interval is considered to be in a cluster when the number of points it contains is larger than a threshold

$$\eta = \frac{\alpha s N}{d_i}, \tag{9.43}$$

where α is a parameter called the cluster dominance factor, s is the size of the interval, d_i is the size of the dimension, and N is the number of data points. The parameter α provides a magnitude of deviation of the histogram values from the uniform distribution, which suggests a low likelihood of the presence of a cluster in the corresponding dimensions.

Similar to CLIQUE, MAFIA also uses a bottom-up algorithm to generate dense units, starting from one dimension. However, for MAFIA, candidate dense units in k-dimension are built from (k – 1)-dimensional units as long as they share any of the k – 2 dimensions, not just the first k – 2 dimensions. This change allows MAFIA to examine many more candidate dense units and cover all possible subspaces for exploring the embedded clusters.

ENCLUS (ENtropy-based CLUStering) is similar to CLIQUE but uses an entropy-based method to select subspaces (Cheng et al., 1999). In addition to the coverage criterion defined in CLIQUE, ENCLUS incorporates two more criteria on the data density and correlation of dimensions for goodness of clustering in subspaces. All three criteria are related to the concept of entropy in information theory. Given $p(\mathbf{u})$ as the density for a unit $\mathbf{u}$, the entropy for the data set $\mathbf{X}$ can be defined as

$$H(\mathbf{X}) = -\sum_{\mathbf{u}} p(\mathbf{u}) \log p(\mathbf{u}). \tag{9.44}$$

Hence, clusters with densely populated data points in some subspaces correspond to low entropy, while the uniform distribution leads to the highest entropy for the data set. A subspace with entropy below some threshold is then regarded as having good clustering.

9.4.2 PROCLUS and ORCLUS

PROCLUS (PROjected CLUstering) is a K-medoids-based (Ng and Han, 2002) projected clustering method (Aggarwal et al., 1999). Given two user-specified parameters, the number of clusters K and the average number of dimensions l of each cluster, PROCLUS identifies a set of K clusters, with each cluster associated with a subset of dimensions, through three steps: initialization, iterative, and cluster refinement.

The goal of the initialization step is to find a good superset of a piercing set of medoids, in which each point belongs to a different cluster. The superset is constructed with a greedy strategy that first generates a random sample of data points with its size a few times larger than the number of clusters. Candidate medoids are then drawn from the sample set in an iterative way, with the requirement that the distance between current medoid and medoids already selected is large enough. The initialization process significantly reduces the number of data points that are used for further analysis.

The iterative step replaces bad medoids, whose corresponding clusters have sizes below a threshold $\alpha(N/K)$, where α is a constant less than 1, with candidates in the subset formed in the initialization step until a good set of medoids is obtained. In order to calculate an appropriate set of dimensions for each medoid, a locality analysis is performed, which uses the information contained in the neighbor points of medoids to determine the subspaces of the clusters. Each point is then assigned to the cluster that has the nearest medoid. The distance used here is called Manhattan segmental distance, defined with respect to a set of dimensions $\mathbf{\Lambda}$,

$$D(\mathbf{x}_i, \mathbf{x}_j) = \frac{\sum_{k \in \Lambda} |x_{ik} - x_{jk}|}{|\Lambda|}. \tag{9.45}$$

The cluster refinement step recalculates the dimensions for each medoid using information from all points in the clusters rather than points in the locality. The final clusters are formed based on the new dimensions.

ORCLUS (arbitrarily ORiented projected CLUster generation) extends PROCLUS by allowing the detection of arbitrary oriented subspaces, not just the directions parallel to the original axes (Aggarwal and Yu, 2002). ORCLUS begins with a set of randomly selected K_0 seeds with the full dimension. Both the dimension of clusters and the number of clusters are iteratively decayed until the pre-specified number of clusters is reached. By choosing appropriate decay factors, the dimension is also reduced to the predefined value at the same time. ORCLUS consists of three main steps, known as assignment, vector finding, and merge. The goal of the assignment step is to place each point into a current cluster based on its distance to the seeds. The seeds are then updated using the centroids of the formed cluster. Each cluster is associated with a subspace of dimension l_c, which is obtained by calculating the covariance matrix of the cluster and choosing a set of orthonormal eigenvectors corresponding to the l_c smallest eigenvalues. A pair of seeds is merged if the projected energy of the union of the corresponding clusters in the least spread subspace is minimal.

In order to scale to very large data sets, ORCLUS uses the extended cluster feature (ECF) vector, similar to CF used in BIRCH (Zhang et al., 1996), to represent the cluster summary. Each ECF vector consists of information on the number of points in the cluster, the sum of the ith components for points in the cluster, and the sum of the products of the ith and j^{th} components for points in the cluster. ORCLUS has overall time complexity of $O(K_0^3 + K_0 Na + K_0^2 d^3)$ and space complexity of $O(K_0 d^2)$, where K_0 is the number of initial seeds, N is the number of data points, and d is the number of features. Obviously, the scalability of ORCLUS largely relies on K_0.

9.4.3 Other Methods

OptiGrid (Hinneburg and Keim, 1999) is designed to obtain an optimal grid-partitioning of data, which is achieved by constructing the best cutting hyperplanes through a set of contracting projections. Given a d-dimensional data space $\mathbf{X}$, the contracting projection is defined as a liner transformation P for all points $\mathbf{x} \in \mathbf{X}$,

$$P(\mathbf{x}) = A\mathbf{x}, \|A\| = \max_{\mathbf{x} \in \mathbf{X}} \left(\frac{\|A\mathbf{x}\|}{\|\mathbf{x}\|} \right) \leq 1. \tag{9.46}$$

The cutting hyperplanes are determined while following two constraints: (1) they go through regions with low density; (2) they are capable of detecting as many clusters as possible. OptiGrid recursively partitions the data into subsets, each of which includes at least one cluster and is operated recursively, too. The recursion ends when no more good cutting hyperplanes can be obtained.

The time complexity for OptiGrid is in the interval of $O(Nd)$ and $O(Nd\log N)$.

pCluster (Wang et al., 2002) is a generalized subspace clustering model focusing on the detection of shifting or scaling patterns of original data, in addition to the measurement of value proximity using similarity or distance functions such as Euclidean or city block distance. The defined pClusters are generated by a depth-first clustering algorithm that is capable of finding multiple clusters simultaneously. With this method, clusters are first generated in the high-dimensional space, and in the lower-dimensional space, the clusters that are not covered by the high-dimensional clusters are further identified. Compared with the approaches that merge low-dimensional clusters to obtain higher-dimensional clusters, the pCluster algorithm achieves more efficient and effective computational cost.

IPCLUS (Interactive Projected CLUStering) incorporates human interaction into high-dimensional data clustering (Aggarwal, 2004). The inclusion of human prior knowledge, intuition, and judgment into the clustering process could provide meaningful instructions for cluster discovery and make the clustering results easy to understand. In the framework of IPCLUS, subspaces are selected by computational algorithms, and the clusters in these subspaces are then visually identified by the users. The user interactions are stored in the form of a set of identity strings that are processed to generate the final clusters.

Unlike PROCLUS and ORCLUS, which require users to specify the number of clusters and the average dimension of subspaces in advance, EPCH (Efficient Projective Clustering by Histograms) explores clusters in subspaces with varied dimensions (Ng et al., 2005). D_0-dimensional histograms are constructed in the beginning to model data distributions in the corresponding subspaces. A "signature" for a data point records whether or not the point is located in the dense regions, and if so, where it is located in the subspace for each histogram. Signatures with lots of data points suggest clusters in subspaces, and similar signatures are merged until a prespecified maximum number of clusters K_{max} is reached or the highest similarity is below some threshold. For the latter case, only the top K_{max} ranked clusters are kept.

HARP (a Hierarchical approach with Automatic Relevant dimension selection for Projected clustering) decreases the dependence on user-specified parameters by dynamically adjusting the internal thresholds (Yip et al., 2004). HARP is based on agglomerative hierarchical clustering and iteratively merges pairs of clusters to form larger clusters based on the calculated merge score. A merge that results in a cluster that has the selected dimensions less than a threshold d_{min} is forbidden. Similarly, a dimension is only selected for a cluster if its relevance index is above another threshold R_{min}. Both thresholds start with the highest values and are updated with a linear loosening method.

DOC (Density-based Optimal projective Clustering) is a Monte Carlo algorithm that performs a greedy search process to identify one projected cluster at a time (Procopiuc et al., 2002). At each step, a data point $\mathbf{x}$ is randomly selected as a medoid, and a discriminating set with respect to **x** is also constructed via random sampling from the data set. A dimension is considered to be associated with the cluster when the distances between **x** and each point of the data set in this dimension are less than a threshold. This process is repeated many times, and the best cluster is kept. FASTDOC is a variant of DOC attempting to reduce the search time (Procopiuc et al., 2002). Proposed by Yiu and Mamoulis (2005), a modification of DOC, called FPC (Frequent-Pattern-based Clustering), combines branch and bound methods to find the projected clusters.

9.5. APPLICATIONS

9.5.1 Triplet Distribution Visualization

As discussed in Chapter 7, genetic text of DNA is a linear sequence of the letters A, C, G, and T, representing four different types of nucleotides. Proteins, whose primary structure is also linear, are encoded from genes. Areas between genes are called intergenic regions or junk (though it is not exactly true). The information that defines the order of amino acids in protein is coded in DNA by triplets of nucleotides, called codons. If we take an arbitrary window of coding sequence and divide it into successive, non-overlapping triplets starting from the first base pair in the window, then this decomposition and arrangement of the real codons may not be in phase. We can divide the window into triplets in three ways, shifting every time by one base pair from the beginning. So, we have three possible triplet distributions, and one of them coincides with the real codon distribution. The coding regions are characterized by the presence of a distinguished phase. Junk evidently has no such feature because neither inserting nor deleting a letter in junk changes properties of DNA considerably; thus, this kind of mutation is allowed in the process of evolution. But every such mutation breaks the phase, so we can expect that distributions of triplets in junk will be similar for all three phases.

Elastic maps were applied to analyze the distribution of triplet frequencies for several real genomes and model genetic sequences, with window size 300, sliding along the whole sequence (Gorban et al., 2003). Particularly, the results of visualization for the *Caulobacter crescentus* complete genome (GenBank accession code is NC_002696) are shown in Fig. 9.7.

Figure 9.7 depicts the distribution of data point projections on the elastic map. Coloring is made by estimation of point density. It is clear that the distribution has six well-defined clusters and a sparse cloud of points between these clusters. For clarity, non-coding regions are marked by black circles points, while squares and triangles correspond to the coding regions, but in different strands of the genome (in bacterial genomes, a gene can be positioned in a forward strand or a complementary strand; in the latter case

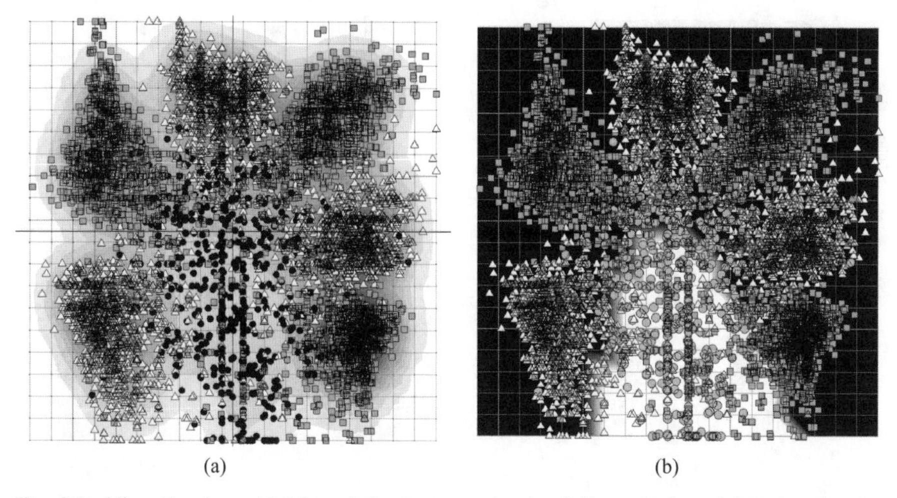

Fig. 9.7. Visualization of DNA triplet frequencies in sliding window. (a) Point density coloring; (b) Coloring by linear discrimination function. (From A. Gorban, A. Zinovyev, and D. Wunsch, II, Application of the method of elastic maps in analysis of genetic texts, In Proceedings of International Joint Conference on Neural Networks—IJCNN'03, vol. 3, pp. 1826–1831, Copyright © 2003 IEEE.)

it is red in the opposite direction and consists of complementary letters G ⇔ C, T ⇔ A).

In the early stages of computational gene recognition, many statistics, locally defined on DNA text, were compared by means of application of linear discrimination analysis (two classes, coding and non-coding, were separated). In Fig. 9.7b, coloring is made by value of linear discriminate function. White color corresponds to the non-coding regions. It can be seen that linear discrimination in this case results in many false positives (many non-coding regions are predicted to be coding). More subtle analysis shows that linear discrimination is not appropriate in this case.

Figures 9.8a and b show coloring by the value of two triplet frequencies: ATG and CAT codon. It is known that most bacterial genes start with an ATG start codon and cannot be in the middle of the gene. The pictures show that the right bottom and top middle clusters correspond to those windows, where triplet decomposition coincides with real codons, and the other four clusters are windows with "shifted" phase.

9.5.2 Face Recognition

Face recognition is one important application in image processing, taking images as points in a vector space usually accompanied by high-dimensional data, particularly when the resolution of the images is high. For example, even for a low-resolution image of 64×64 pixels, the dimension will be as high as

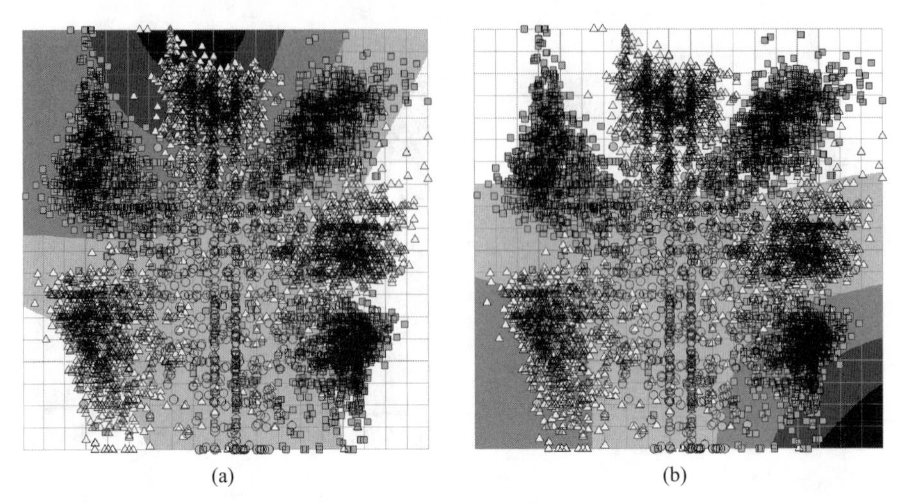

Fig. 9.8. Visualization of DNA triplet frequencies in sliding window. (a) Coloring by ATG (start codon) frequency; (b) coloring by CAT (complemented start codon) frequency. (From A. Gorban, A. Zinovyev, and D. Wunsch, II, Application of the method of elastic maps in analysis of genetic texts, In Proceedings of International Joint Conference on Neural Networks—IJCNN'03, vol. 3, pp. 1826–1831, Copyright © 2003 IEEE.)

4,096. An application of LLE for human face expression recognition is illustrated in Fig. 9.9 (Roweis and Saul, 2000). The data set includes 2,000 face images from the same individual but with different expressions. Each input pattern is a 560-dimensional vector, corresponding to the 20×28 gray scale of the images. The faces are mapped into a two-dimensional space, constituted of the first two constructed coordinates of LLE. The K-nearest neighbors method is used with K set at 12. As can be seen from the figure, the coordinates of the embedding space are tightly related to the feature of face pose and expression.

Detection of rotated human faces by DOC was illustrated in Procopiuc et al. (2002). Tenenbaum et al. (2000) used ISOMAP to project face images into a three-dimensional manifold with coordinates representing up-down pose, left-right pose, and lighting direction. More applications on PCA, ICA, nonlinear PCA, and kernel PCA in face recognition can be found in Moghaddam (2002) and Bartlett et al. (2002).

9.6. SUMMARY

Data sets that display high dimensions result from the employment of certain measurement devices or selection of sensors. Not all of these measured features are important or even related to the representation of the data structure.

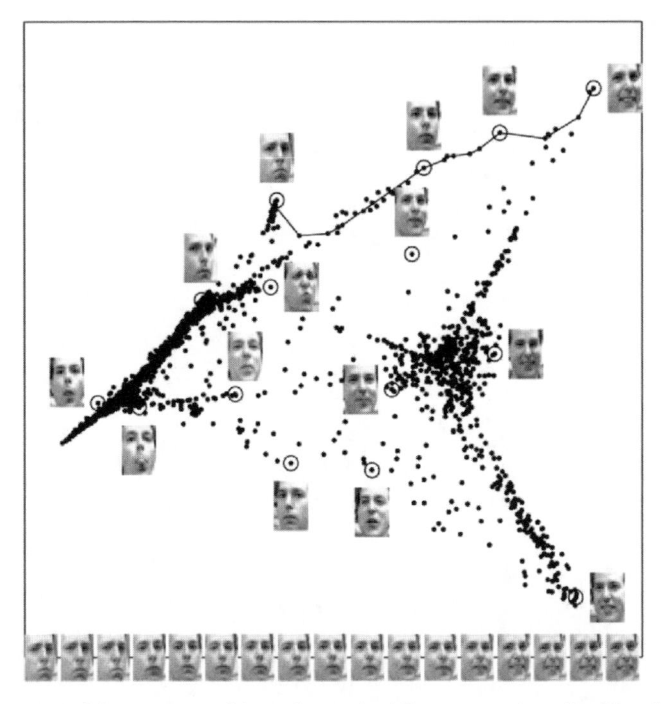

Fig. 9.9. Images of faces mapped into the embedding space described by the first two coordinates of LLE. Representative faces are shown next to circled points in different parts of the space. The bottom images correspond to points along the top-right path (linked by a solid line), illustrating one particular mode of variability in pose and expression. (From S. Roweis and L. Saul. Nonlinear dimensionality reduction by locally linear embedding. Science, vol. 290, pp. 2323–2326, 2000. Reprinted with permission from AAAS.)

Many variables in practical cases are highly correlated, which implies a lower intrinsic dimensionality that provides effective representation of the data. Methods that either explicitly build a set of clusters on some subspaces, or that provide visualizations of high-dimensional data on a two- or three-dimensional space, represent data with fewer degrees of freedom while decreasing the information loss as much as possible.

CHAPTER 10

CLUSTER VALIDITY

10.1. INTRODUCTION

In the previous chapters, we discussed many clustering algorithms, whose goal is to expose the inherent partitions in the underlying data. Each algorithm can partition data, but different algorithms or input parameters cause different clusters, or reveal different clustering structures. Thus, the problem of objectively and quantitatively evaluating the resulting clusters, or whether the clustering structure derived is meaningful, which is referred to as cluster validation, is particularly important (Dubes, 1993; Gordon, 1998; Halkidi et al., 2002; Jain and Dubes, 1988). For example, if there is no clustering structure in a data set, the output from a clustering algorithm becomes meaningless and is just an artifact of the clustering algorithm. In this case, it is necessary to perform some type of tests to assure the existence of the clustering structure before performing any further analysis. Such problems are called clustering tendency analysis and are reviewed in Jain and Dubes (1988) and Gordon (1998).

With respect to three types of clustering structures, i.e., partitional clustering, hierarchical clustering, and individual clusters, there are three categories of testing criteria, known as external criteria, internal criteria, and relative criteria (Jain and Dubes, 1988; Theodoridis and Koutroumbas, 2006). Given a data set $\mathbf{X}$ and a clustering structure $\boldsymbol{C}$ derived from the application of a certain clustering algorithm on $\mathbf{X}$, external criteria compare the obtained clustering structure $\boldsymbol{C}$ to a prespecified structure, which reflects *a priori* information on the clustering structure of $\mathbf{X}$. For example, an external criterion can

Clustering, by Rui Xu and Donald C. Wunsch, II

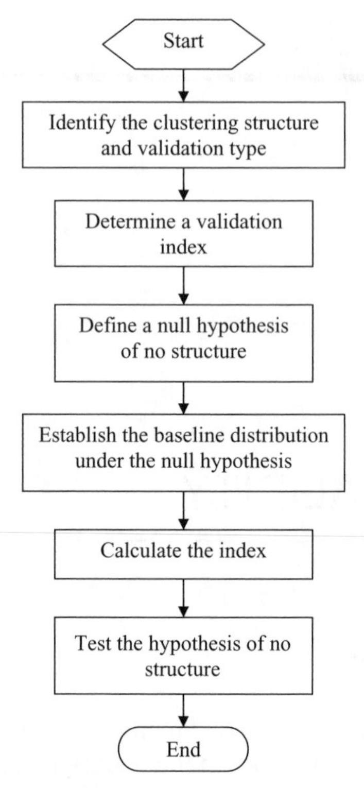

Fig. 10.1. Flowchart of the validity paradigm for clustering structures. The null hypothesis is a statement of randomness regarding to the structure on the data set. The values of the appropriate statistic indices are then compared with the values obtained from the baseline distribution under the null hypothesis.

be used to examine the match between the cluster labels with the category labels based on *a priori* information. In contrast to external criteria, internal criteria evaluate the clustering structure exclusively from **X**, without any external information. For example, an internal criterion would use the proximity matrix of **X** to assess the validity of ***C***. Relative criteria compare ***C*** with other clustering structures, obtained from the application of different clustering algorithms or the same algorithm but with different parameters on **X**, and determine which one may best represent **X** in some sense. For example, a relative criterion would compare a set of values of K for the K-means algorithm to find the best fit of the data.

Both external criteria and internal criteria are closely related to statistical methods and hypothesis tests (Jain and Dubes, 1988). The validity paradigm for a clustering structure is depicted in Fig. 10.1 (Dubes, 1993). Particularly, in the framework of cluster validation, a null hypothesis H_0 states that there is no structure on the data set **X**, or the structure of **X** is random. Generally, there are three commonly used null hypotheses defined below (Jain and Dubes, 1988):

- Random position hypothesis:
  H_0: All the locations of N data points in some specific region of a d-dimensional space are equally likely.
- Random graph hypothesis:
  H_0: All $N \times N$ rank order proximity matrices are equally likely.
- Random label hypothesis:
  H_0: All permutations of the labels on N data objects are equally likely.

Typically, the random position hypothesis is appropriate for ratio data, the random graph hypothesis works for ordinal proximities between pairs of data objects, and the random label hypothesis has applications to all data types.

In order to obtain the baseline distribution under the null hypothesis, statistical sampling techniques like Monte Carlo analysis and bootstrapping are used (Jain and Dubes, 1988). Based on the baseline distribution, a threshold that represents a certain level of statistical significance can then be obtained, which is further compared to the calculated validity index to determine whether the index meets the requirement to be considered as unusual.

In the following sections, we discuss cluster validity approaches based on external criteria, internal criteria, and relative criteria. In particular, under the discussion of relative criteria, we focus on the problem of how to estimate the number of clusters, which is called "the fundamental problem of cluster validity" (Dubes, 1993).

10.2. EXTERNAL CRITERIA

If $\boldsymbol{P}$ is a prespecified partition of data set $\mathbf{X}$ with N data points and is independent from the clustering structure $\boldsymbol{C}$ resulting from a clustering algorithm, then the evaluation of $\boldsymbol{C}$ by external criteria is achieved by comparing $\boldsymbol{C}$ to $\boldsymbol{P}$. Considering a pair of data points $\mathbf{x}_i$ and $\mathbf{x}_j$ of $\mathbf{X}$, there are four different cases based on how $\mathbf{x}_i$ and $\mathbf{x}_j$ are placed in $\boldsymbol{C}$ and $\boldsymbol{P}$.

- Case 1: $\mathbf{x}_i$ and $\mathbf{x}_j$ belong to the same clusters of $\boldsymbol{C}$ and the same category of $\boldsymbol{P}$.
- Case 2: $\mathbf{x}_i$ and $\mathbf{x}_j$ belong to the same clusters of $\boldsymbol{C}$ but different categories of $\boldsymbol{P}$.
- Case 3: $\mathbf{x}_i$ and $\mathbf{x}_j$ belong to different clusters of $\boldsymbol{C}$ but the same category of $\boldsymbol{P}$.
- Case 4: $\mathbf{x}_i$ and $\mathbf{x}_j$ belong to different clusters of $\boldsymbol{C}$ and different category of $\boldsymbol{P}$.

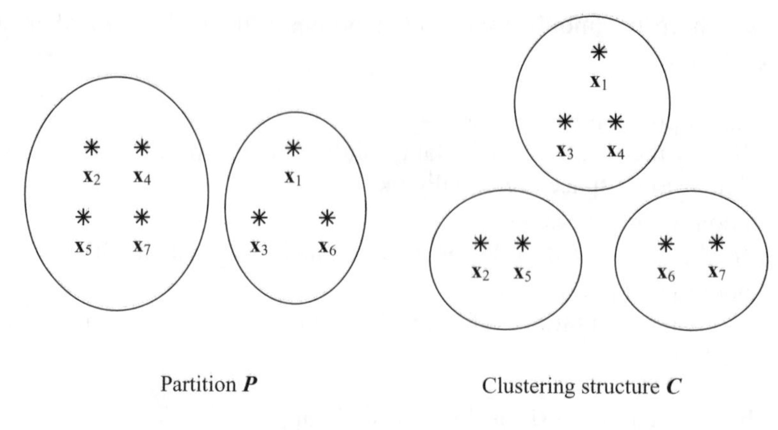

Case	Pairs of data points	Total
1	$\mathbf{x}_1$ and $\mathbf{x}_3$; $\mathbf{x}_2$ and $\mathbf{x}_5$	2
2	$\mathbf{x}_1$ and $\mathbf{x}_4$; $\mathbf{x}_3$ and $\mathbf{x}_4$; $\mathbf{x}_6$ and $\mathbf{x}_7$	3
3	$\mathbf{x}_1$ and $\mathbf{x}_6$; $\mathbf{x}_2$ and $\mathbf{x}_4$; $\mathbf{x}_2$ and $\mathbf{x}_7$; $\mathbf{x}_3$ and $\mathbf{x}_6$; $\mathbf{x}_4$ and $\mathbf{x}_5$; $\mathbf{x}_4$ and $\mathbf{x}_7$; $\mathbf{x}_5$ and $\mathbf{x}_7$	7
4	$\mathbf{x}_1$ and $\mathbf{x}_2$; $\mathbf{x}_1$ and $\mathbf{x}_5$; $\mathbf{x}_1$ and $\mathbf{x}_7$; $\mathbf{x}_2$ and $\mathbf{x}_3$; $\mathbf{x}_2$ and $\mathbf{x}_6$; $\mathbf{x}_3$ and $\mathbf{x}_5$; $\mathbf{x}_3$ and $\mathbf{x}_7$; $\mathbf{x}_4$ and $\mathbf{x}_6$; $\mathbf{x}_5$ and $\mathbf{x}_6$	9

Fig. 10.2. Illustration of four different cases on how a pair of data points is placed in a prespecified partition $\boldsymbol{P}$ and a resulting clustering structure $\boldsymbol{C}$. The data set consists of seven data points. The values of a, b, c, and d are 2, 3, 7, and 9, respectively.

Correspondingly, the numbers of pairs of points for the four cases are denoted as a, b, c, and d, respectively (see Fig. 10.2). Because the total number of pairs of points is $N(N-1)/2$, denoted as M, we have $a + b + c + d = M$.

We now can give some commonly used external indices for measuring the match between $\boldsymbol{C}$ and $\boldsymbol{P}$ as follows:

1. Rand index (Rand, 1971)

$$R=(a+d)/M; \tag{10.1}$$

2. Jaccard coefficient

$$J=a/(a+b+c); \tag{10.2}$$

3. Fowlkes and Mallows index (Fowlkes and Mallows, 1983)

$$FM=\sqrt{\frac{a}{a+b}\frac{a}{a+c}}; \tag{10.3}$$

4. Γ statistics

$$\Gamma = \frac{Ma - m_1 m_2}{\sqrt{m_1 m_2 (M - m_1)(M - M_2)}}, \quad (10.4)$$

where $m_1 = a + b$ and $m_2 = a + c$.

As can be seen from the definition, the larger the values of the first three indices, the more similar are $\boldsymbol{C}$ and $\boldsymbol{P}$. Specifically, the value of Γ statistics is between −1 and 1, while the values of both the Rand index and the Jaccard coefficient are in the range of [0, 1]. The major difference between the latter two statistics is that the Rand index emphasizes the situation that pairs of points belong to the same group or different groups in both $\boldsymbol{C}$ and $\boldsymbol{P}$, but the Jaccard coefficient excludes d in the similarity measure.

The above indices, denoted as *Ind*, can be corrected to provide a normalized formula (Jain and Dubes, 1988), written as

$$Ind' = \frac{Ind - E(Ind)}{\max(Ind) - E(Ind)}, \quad (10.5)$$

where $E(Ind)$ is the expected value of *Ind* under the baseline distribution and max(*Ind*) is the maximum possible value of *Ind*. For example, the corrected Rand index (Hubert and Arabie, 1985) can be formulated as,

$$R' = \frac{R - E(R)}{1 - E(R)}. \quad (10.6)$$

10.3. INTERNAL CRITERIA

A Cophenetic correlation coefficient (CPCC) (Jain and Dubes, 1988; Rohlf and Fisher, 1968) is an index used to validate hierarchical clustering structures. Given the proximity matrix $\mathbf{P} = \{p_{ij}\}$ of $\mathbf{X}$, CPCC measures the degree of similarity between $\mathbf{P}$ and the cophenetic matrix $\mathbf{Q} = \{q_{ij}\}$, whose elements record the proximity level where pairs of data points are grouped in the same cluster for the first time. Let μ_P and μ_Q be the means of $\mathbf{P}$ and $\mathbf{Q}$, i.e.,

$$\mu_P = \frac{1}{M} \sum_{i=1}^{N-1} \sum_{j=i+1}^{N} p_{ij}, \quad (10.7)$$

$$\mu_Q = \frac{1}{M} \sum_{i=1}^{N-1} \sum_{j=i+1}^{N} q_{ij}, \quad (10.8)$$

where $M = N(N-1)/2$, CPCC is defined as

$$CPCC = \frac{\frac{1}{M}\sum_{i=1}^{N-1}\sum_{j=i+1}^{N} p_{ij}q_{ij} - \mu_P\mu_Q}{\sqrt{\left(\frac{1}{M}\sum_{i=1}^{N-1}\sum_{j=i+1}^{N} p_{ij}^2 - \mu_P^2\right)\left(\frac{1}{M}\sum_{i=1}^{N-1}\sum_{j=i+1}^{N} q_{ij}^2 - \mu_Q^2\right)}}. \tag{10.9}$$

The value of CPCC lies in the range of [−1, 1], and an index value close to 1 indicates a significant similarity between **P** and **Q** and a good fit of hierarchy to the data. However, for the group average linkage (UPGMA) method, even large CPCC values (such as 0.9) cannot assure sufficient similarity between the two matrices (Rohlf, 1970).

When matrices **P** and **Q** are on an ordinal scale, the Goodman-Kruskal γ statistic (Hubert, 1974) and Kendall's τ statistic (Cunningham and Ogilvie, 1972) can be used for cluster validity. Both statistics are further discussed with examples in Jain and Dubes (1988). Moreover, an investigation of a comparison of 30 different internal indices was performed in Milligan (1981).

10.4. RELATIVE CRITERIA

External and internal criteria require statistical testing, which could become computationally intensive. Relative criteria eliminate such requirements and concentrate on the comparison of clustering results generated by different clustering algorithms or the same algorithm but with different input parameters. As we have discussed in the previous chapters, one important issue is to determine the true number of clusters, K. For hierarchical clustering algorithms, a cutting point must be determined to cut the dendrogram at certain levels in order to form a set of clusters. For many partitional clustering algorithms, K is required as a user-specified parameter. Although in some cases, K can be estimated in terms of the user's expertise or *a priori* information for more applications, K still needs to be estimated exclusively from the data themselves. Either over-estimation or under-estimation of K will affect the quality of resulting clusters. A partition with too many clusters complicates the true clustering structure, therefore making it difficult to interpret and analyze the results. On the other hand, a partition with too few clusters causes the loss of information and misleads the final decision. In the following section, we focus on the methods, indices, and criteria used to address this fundamental problem.

10.4.1 Data Visualization

Perhaps the most direct method to estimate the value of K is to project data points onto a two or three-dimensional Euclidean space, using data visualization methods, such as those introduced in Chapter 9. In that way, visual inspections can provide some useful insight on the number of clusters, and for that matter, even the assignment of data points to different clusters could be determined. However, the complexity of many real data sets makes the two or three-dimensional projections far from sufficient to represent the true data structures; therefore, the visual estimation of K can only be restricted to a very small scope of applications.

10.4.2 Validation Indices and Stopping Rules

For a clustering algorithm that requires the input of K from users, a sequence of clustering structures can be obtained by running the algorithm several times from the possible minimum K_{min} to the maximum K_{max}. These structures are then evaluated based on the constructed indices, and the expected clustering solution is determined by choosing the one with the best index. In the case of hierarchical clustering structures, the indices are also known as stopping rules, which tell where the best level is in order to cut the dendrogram. As the standard for assessing clusters, these indices combine the information on the compactness of intra-cluster and the isolation of inter-cluster and are the functions of certain factors, such as the defined squared error, the geometric or statistical properties of the data, the number of data objects, the dissimilarity or similarity measurement, and of course, the number of clusters. Milligan and Cooper (1985) compared and ranked 30 indices according to their performance over a series of artificial data sets.

Among these, the best-performing was the Caliński and Harabasz index (Caliński and Harabasz, 1974), which is defined as,

$$CH(K) = \frac{Tr(\mathbf{S_B})}{K-1} \Big/ \frac{Tr(\mathbf{S_W})}{N-K}, \tag{10.10}$$

where N is the number of objects, and $Tr(\mathbf{S}_B)$ and $Tr(\mathbf{S}_W)$ are the trace of the between and within-class scatter matrix (see Chapter 4), respectively. The value of K that maximizes $CH(K)$ suggests an estimation of K.

The Davies-Bouldin index (Davies and Bouldin, 1979), another index examined by Milligan and Cooper (1985), attempts to maximize the between-cluster distance while minimizing the distance between the cluster centroid and the other points. By defining each individual cluster index R_i as the maximum comparison between cluster i and other clusters in the partition,

$$R_i = \max_{j \neq i} \left(\frac{e_i + e_j}{D_{ij}} \right), \tag{10.11}$$

where D_{ij} is the distance between the centroids of cluster i and j, and e_i and e_j are the average errors for cluster i and j, respectively, the Davies-Bouldin index can be written as

$$DB(K) = \frac{1}{K}\sum_{i=1}^{K} R_i. \tag{10.12}$$

The minimum $DB(K)$ indicates the potential number of clusters in the data.

The Dunn index (Dunn, 1974b) attempts to identify clusters that are compact and well separated. Defining the distance $D(C_i, C_j)$ between two clusters C_i and C_j as the minimum distance between a pair of points **x** and **y** belonging to C_i and C_j, respectively,

$$D(C_i, C_j) = \min_{\mathbf{x}\in C_i, \mathbf{y}\in C_j} D(\mathbf{x}, \mathbf{y}), \tag{10.13}$$

and the diameter $diam(C_i)$ of the cluster C_i as the maximum distance between two of its members

$$diam(C_i) = \max_{\mathbf{x},\mathbf{y}\in C_i} D(\mathbf{x}, \mathbf{y}), \tag{10.14}$$

the Dunn index is constructed as

$$Du(K) = \min_{i=1,\ldots,K}\left(\min_{j=i+1,\ldots,K}\left(\frac{D(C_i, C_j)}{\max\limits_{l=1,\ldots,K} diam(C_l)}\right)\right). \tag{10.15}$$

A large value of $Du(K)$ suggests the presence of compact and well-separated clusters, and the corresponding K provides an estimation of the number of clusters.

The Dunn index has the disadvantage of over-sensitivity to noise, for which a family of 18 cluster validation indices is proposed based on the different definitions of cluster distance and cluster diameter (Bezdek and Pal, 1998). For example, the distance between two clusters can be calculated as the maximum or average distance between pairs of points taken from each cluster, which correspond to the concept of complete linkage and average linkage used in hierarchical clustering (Chapter 3), respectively. On the other hand, the cluster diameter can take the following forms that are more robust to noise:

$$diam(C_i) = \frac{1}{N_{C_i}(N_{C_i} - 1)} \sum_{\mathbf{x}, \mathbf{y}\in C_i} D(\mathbf{x}, \mathbf{y}), \tag{10.16}$$

where N_{C_i} is the number of data points in the cluster C_i, and

$$diam(C_i) = 2\frac{\sum_{\mathbf{x}\in C_i} D(\mathbf{x}, \bar{\mathbf{x}})}{N_{C_i}}, \tag{10.17}$$

where

$$\bar{\mathbf{x}} = \frac{1}{N_{C_i}} \sum_{\mathbf{x} \in C_i} \mathbf{x}. \quad (10.18)$$

More discussions of cluster validation indices or stopping rules are given in Bandyopadhyay and Maulik (2001), Halkidi et al. (2002), Milligan and Cooper (1985), and Theodoridis and Koutroumbas (2006). It is worth noting that these indices are data dependent, and a good performance of an index for certain data does not guarantee the same behavior when applied to different data. As pointed out by Everitt et al. (2001), "it is advisable not to depend on a single rule for selecting the number of groups, but to synthesize the results of several techniques."

10.4.3 Fuzzy Validation Indices

In fuzzy clustering, a data point $\mathbf{x}_j$ is associated with each cluster C_i according to a membership coefficient u_{ij}, which is defined as an element in the $K \times N$ fuzzy partition matrix $\mathbf{U}$. Because the fuzzy c-means algorithm is the most commonly used fuzzy clustering technology, the following discussions on fuzzy validation indices are based on it, although these indices can be used for more generalized cases.

Both partition coefficient (PC) (Bezdek, 1974) and partition entropy (PE) (Bezdek, 1975) use the information on the partition matrix exclusively without considering data themselves. Specifically, PC is defined as

$$PC = \frac{1}{N} \sum_{j=1}^{N} \sum_{i=1}^{K} u_{ij}^2, \quad (10.19)$$

and PE is given as

$$PE = -\frac{1}{N} \sum_{j=1}^{N} \sum_{i=1}^{K} (u_{ij} \log_a u_{ij}), \quad (10.20)$$

where $a \in (1, \infty)$ is the logarithmic base.

The value of PC is in the range of $[1/K, 1]$, while the range of values of PE is $[0, \log_a K]$. If PC takes its maximum or PE takes its minimum, the obtained partition corresponds to a hard partition. The opposite extreme is obtained, where each data point is assigned to all clusters with the same membership, when PC takes its minimum or PE takes its maximum. In this case, it is indicated that either the clustering algorithm is incapable of finding the clustering structure or there is no such structure underlying the data at all (Pal and Bezdek, 1995). Because both the values of PC and PE are dependent on K,

K can be estimated by seeking the significant knees of increase, in the case of PC, or decrease, in the case of PE.

The performances of PC and PE are both influenced by the fuzzification parameter m (Pal and Bezdek, 1995). When m approximates to 1 from the right, the limits of PC and PE are equal to 1 and 0, respectively, no matter what the values of K are. Similarly, when m becomes large, PC and PE also lose their capability to choose appropriate values of K because $K = 2$ always maximizes the indices at those cases.

The fuzzy clustering validity by Gath and Geva (1989) integrates the information on data and clusters, in addition to the membership coefficients. Gath and Geva (1989) summarize three criteria for an optimal partition of data: 1) The clusters are clearly separated; 2) The clusters have the minimal volumes; and 3) The clusters have the maximal concentration of data points around the cluster centroids. Correspondingly, three indices were defined to measure whether the derived clustering structure meets these requirements:

- The total fuzzy hypervolume

$$FHV = \sum_{i=1}^{K} [\det(\Sigma_i)]^{1/2}, \tag{10.21}$$

  where Σ_i is the fuzzy covariance matrix of the cluster C_i and is defined as

$$\Sigma_i = \frac{\sum_{j=1}^{N} u_{ij}^m (\mathbf{x}_j - \mathbf{m}_i)(\mathbf{x}_j - \mathbf{m}_i)^T}{\sum_{j=1}^{N} u_{ij}^m}. \tag{10.22}$$

  Small values of FHV are indicative of compact clusters.
- The average partition density

$$APD = \frac{1}{K} \sum_{i=1}^{K} \frac{S_i}{[\det(\Sigma_i)]^{1/2}}, \tag{10.23}$$

  where S_i is the sum of central members of the cluster C_i and considers of only those points that are inside the defined hyperellipsoid,

$$S_i = \sum_{j=1}^{N} u_{ij}, \forall \mathbf{x}_j \in \{\mathbf{x}_j : (\mathbf{x}_j - \mathbf{m}_i)^T \Sigma_i^{-1} (\mathbf{x}_j - \mathbf{m}_i) < 1\}. \tag{10.24}$$

Large values of APD are indicative of compact clusters.

- The partition density

$$PD = \frac{S}{FHV}, \tag{10.25}$$

where

$$S = \sum_{i=1}^{K} S_i. \tag{10.26}$$

Again, large values of PD are indicative of compact clusters.

The Xie-Beni index (Xie and Beni, 1991) also considers data-related information, attempts to measure the compactness and separation of a fuzzy partition, and is defined as,

$$XB = \frac{\left(\sum_{i=1}^{K}\sum_{j=1}^{N} u_{ij}^2 \|\mathbf{x}_j - \mathbf{m}_i\|^2\right)/N}{\left(\min_{i \neq j} \|\mathbf{m}_i - \mathbf{m}_j\|^2\right)}. \tag{10.27}$$

The separation of the clusters is represented via the minimum distance between pairs of cluster prototypes (numerator of the above equation), while the compactness is measured by the ratio of the total variance to the number of data points (denominator). A smaller value of XB indicates a compact and well-separated clustering partition. However, as observed by Xie and Beni (1991), when K is close to N, the XB index decreases monotonically. Xie and Beni (1991) suggested three heuristics to address this problem. For example, they recommended plotting the Xie-Beni index as a function of $K = 2$ to $N - 1$ and determining the starting point of the monotonically decreasing tendency as the maximum K for further analysis.

Consider the cost function J of the fuzzy c-means algorithm (Eq. 4.59, Chapter 4). When the squared Euclidean distance function is used and the fuzzification parameter m is set to 2, it is clear that the Xie-Beni index can be represented in terms of J. By removing the restriction of m, a generalized Xie-Beni index, called the extended Xie-Beni index, can be defined as,

$$EXB = \frac{\left(\sum_{i=1}^{K}\sum_{j=1}^{N} u_{ij}^m \|\mathbf{x}_j - \mathbf{m}_i\|^2\right)/N}{\left(\min_{i \neq j} \|\mathbf{m}_i - \mathbf{m}_j\|^2\right)}. \tag{10.28}$$

Another index for measuring compact and well-separated fuzzy clusters is the Fukuyama-Sugeno index (Fukuyama and Sugeno, 1993), written as,

$$FS = \sum_{i=1}^{K}\sum_{j=1}^{N} u_{ij}^{m}\left(\|\mathbf{x}_j - \mathbf{m}_i\|_{\mathbf{A}}^2 - \|\mathbf{m}_i - \bar{\mathbf{m}}\|_{\mathbf{A}}^2\right), \tag{10.29}$$

where $\bar{\mathbf{m}}$ is the mean of the cluster prototypes and $\mathbf{A}$ is a positive definite and symmetric matrix. Similar to the Xie-Beni index, small values of FS suggest a good partition with compact and well-separated clusters.

The fuzzification parameter m also affects the Xie-Beni index, extended Xie-Beni index, and Fukuyama-Sugeno index (Pal and Bezdek, 1995). When m approaches infinity, the limit of the Xie-Beni index also tends to infinity and therefore causes instability. As to the extended Xie-Beni index or Fukuyama-Sugeno index, the above case leads to an indeterminate form or a zero index value. Again, the indices are no longer capable of discriminating the number of clusters. At the other extreme, when $m \to 1$ from the right, the Xie-Beni index and extended Xie-Beni index still work well while, but the Fukuyama-Sugeno index behaves like the trace of the within-cluster scatter matrix, which is not desirable as a validity index.

Numerical simulation on PE, PC, the Xie-Beni index, the extended Xie-Beni index, and the Fukuyama-Sugeno index has been performed, and the Xie-Beni index was shown to be the most reliable upon the testing data (Pal and Bezdek, 1995). Also, an interval of [1.5, 2.5] is recommended for the values of m. The modified versions of the fuzzy hypervolume, average partition density, partition density, Xie-Beni index, and Fukuyama-Sugeno index to discontinuity data clustering are given in Hammah and Curran (2000).

10.4.4 Model Selection Criteria

For probabilistic mixture model-based clustering (Chapter 4), finding the correct number of clusters (components) K is equivalent to fitting a model with observed data and optimizing some criterion (McLachlan and Peel, 2000). Usually, the EM algorithm is used to estimate the model parameters for a given K, which also goes through a predefined range of values. The value of K that maximizes or minimizes the defined criterion is considered optimal. Smyth (1996) presented a Monte Carlo cross validation (MCCV) method, which randomly divides data into training and test sets a certain number of times using a fraction β (e.g., a half-half division of training and test set works well based on the empirical results). K can be selected either directly based on the criterion function or through the calculated posterior probabilities.

Assuming that N is the number of data objects, N_k is the number of parameters for each cluster, N_p is the total number of independent parameters to be estimated in the model, and $l(\hat{\boldsymbol{\theta}})$ is the maximum log-likelihood, some commonly used criteria are defined as follows.

- Akaike's information criterion (AIC) (Akaike, 1974; Windham and Culter, 1992),

$$AIC(K)=\frac{-2(N-1-N_k-K/2)l(\hat{\boldsymbol{\theta}})}{N}+3N_p, \tag{10.30}$$

  K is selected with the minimum value of $AIC(K)$.
- Bayesian information criterion (BIC) (Pelleg and Moore, 2000; Schwarz, 1978),

$$BIC(K)=l(\hat{\boldsymbol{\theta}})-(N_p/2)\log(N). \tag{10.31}$$

K is selected with the maximum value of $BIC(K)$.
- Minimum description length (MDL) (Rissanen, 1978, 1996),

$$MDL(K)=-l(\hat{\boldsymbol{\theta}})+(N_p/2)\log(N). \tag{10.32}$$

  MDL is equivalent to BIC except for the sign. K is selected with the minimum value of $MDL(K)$.
- Non-coding information theoretic criterion (ICOMP) (Bozdogan, 1994),

$$ICOMP(K)=-l(\hat{\boldsymbol{\theta}})+C_1(\mathbf{I}(\boldsymbol{\theta})^{-1}), \tag{10.33}$$

  where $\mathbf{I}(\theta)$ is the observed Fisher information matrix and is defined as

$$C_1(\mathbf{I}(\boldsymbol{\theta})^{-1})=\frac{N_p}{2}\log\left(\frac{Tr(\mathbf{I}(\boldsymbol{\theta})^{-1})}{N_p}\right)-\frac{1}{2}\log(\det(\mathbf{I}(\boldsymbol{\theta})^{-1})). \tag{10.34}$$

More criteria, such as minimum message length (MML) (Grünwald et al., 1998; Oliver et al., 1996; Wallace and Boulton, 1968), approximate weight of evidence criterion (AWE) (Banfield and Raftery, 1993), cross validation-based information criterion (CVIC) (Smyth, 1998), Bayesian Ying-Yang criterion (Guo et al., 2002; Xu, 1997) and covariance inflation criterion (CIC) (Tibshirani and Knight, 1999), were also published. McLachlan and Peel (2000) offer a further discussion of the characteristics of these criteria. Similar to the previous discussion for validation index, no criterion is superior to others in general cases. The selection of different criteria remains data-dependent.

10.4.5 Heuristic Approaches

A large number of heuristic methods have also been proposed to estimate the number of clusters in data. These methods are based on a wide variety of theories and provide a supplement to the schemes above.

- Girolami (2002) performed eigenvalue decomposition on the kernel matrix $\mathbf{K} = \{k(\mathbf{x}_i, \mathbf{x}_j)\}$ (Chapter 6) in the high-dimensional feature space and used the dominant K components in the decomposition summation as an indication of the possible existence of K clusters. Given a data set $\mathbf{X}$ with N data objects, when a Gaussian kernel is used, we will have the following approximation,

$$\begin{aligned}\int_{\mathbf{x}} p(\mathbf{x})^2 d\mathbf{x} &= \frac{1}{N^2}\sum_{i=1}^{N}\sum_{j=1}^{N} k(\mathbf{x}_i, \mathbf{x}_j) \\ &= \mathbf{b}_N^T \mathbf{K} \mathbf{b}_N, \end{aligned} \tag{10.35}$$

  where $\mathbf{b}_N$ is an $N \times 1$ dimensional vector including elements of value $1/N$. By performing an eigenvalue decomposition on $\mathbf{K}$, we further derive the formula as shown below,

$$\mathbf{b}_N^T \mathbf{K} \mathbf{b}_N = \sum_{i=1}^{N} \lambda_i \{\mathbf{b}_N^T \mathbf{v}_i\}^2, \tag{10.36}$$

  where λ_i are the eigenvalues of $\mathbf{K}$ and $\mathbf{v}_i$ are the corresponding eigenvectors. Theoretically, if the data is composed of K clusters, there will also be K dominant components in the summation of Eq. 10.36.
- Kothari and Pitts (1999) described a scale-space, theory-based method, in which the sum-of-squared distances from a cluster centroid to the clusters in its neighborhood are added as a regularization term in the original squared error criterion (Chapter 4). The smaller the neighborhood is, the more clusters tend to form, and vice versa. The neighborhood of clusters then works as a scale parameter, and we can plot the change of K under a range of the scale parameter. Finally, the K that is persistent in the largest interval of the neighborhood parameter is considered optimal.
- Constructive clustering algorithms adaptively and dynamically adjust the number of clusters rather than asking for a pre-specified and fixed number (Boujemaa, 2000). In this context, a cluster could be created when some certain condition is triggered, such as ART networks, which generate a new cluster only when all existing clusters cannot represent the given input pattern according to the vigilance test (Carpenter and Grossberg, 1987a). A similar mechanism can also be found in the growing cell structures (Fritzke, 1994) and growing neural gas (Fritzke, 1995). On the other

hand, the number of clusters could be adjusted from an initial value by merging and splitting the corresponding clusters based on some criteria. A well-known example is the ISODATA algorithm (Ball and Hall, 1967) that splits a cluster if the within-cluster variability is above a threshold, or combines two clusters if their prototypes are close enough (judged by another threshold). The disadvantage of this approach is the need for the user to select the parameters. As another example, the robust competitive clustering algorithm (RCA) that describes a competitive agglomeration process progressing in stages, begins with an over-specified number of clusters. The clusters of RCA that lose in the competition are then discarded, and their members are moved into other clusters (Frigui and Krishnapuram, 1999). This process is generalized in Boujemaa (2000), which attains the number of clusters by balancing the effect between complexity and fidelity. Another learning scheme, the self-splitting competitive learning network, iteratively chooses and divides cluster prototypes from a single and random prototype until no more prototypes satisfy the split criterion (Zhang and Liu, 2002). The divisibility of a prototype is based on the consideration that each prototype represents only one natural cluster instead of combinations of several clusters. More examples of the constructive clustering algorithms, including the fully automatic clustering system, the plastic neural gas, and the cluster detection and labeling network, can be accessed in Patanè and Russo (2002), Ridella et al. (1998), and Eltoft and deFigueiredo (1998), respectively. Thus, the problem of determining the number of clusters is converted into a parameter selection problem, and the resulting number of clusters is largely dependent on parameter tweaking.

10.5. SUMMARY

Justifying clustering results and assuring a correct understanding of the underlying structure of data is an indispensable step in cluster analysis. Although a clustering structure resulting from a certain algorithm could be assessed by domain knowledge and expert experience, cluster validity emphasizes the evaluation of the clustering result in an objective and quantitative way, which is usually statistically based. Cluster validity provides meaningful insights in answering questions such as "among a set of clustering results, which one fits the data best?", "how many clusters are there in the data?", or "from a practical point of view, are the resulting clusters meaningful or are they just the artifacts of the clustering algorithm?". Validity criteria are used to examine hierarchical clustering structures, partitional clustering structures, or individual clusters. Particularly, estimating the number of clusters in underlying data is a central task in cluster validation, attracting efforts from a wide variety of disciplines and leading to a large number of indices, criteria, and ad hoc methods. With all these validation criteria, indices, and methods, it is important to keep in mind that no criterion, index, or method is superior to any other for all problems encountered. It is reasonable to examine the clustering results with different validation methods before making any final conclusions. Altogether, as noted by Jain and Dubes (1988), cluster validity is the "most difficult and frustrating part of cluster analysis," and we cannot overemphasize its importance.

CHAPTER 11

CONCLUDING REMARKS

Cluster analysis, without prior information, expert knowledge, or category labels, organizes data either into a set of groups with a pre-specified number, or more preferably, dynamically, or in a hierarchical way, represented as a dendrogram. The absence of prior information usually makes cluster analysis more difficult than supervised classification. Therefore, the object of cluster analysis is to unveil the underlying structure of the data, rather than to establish classification rules for discrimination.

Intuitively, data objects that belong to the same cluster should be more similar to each other than to the ones outside. Such similarity or dissimilarity is measured through the defined proximity methods, which is also an important step in cluster analysis. Basically, cluster analysis consists of a series of steps, ranging from preprocessing (such as feature selection and extraction), proximity definition, and clustering algorithm development or selection, to clustering result validity and evaluation and knowledge extraction. Each step is tightly related to all others and could have a large influence on the performance of other steps. For example, a good representation of data objects with appropriate features makes it easier to find the clustering structures, while data sets with many redundant or irrelevant features make the clustering structure vague and the subsequent analysis more complicated. In this sense, they all are equally important in cluster analysis and deserve the same efforts from the scientific disciplines.

In essence, clustering is also a subjective process, which asks for extra attention when performing a cluster analysis on the data. The assumption on the

Clustering, by Rui Xu and Donald C. Wunsch, II

data, the definition of the proximity measure, the construction of the optimum criterion, the selection of the clustering algorithm, and the determination of the validation index all have subjectivity. Moreover, given the same data set, different goals usually lead to different partitions. A simple and direct example is on the partition of animals: an eagle, a cardinal, a lion, a panther, and a ram. If they are divided based on the criterion of whether or not they can fly, we have two clusters with the eagle and the cardinal in one cluster and the rest in the other cluster. However, if the criterion changes to whether or not they are carnivores, we have a completely different partition with the cardinal and the ram in one cluster and the other three in the second cluster.

We have discussed a wide variety of clustering algorithms. These evolve from different research communities, aim to solve different problems, and have their own pros and cons. Though we have already seen many examples of successful applications of cluster analysis, there still remain many open problems due to the existence of many inherent, uncertain factors. These problems have already attracted and will continue to attract intensive efforts from broad disciplines. It will not be surprising to see the continuous growth of clustering algorithms in the future.

In conclusion, we summarize and emphasize several important issues and research trends for cluster analysis:

- At the pre-processing and post-processing phases, feature selection/extraction (as well as standardization and normalization) and cluster validation are as important as the clustering algorithms. If the data sets include too many irrelevant features, this not only increases the computational burden for subsequent clustering, but, more negatively, affects the effective computation of proximity measures and, consequently, cluster formation. On the other hand, clustering result evaluations and validation reflect the degree of confidence to which we can rely on the generated clusters, which is critically important because of the fact that clustering algorithms can always provide us with a clustering structure, which may be just an artifact of the algorithm or may not exist at all. Both processes lack universal guidance and depend on the data. Ultimately, the tradeoff among different criteria and methods remains dependent on the applications themselves.
- There is no clustering algorithm that universally solves all problems. Usually, clustering algorithms are designed with certain assumptions about cluster shapes or data distributions and inevitably favor some type of bias. In this sense, it is not accurate to say "best" in the context of clustering algorithms, although some comparisons are possible. These comparisons are mostly based on some specific applications, under certain conditions, and the results may become quite different if the conditions change.
- As a result of the emergence of more new technologies, more complicated and challenging tasks have appeared, which require more powerful

clustering algorithms. The following properties are important to the efficiency and effectiveness of a novel algorithm:

- Generate arbitrary shapes of clusters rather than being confined to some particular shape. In practice, irregularly-shaped clusters are much more common than shapes like hyperspheres or hyperrectangles.
- Handle a large volume of data as well as high-dimensional features with reasonable time and storage complexity. It is not unusual to see data sets including millions of records with up to tens of thousands of features. Linear or near linear complexity is highly desirable for clustering algorithms to meet such requirements. At the same time, high dimensions with many features irrelevant to the resulting clusters make the algorithms working in low dimensions no longer effective. It is important to find and use the real dimensions to represent the data.
- Detect and remove possible outliers and noise. Noise and outliers are inevitably present in the data due to all kinds of different factors in the measurement, storage, and processing of the data. Their existence can affect the form of the resulting clusters and distort the clustering algorithm.
- Decrease the reliance of algorithms on user-dependent parameters. Most current clustering algorithms require users to specify several parameters. Such parameters are usually hard to determine due to a lack of effective guidance. Furthermore, the formed clusters may be sensitive to the selection of these parameters.
- Have the capability to deal with newly-occurring data without re-learning from scratch. This property could save lots of computational time and increase the clustering efficiency.
- Be immune to the effects of the order of input patterns, or at least provide guidance about what to expect regarding order-dependency. Most current online clustering algorithms suffer from this problem. Different orders of the presentation of input patterns leads to different resulting clusters, which may just be artifacts of the algorithms, making the results questionable.
- Provide some insight for the number of potential clusters without prior knowledge. As discussed in Chapter 10, estimating the number of clusters is one of the most fundamental and important problems in cluster analysis. Many current algorithms require that number as a user-specified parameter; however, without prior knowledge, users usually do not have such information. Over-estimation or under-estimation both could cause the incorrect explanation of the clustering structure.
- Show good data visualization and provide users with results that can simplify further analysis. The ultimate goal of cluster analysis is to find the potential data structure that can be further used to solve a

more complicated problem. An understandable and visualized result is very helpful in this regard.

- Be capable of handling both numerical and nominal data or be easily adaptable to some other data types. More commonly, data are described by features with different types, which requires the clustering algorithm to be flexible in this respect.

Of course, detailed requirements for specific applications will affect these properties.

Although we have endeavored to provide a broad overview of clustering as it now stands, we fully expect (and hope to contribute to) a rapid evolution of this important field. We are eager to see not only the advances on the important challenges to clustering as discussed above, but also the unpredictable array of forthcoming applications of this singularly fascinating technology.

PROBLEMS

CHAPTER 2

1. Discuss the possible conversion methods between the four scales in Section 2, i.e., nominal, ordinal, interval, and ratio.
2. Prove that the Mahalanobis distance is scale invariant, but the Euclidean distance is not.
3. The regularized Mahalanobis distance is defined as, $D(\mathbf{x}_i,\mathbf{x}_j) = (\mathbf{x}_i - \mathbf{x}_j)^T [(1-\lambda)(\mathbf{S}+\varepsilon\mathbf{I})^{-1} + \lambda\mathbf{I}](\mathbf{x}_i - \mathbf{x}_j)$, where $\mathbf{S}$ is the covariance matrix and $\mathbf{I}$ is the identity matrix. Discuss the effect of the parameters λ and ε.
4. Verify whether or not the measures summarized in Table 2.3 are metric.
5. Suppose D is a metric; it is called an ultrametric if the following formula holds: $D(\mathbf{x}_i,\mathbf{x}_j) \le \max(D(\mathbf{x}_i,\mathbf{x}_l), D(\mathbf{x}_l,\mathbf{x}_j)), \forall \mathbf{x}_l$. Verify whether or not the measures summarized in Table 2.3 are ultrametric.
6. Given a dissimilarity matrix $\mathbf{D} = \{D_{ij}\}$ that is a non-metric, prove that the matrix $\mathbf{D'} = \{D_{ij} + b\}(i \ne j)$ is metric, where $b \ge \max_{l,m,n} |D_{lm} + D_{ln} - D_{mn}|$.
7. Suppose the dissimilarity matrix $\mathbf{D} = \{D_{ij}\}$ is a metric; prove that the following matrices derived from $\mathbf{D}$ are also metric:
   a) $\mathbf{D}^{(1)} = \{D_{ij} + b^2\}$ $(i \ne j)$, where b is a real constant;
   b) $\mathbf{D}^{(2)} = \{D_{ij}^{1/r}\}(i \ne j)$, where $r \ge 1$;
   c) $\mathbf{D}^{(3)} = \{D_{ij}/(D_{ij} + b^2)\}$ $(i \ne j)$, where b is a real constant.

8. Verify the generalized string property (Sung and Jin, 2000): Given a data set $\mathbf{X}$ defined in a Euclidean space, if $\mathbf{x}_i, \mathbf{x}_j \in \mathbf{X}$ belong to the same cluster, then every data point $\mathbf{x}_l \in \mathbf{X}$ that satisfies the inequality $D(\mathbf{x}_l, \mathbf{x}_j) < D(\mathbf{x}_i, \mathbf{x}_j)$ or $D(\mathbf{x}_l, \mathbf{x}_i) < D(\mathbf{x}_i, \mathbf{x}_j)$, where $D(\cdot,\cdot)$ is the Euclidean distance, is also in the cluster.
9. For all data points $\mathbf{x}_i$ and $\mathbf{x}_j$ in the feature space, the L_p norm is defined in Eq. 2.19. Let $D_1(\mathbf{x}_i, \mathbf{x}_j)$ denote L_1 norm, $D_2(\mathbf{x}_i, \mathbf{x}_j)$ denote L_2 norm, and so on; prove that $D_\infty(\mathbf{x}_i, \mathbf{x}_j) \le D_2(\mathbf{x}_i, \mathbf{x}_j) \le D_1(\mathbf{x}_i, \mathbf{x}_j)$.

CHAPTER 3

1. Monotonicity. A clustering algorithm is said to be monotone if the two clusters C_i and C_j are merged into a cluster C_{ij}, then for all other clusters C_l, $l \ne i, j$,

$$D(C_l, C_{ij}) \ge D(C_i, C_j).$$

   Prove that
   a) if the coefficients in Eq. 3.1 satisfy $\alpha_i + \alpha_j + \beta \ge 1$, and $\gamma \ge 0$, the corresponding hierarchical clustering method is monotone;
   b) if $\alpha_i + \alpha_j + \beta \ge 1$, and $0 > \gamma \ge \max(-\alpha_i, -\alpha_j)$, the hierarchical clustering method is monotone.
2. Verify that the hierarchical clustering algorithms summarized in Table 3.1 can all be obtained from the Lance and Williams' updating formula by selecting appropriate parameters.
3. Show that the time and space complexities for the general agglomerative hierarchical clustering algorithm are $O(N^2\log N)$ and $O(N^2)$, respectively.
4. Derive Eq. 3.8 from Eq. 3.7.
5. Derive Eq. 3.10 in Ward's algorithm.
6. Derive the time complexity of the algorithm DIANA.
7. Prove the CF Additivity Theorem in Eq. 3.18.
8. Show that link$(\mathbf{x}_i, \mathbf{x}_j)$ in the algorithm ROCK is the number of distinct paths with length 2 between $\mathbf{x}_i$ and $\mathbf{x}_j$ so that each pair of consecutive points on the path are neighbors (Guha et al., 2000).
9. a) In Eq. 3.21, $EC(C_i, C_j)$ is a measure of the absolute interconnectivity. Discuss the advantage of the relative interconnectivity over the absolute interconnectivity.
   b) The absolute closeness is defined as the average weight of the edges connecting vertices between a pair of clusters. Similarly, discuss the advantage of the relative closeness defined in Eq. 3.22 over the absolute closeness.

CHAPTER 4

1. Derive Eq. 4.1 and calculate the number of clusters for $N = 50$ and $K = 5$.
2. Show that the computational complexity of the K-means algorithm is $O(NKdT)$.
3. Verify that the measures in Eqs. 4.7 and 4.8 are invariant to linear transformations.
4. Bradley et al. (2000) proposed a constrained K-means algorithm by adding an additional constraint to the criterion function in Eq. 4.1, which requires that clusters C_i, $i = 1, \ldots, K$ consist of at least N_i data points. Derive the updating equation of the cluster centroid and draw the flowchart of this modified K-means algorithm. Discuss the advantage of this variant.
5. Consider that the component densities are multivariate Gaussian densities,
   a) fill in the gap of deriving Eqs. 4.39–4.42;
   b) derive the ML estimation if only the mean vector $\boldsymbol{\mu}$ is unknown.
6. Suppose that data vector $\mathbf{x}$ is a d-dimensional binary vector and the component densities are multivariate Bernoulli distributions, written as

$$p(\mathbf{x}|\boldsymbol{\theta}_i) = \prod_{j=1}^{d} \theta_{ij}^{x_j}(1-\theta_{ij})^{1-x_j};$$

   derive the ML estimation of $\boldsymbol{\theta}_i$.
7. Refer to Celeux and Govaert (1992). Show that K-means is equivalent to the CEM algorithm under a spherical Gaussian mixture, where the dimensions have the same variance.
8. Given a set of 2-dimensional data points $\mathbf{X} = \{\mathbf{x}_i, i = 1, \ldots, 9\}$, where $\mathbf{x}_1 = [1, 4]^T$, $\mathbf{x}_2 = [2, 3]^T$, $\mathbf{x}_3 = [3, 3.5]^T$, $\mathbf{x}_4 = [2.5, 2.5]^T$, $\mathbf{x}_5 = [4, 3]^T$, $\mathbf{x}_6 = [7, 7.5]^T$, $\mathbf{x}_7 = [6, 7]^T$, $\mathbf{x}_8 = [8, 6]^T$, $\mathbf{x}_9 = [6.5, 5]^T$,
   a) Construct a weighted graph G based on the Euclidean distance;
   b) Create a minimum spanning tree of G;
   c) Identify the inconsistent edges;
   d) Remove such edges and examine the formed clusters;
   e) Write a program to implement the above steps to general situations, with $\mathbf{X} = \{\mathbf{x}_i, i = 1, \ldots, N\}$, $\mathbf{x}_i \in \Re^d$.
9. Assume that there is no prior information available, and show how to estimate the parameters μ_B, σ_B^2, μ_W, σ_W^2, and P_0 in Eqs 4.51–4.53 using the EM algorithm.
10. Given a complete graph G, prove that Eq. 4.56 holds for any cut E_c of G. Extend such discussion to the case that G is incomplete.

11. Derive the updating equations of the membership and prototype in Eqs. 4.62 and 4.63 via the optimization of the criterion function in Eq. 4.59.
12. It is important for a clustering algorithm to be capable of handling outliers. Keller (2000) modified the criterion function in Eq. 4.59 by introducing an additional weighting factor ω for each data point, written as

$$J(\mathbf{U},\mathbf{M})=\sum_{i=1}^{c}\sum_{j=1}^{N}(u_{ij})^m\frac{1}{\omega_j^q}D_{ij}^2,$$

where q is a constant to adjust the effect of ωs, and the weighting factors follow the constraint $\sum_{j=1}^{N}\omega_j=\Omega$, where Ω is a constant. It can be seen that data points that fit well with at least one cluster will have small weighting factors, while outliers will be associated with large weighting factors. Show the steps for the corresponding fuzzy clustering algorithm.
13. Discuss the behavior of fuzzy c-means clustering when L_1 and L_∞ norms are used.
14. Compare the possibilistic c-means clustering algorithm and the fuzzy c-means clustering algorithm. Discuss the advantages and disadvantages of both algorithms.
15. The distance function in the fuzzy c-spherical shells algorithm is defined as

$$D_{ij}^2=\left(\|\mathbf{x}_j-\mathbf{m}_i\|^2-r_i^2\right)^2,$$

where $\mathbf{m}_i$ is the center of the cluster representing a hyperspherical shell, and r_i is the radius. The distance function can also be written as

$$D_{ij}^2=\mathbf{p}_i^T\mathbf{\Phi}_j\mathbf{p}_i+\mathbf{\Psi}_j^T\mathbf{p}_i+b_j,$$

where $\mathbf{p}_i=\left[-2\mathbf{m}_i \quad \mathbf{m}_i^T\mathbf{m}_i-r_i^2\right]^T$, $\mathbf{q}_j=[\mathbf{x}_j \ 1]^T$, $\mathbf{\Phi}_j=\mathbf{q}_j\mathbf{q}_j^T$, $\mathbf{\Psi}_j=2(\mathbf{x}_j^T\mathbf{x}_j)\mathbf{q}_j$, and $b_j=(\mathbf{x}_j^T\mathbf{x}_j)^2$. Derive the updating equation of $\mathbf{p}_i$ via the minimization of the criterion function in Eq. 4.59.
16. Following the steps for GGA, design an ES or EP-based clustering algorithm and implement the derived algorithm.
17. Modify GGA in terms of coding strategy, selection operator, crossover operator, and mutation operator. Discuss the pros and cons of the proposed modification compared with the original methods.
18. Based on Eq. 4.85,

a) Derive the updating equation of prototype vectors $\mathbf{m}_j$, $j = 1, \ldots, K$, when the Euclidean distance is used;
b) Show that Eq. 4.85 can be approximated as

$$\frac{1}{N}\sum_i p(\mathbf{m}_j|\mathbf{x}_i)\frac{d}{d\mathbf{m}_j}D(\mathbf{x}_i, \mathbf{m}_j) = 0.$$

19. Consider the velocity updating equation of PSO in Eq. 4.86; the inertia weight W_I is fixed as a constant. Design a reasonable method to dynamically adjust W_I during the search process.
20. Implement a TS-based clustering algorithm based on Fig. 4.7. Investigate the effects of the tabu list size on the performance of the clustering algorithm.

CHAPTER 5

1. Show that when the instar rule in Eq. 5.4 is used, if the input pattern is normalized, the weight vector will also be normalized.
2. Derive the relation in Eq. 5.40.
3. For both ART1 and Fuzzy ART networks, prove that no two clusters have the same weight vectors.
4. Given an ART1 network, design an example to show that the resulting clusters are dependent on the order of the presentation of the input patterns.
5. For the ART1 network, prove that the maximum number of learned clusters is 2^d, where d is the number of features of the input patterns.
6. Assuming that a list of input patterns is repeatedly presented to a Fuzzy ART network with fast learning, prove that no uncommitted node will be selected after the first presentation of the input patterns.
7. Assuming that a list of binary input patterns is repeatedly presented to a Fuzzy ART network with fast learning and enough nodes in the F_2 layer, prove that if $\alpha \leq \rho/(1 - \rho)$, then the weights will stabilize in one presentation of the input patterns.
8. Derive the updating equations (Eqs. 5.69–5.71) of prototype vectors by calculating the gradient of the cost function in Eq. 5.66.
9. Replace the weight factors v_{ji} in Eq. 66 with the fuzzy membership coefficient u_{ji} in Eq. 5.74. Calculate the gradient of the new cost function with respect to the prototype vectors.
10. Verify that FLVQ is equivalent to FCM when $m_0 = m_f$.
11. Consider the projection of 500 four-dimensional input patterns into a two-dimensional lattice consisting of 8×8 neurons. Each dimension of the input data follows a uniform distribution as presented here:

x_1: $0 < x_1 < 2$, x_2: $0 < x_2 < 1$, x_3: $0 < x_3 < 1$, and x_4: $0 < x_4 < 0.8$. Generate such a data set and write a program to illustrate the performance of SOFM after 10, 50, 100, 500, 1,000, 5,000, and 10,000 iterations.

12. Based on the work by Cottrell et al. (2006), write the pseudo-code for the batch version NG algorithm.
13. Compare ART, SOFM, and GNG in terms of learning mechanism, updating rule, and other properties of interest. List the pros and cons of each algorithm.

CHAPTER 6

1. Construct a kernel-based
   a) *K*-means algorithm;
   b) LVQ algorithm.

   Consider the possibility of reformulating nonlinear variants of other linear algorithms.
2. Positive definite kernel. A kernel function k is called a positive definite function if, and only if, the following conditions are satisfied:
   a) $k(\mathbf{x}_i, \mathbf{x}_j) = k(\mathbf{x}_j, \mathbf{x}_i)$, for $\forall \mathbf{x}_i, \mathbf{x}_j \in \mathbf{X}$, and
   b) $\sum_{i,j=1}^{N} b_i b_j k(\mathbf{x}_i, \mathbf{x}_j) \geq 0$, for all $N > 1$, $\mathbf{x}_1, \ldots, \mathbf{x}_N \subseteq \mathbf{X}$, and $b_1, \ldots, b_N \subseteq \Re$.

   Verify that the polynomial kernel and Gaussian radial basis function kernel are both positive definite functions.
3. Show that the data can be centered in the feature space by using the new kernel matrix defined in Eq. 6.7.
4. Show that the principal components in kernel PCA are uncorrelated.
5. Derive Eq. 6.18 from Eq. 6.15 using the definition in Eq. 6.16 and the assumption that a Gaussian kernel is used.
6. Complete the steps to solve the nonlinear optimization problem in Eq. 6.18.
7. Derive the Wolfe dual form in Eq. 6.22 based on the Lagrangian form in Eq. 6.21.
8. Prove that the number of bounded support vectors satisfies $N_{bsv} < 1/C$.
9. Write a program to investigate the effect of the Gaussian kernel width parameter σ on the number of clusters.

CHAPTER 7

1. Find the global and local alignments for the DNA sequences AGGCTAG and AGTCTG using the Needleman-Wunsch and Smith-Waterman algorithms, respectively. Score +2 for a match, –1 for a mismatch, and –2 for a gap.

2. Discuss the computational complexity of the Needleman-Wunsch algorithm in terms of both space and time.
3. Verify that
   a) the computational complexity of $P(\mathbf{o}|\boldsymbol{\lambda})$ based on Eq. 7.14 is $O(N^T T)$;
   b) the computational complexity of $P(\mathbf{o}|\boldsymbol{\lambda})$ based on the Forward or Backward algorithm is $O(N^2 T)$.
4. Derive Eq. 7.28.
5. Complete the steps in deriving Eqs. 7.34–7.36. Explain the meaning of these estimations.
6. Apply an HMM-based clustering method to the following artificial data set as in Smyth (1997). The data are generated from two two-state HMMs with transition probability distributions $\mathbf{A}_1 = \begin{bmatrix} 0.6 & 0.4 \\ 0.4 & 0.6 \end{bmatrix}$ and $\mathbf{A}_2 = \begin{bmatrix} 0.4 & 0.6 \\ 0.6 & 0.4 \end{bmatrix}$. The first state emits observance obeying a Gaussian density with mean 0 and variance 1, while the second state emits observance from Gaussian density with mean 3 and variance 1. Twenty sequences are generated from each model, with length equal to 200.
7. Obtain the M-step parameter estimation in the EM algorithm for ARMA mixtures, using the log-likelihood function defined in Eq. 7.47.
8. Write a program to implement the HMM architecture in Fig. 7.11.

CHAPTER 8

1. Prove Eq. 8.3.
2. Study the BIRCH algorithm. Write the pseudo-code on the insertion of an entry into a CF tree and perform the computational analysis of the step 1 summarized in Fig. 3.4 in Chapter 3.
3. Compare the clustering algorithm developed by Chiu et al. (2001) for mixed data types with the BIRCH algorithm. Show how the modification is made in order to deal with both continuous and categorical data.
4. According to the DBSCAN algorithm, a point $\mathbf{x}$ is directly density-reachable from a point $\mathbf{y}$ if 1) $\mathbf{x} \in N_{Eps}(\mathbf{y})$ and 2) $|N_{Eps}(\mathbf{y})| \geq Minpts$. Show that direct density-reachability is symmetric for pairs of core points, but not for pairs consisting of one core point and one border point, generally.
5. Using the cluster definition of the DBSCAN algorithm, prove that,
   a) given a data point $\mathbf{x} \in \mathbf{X}$ and $|N_{Eps}(\mathbf{x})| \geq Minpts$, the set $C = \{\mathbf{a} \in \mathbf{X} | \mathbf{a}$ is density-reachable from $\mathbf{x}$ with respect to Eps, and $Minpts\}$ is a cluster with respect to Eps and $Minpts$;

b) given $\boldsymbol{C}$ as a cluster with respect to *Eps* and *Minpts* and $\mathbf{x}$ as any point in $\boldsymbol{C}$ with $|N_{Eps}(\mathbf{x})| \geq Minpts$, $\boldsymbol{C}$ is equal to the set $\boldsymbol{A} = \{\mathbf{a} | \mathbf{a}$ is density-reachable from $\mathbf{x}$ with respect to *Eps* and *Minpts*$\}$.

6. Discuss the generality of the DENCLUE algorithm to hierarchical clustering algorithms.
7. Following Eqs. 8.13 and 8.14, calculate the parameters—standard deviation, minimum, maximum—at higher level cells using the corresponding parameters at lower levels.
8. Discuss the relation between the algorithms STING and DBSCAN.
9. Clustering problems can be constructed as traveling salesman problems. Given a d-dimensional data set with N objects, denoted as a matrix $\mathbf{X} = \{x_{ij}\}_{N \times d}$, where each row corresponds to an object and each column represents a feature of the object, the goal of clustering is to find an optimal permutation P_0 of the rows that maximizes (minimizes) the sum of the similarities (distances) between adjacent rows. Considering the fact that, in natural clusters, the intra-cluster distances are usually much less than the inter-cluster distances, it is more appropriate not to include the inter-cluster distances during the optimization computation, written as,

   $J(P_0) = \min\left(\sum_{i=1}^{K} \sum_{j=n_i(1)}^{n_i(L_i)-1} D(j, j+1)\right)$, where K is the number of clusters, $n_i(1)$ is the first element of cluster i, $n_i(L_i)$ is the last element of cluster i, and $D(\cdot)$ is the distance function. Map this problem into the traveling salesman problem and give the steps of your algorithm.

CHAPTER 9

1. Derive Eq. 9.10 by minimizing Eq. 9.8 using the Lagrange optimization method. Using the result in Eq. 9.10, fill the gaps in calculating the value of the error criterion function as shown in Eq. 9.11.
2. Prove that even non-linear activation functions are used for the hidden layer in Fig. 9.1; the three-layer neural network cannot be used to realize nonlinear PCA.
3. Show the detailed steps in the training of the auto-associative neural network for PCA as depicted in Fig. 9.1.
4. Given a Gaussian density function with mean μ and variance σ^2, calculate the corresponding differential entropy. Consider other continuous density functions that have the same mean and variance. Are the resulting entropies larger than that for the Gaussian distribution?
5. Based on the results shown in Fig. 9.3, discuss the relations between PCA, ICA, and projection pursuit in detail.
6. In Fig. 9.4, the bottleneck layer consists of linear nodes. Consider the creation of a novel four-layer neural network by combing the weights

of the pre- and post-bottleneck layers. What is the difference between this new network and the original five-layer network? Why not omit the bottleneck layer?

7. Find the optimal configuration of nodes in elastic maps by minimizing the energy function in Eq. 9.30.
8. Prove the monotonicity: if a set of data points $\boldsymbol{U}$ is a cluster in a K-dimensional space, then $\boldsymbol{U}$ is also part of a cluster in any (K-1)-dimensional projection of this space.
9. Discuss the relation between kernel PCA and the locally linear embedding algorithm.
10. Draw the flowchart of the algorithm ORCLUS. Discuss the computation complexity for each of the three main steps.

CHAPTER 10

1. Consider the Rand index. Show that if the number of clusters in a revealed clustering structure of the data set $\mathbf{X}$ is not equal to the number in a pre-specified partition, then the maximum value of the Rand index is less than one.
2. Show that the value of CPCC is in the range of [–1, 1].
3. Given $\mathbf{U}$ as the fuzzy partition matrix, verify the following relations:
   a) $PC(\mathbf{U}) = 1 \Leftrightarrow PE(\mathbf{U}) = 0 \Leftrightarrow \mathbf{U}$ is a hard K-partition of the given data set;
   b) $PC(\mathbf{U}) = 1/K \Leftrightarrow PE(\mathbf{U}) = \log_a K \Leftrightarrow \mathbf{U}$ is the fuzziest partition with each point assigned to all clusters with equal membership values $1/K$.
4. Given $\mathbf{U}$ as the fuzzy partition matrix and $\mathbf{M}$ as the cluster prototype matrix, verify the following results:
   a) $\lim_{m \to \infty} XB(\mathbf{U}, \mathbf{M}) = \infty$;
   b) $\lim_{m \to \infty} EXB(\mathbf{U}, \mathbf{M}) = \frac{0}{0}$;
   c) $\lim_{m \to \infty} FS(\mathbf{U}, \mathbf{M}) = 0$.
5. Calculate the limit of $FS(\mathbf{U}, \mathbf{M})$ when m approaches one from above.
6. The IRIS data set is one of the most popular data sets in the fields of pattern recognition and machine learning. It can be downloaded from the UCI Machine Learning Repository at http://www.ics.uci.edu/~mlearn/MLRepository.html. There are three categories in the data set (i.e. iris setosa, iris versicolor and iris virginical), each having 50 patterns with 4 features (i.e., sepal length (SL), sepal width (SW), petal length (PL) and petal width (PW)). Use the method of eigenvalue decomposition on

kernel matrix (Girolami, 2002) to estimate the number of clusters in the IRIS data set.

7. Choose a constructive clustering algorithm to perform a cluster analysis on the IRIS data set. Investigate the effect of the corresponding parameters, such as the vigilance parameter in ART, to the resulting number of clusters.
8. Now apply the mixture model-based clustering methods to the IRIS data set. Estimate the number of clusters using the model selection criteria discussed in Section 4.4. Compare the performance of these criteria.

REFERENCES

Aarts, E. and Korst, J. (1989). *Simulated annealing and Boltzmann machines: A stochastic approach to combinatorial optimization and neural computing*. New York, NY: John Wiley & Sons.

Abascal, F. and Valencia, A. (2002). Clustering of proximal sequence space for the identification of protein families. *Bioinformatics*, 18: 908–921.

Aggarwal, C. (2004). A human-computer interactive method for projected clustering. *IEEE Transactions on Knowledge and Data Engineering*, 16(4): 448–460.

Aggarwal, C., Wolf, J., Yu, P., Procopius, C., and Park, J. (1999). Fast algorithm for projected clustering. In *Proceedings of ACM SIGMOD International Conference on Management of Data*, pp. 61–71.

Aggarwal, C. and Yu, P. (2002). Redefining clustering for high-dimensional applications. *IEEE Transactions on Knowledge and Data Engineering*, 14(2): 210–225.

Agrawal, R., Gehrke, J., Gunopulos, D., and Raghavan, P. (1998). Automatic subspace clustering of high dimensional data for data mining applications. In *Proceedings of ACM SIGMOD International Conference on Management of Data*, pp. 94–105.

Agrawal, R. and Srikant, R. (1995). Mining sequential patterns. In *Proceedings of the 11th IEEE International Conference on Data Engineering*, pp. 3–14.

Ahalt, S., Krishnamurthy, A., Chen, P., and Melton, D. (1990). Competitive learning algorithms for vector quantization. *Neural Networks*, 3: 277–290.

Akaike, H. (1974). A new look at the statistical model identification. *IEEE Transactions on Automatic Control*, AC-19: 716–722.

Aldenderfer, M. and Blashfield, R. (1984). *Cluster analysis*. Newbury Park, CA: Sage Publications.

Clustering, by Rui Xu and Donald C. Wunsch, II

Alizadeh, A., Eisen, M., Davis, R., Ma, C., Lossos, I., Rosenwald, A., Boldrick, J., Sabet, H., Tran, T., Yu, X., Powell, J., Yang, L., Marti, G., Moore, T., Hudson, J., Jr, Lu, L., Lewis, D., Tibshirani, R., Sherlock, G., Chan, W., Greiner, T., Weisenburger, D., Armitage, J., Warnke, R., Levy, R., Wilson, W., Grever, M., Byrd, J., Bostein, D., Brown, P., and Staudt, L. (2000). Distinct types of diffuse large B-cell Lymphoma identified by gene expression profiling. *Nature*, 403: 503–511.

Alon, U., Barkai, N., Notterman, D., Gish, K., Ybarra, S., Mack, D., and Levine, A. (1999). Broad patterns of gene expression revealed by clustering analysis of tumor and normal colon tissues probed by Oligonucleotide arrays. *Proceedings of the National Academy of Science, USA 96*, 96(12): 6745–6750.

Alpert, C. and Kahng, A. (1994). Multi-way partitioning via spacefilling curves and dynamic programming. In *Proceedings of the 31st ACM/IEEE design automation conference*, pp. 652–657.

Alpert, C. and Kahng, A. (1995). Recent directions in netlist partitioning: A survey. *VLSI Journal*, 19: 1–81.

Al-Sultan, K. (1995). A Tabu search approach to the clustering problem. *Pattern Recognition*, 28(9): 1443–1451.

Altschul, S., Madden, T., Schäffer, A., Zhang, J., Zhang, Z., Miller, W., and Lipman, D. (1997). Gapped BLAST and PSI-BLAST: A new generation of protein database search programs. *Nucleic Acids Research*, 25(17): 3389–3402.

Altschul, S., Gish, W., Miller, W., Myers, E., and Lipman, D. (1990). Basic local alignment search tool. *Journal of Molecular Biology*, 215: 403–410.

Amaratunga, D. and Cabrera, J. (2004). *Exploration and analysis of DNA microarray and protein array data*. Hoboken, NJ: John Wiley & Sons.

Anagnostopoulos, G. and Georgiopoulos, M. (2000). Hypersphere ART and ARTMAP for unsupervised and supervised incremental learning. In *Proceedings of the IEEE-INNS-ENNS International Joint Conference on Neural Networks (IJCNN'00)*, Vol. 6, pp. 59–64.

Anagnostopoulos, G. and Georgiopoulos, M. (2001). Ellipsoid ART and ARTMAP for incremental unsupervised and supervised learning. In *Proceedings of the IEEE-INNS-ENNS International Joint Conference on Neural Networks (IJCNN'01)*, Vol. 2, pp. 1221–1226.

Anderberg, M. (1973). *Cluster analysis for applications*. New York, NY: Academic Press.

Ankerst, M., Breunig, M., Kriegel, H., and Sander, J. (1999). OPTICS: Ordering points to identify the clustering structure. In *Proceedings of 1999 ACM SIGMOD International Conference on Management of Data*, pp. 49–60.

Asharaf, S., Shevade, S., and Murty, M. (2005). Rough support vector clustering. *Pattern Recognition*, 38: 1779–1783.

Asuncion, A. and Newman, J. (2007). UCI Machine Learning Repository [http://www.ics.uci.edu/~mlearn/MLRepository.html]. Irvine, CA: University of California, School of Information and Computer Science.

Babu, G. and Murty, M. (1993). A near-optimal initial seed value selection in *K*-Means algorithm using a genetic algorithm. *Pattern Recognition Letters*, 14(10): 763–769.

Babu, G. and Murty, M. (1994). Clustering with evolution strategies. *Pattern Recognition*, 27(2): 321–329.

Bach, F. and Jordan, M. (2002). Kernel independent component analysis. *Journal of Machine Learning Research*, 3: 1–48.

Backer, E. and Jain, A. (1981). A clustering performance measure based on fuzzy set decomposition. *IEEE Transactions on Pattern Analysis and Machine Intelligence*, PAMI-3(1): 66–75.

Bagnall, A. and Janacek, G. (2004). Clustering time series from ARMA models with clipped data. *Proceedings of the International Conference on Knowledge Discovery in Data and Data Mining—ACM SIGKDD 2004*, pp. 49–58.

Bagnall, A., Janacek, G., Iglesia, B., and Zhang, M. (2003). Clustering time series from mixture polynomial models with discretized data. In *Proceedings of the 2nd Australasian Data Mining Workshop*, pp. 105–120.

Bairoch, A. and Apweiler, R. (1999). The SWISS-PROT protein sequence data bank and its supplement TrEMBL in 1999. *Nucleic Acids Research*, 27: 49–54.

Baldi, P. and Brunak, S. (2001). *Bioinformatics: The machine learning approach, 2nd edition*. Cambridge, MA: MIT Press.

Baldi, P. and Hornik, K. (1989). Neural networks and principal component analysis: Learning from examples without local minima. *Neural Networks*, 2: 53–58.

Baldi, P. and Long, A. (2001). A Bayesian framework for the analysis of microarray expression data: Regularized t-test and statistical inferences of gene changes. *Bioinformatics*, 17: 509–519.

Ball, G. and Hall, D. (1967). A clustering technique for summarizing multivariate data. *Behavioral Science*, 12: 153–155.

Bandyopadhyay, S. (2005). Simulated annealing using a reversible jump Markov chain Monte Carlo algorithm for fuzzy clustering. *IEEE Transactions on Knowledge and Data Engineering*, 17(4): 479–490.

Bandyopadhyay, S. and Maulik, U. (2001). Nonparametric genetic clustering: Comparison of validity indices. *IEEE Transactions on Systems, Man, and Cybernetics—Part C: Applications and Reviews*, 31(1): 120–125.

Banerjee, A. and Ghosh, J. (2004). Frequency-sensitive competitive learning for scalable balanced clustering on high-dimensional hyperspheres. *IEEE Transactions on Neural Networks*, 15(3): 702–719.

Banfield, J. and Raftery, A. (1993). Model-based Gaussian and non-Gaussian clustering. *Biometrics*, 49: 803–821.

Baraldi, A. and Alpaydin, E. (2002). Constructive feedforward ART clustering networks—Part I and II. *IEEE Transactions on Neural Networks*, 13(3): 645–677.

Baraldi, A. and Blonda, P. (1999). A survey of fuzzy clustering algorithms for pattern recognition—Part I and II. *IEEE Transactions on Systems, Man, and Cybernetics—Part B: Cybernetics*, 29(6): 778–801.

Baraldi, A., Blonda, P., Parmiggiani, F., Pasquariello, G., and Satalino, G. (1998). Model transitions in descending FLVQ. *IEEE Transactions on Neural Networks*, 9(5): 724–738.

Baraldi, A., Bruzzone, L., and Blonda, P. (2005). Quality assessment of classification and cluster maps without ground truth knowledge. *IEEE Transactions on Geoscience and Remote Sensing*, 43(4): 857–873.

Baraldi, A. and Schenato, L. (1999). Soft-to-hard model transition in clustering: A review. Technical Report, TR-99-010.

Barbará, D. and Chen, P. (2000). Using the fractal dimension to cluster datasets. In *Proceedings of the 6th ACM SIGKDD International Conference on Knowledge Discovery and Data Mining*, pp. 260–264.

Bargiela, A., Pedrycz, W., and Hirota, K. (2004). Granular prototyping in fuzzy clustering. *IEEE Transactions on Fuzzy Systems*, 12(5): 697–709.

Bar-Joseph, Z., Gerber, G., Gifford, D., Jaakkola, T., and Simon, I. (2002). A new approach to analyzing gene expression time series data. In *Proceedings of the 6th Annual International Conference on Research in Computational Molecular Biology—RECOMB*, pp. 39–48.

Barni, M., Cappellini, V., and Mecocci, A. (1996). Comments on "A possibilistic approach to clustering". *IEEE Transactions on Fuzzy Systems*, 4(3): 393–396.

Barreto, G., Mota, J., Souza, L., Frota, R., and Aguayo, L. (2005). Condition monitoring of 3G cellular networks through competitive neural models. *IEEE Transactions on Neural Networks*, 16(5): 1064–1075.

Bartfai, G. and White, R. (1997). ART-based modular networks for incremental learning of hierarchical clusterings. *Connection Science*, 9(1): 87–112.

Bartlett, M., Movellan, J., and Sejnowski, T. (2002). Face recognition by independent component analysis. *IEEE Transactions on Neural Networks*, 13(6): 1450–1464.

Basak, J. and Krishnapuram, R. (2005). Interpretable hierarchical clustering by constructing an unsupervised decision tree. *IEEE Transactions on Knowledge and Data Engineering*, 17(1): 121–132.

Belkin, M. and Niyogi, P. (2002). Laplacian eigenmaps for dimensionality reduction and data representation. *Neural Computation*, 13: 1373–1396.

Bellman, R. (1957). *Dynamic Programming*. Princeton, NJ: Princeton University Press.

Bellman, R. (1961). *Adaptive control processes: A guided tour*. Princeton, NJ: Princeton University Press.

Ben-Dor, A., Shamir, R., and Yakhini, Z. (1999). Clustering gene expression patterns. *Journal of Computational Biology*, 6: 281–297.

Bengio, Y. (1999). Markovian models for sequential data. *Neural Computing Surveys*, 2: 129–162.

Ben-Hur, A., Horn, D., Siegelmann, H., and Vapnik, V. (2000). A support vector clustering method. In *Proceedings of the 15th International Conference on Pattern Recognition*, Vol. 2, pp. 724–727.

Ben-Hur, A., Horn, D., Siegelmann, H., and Vapnik, V. (2001). Support vector clustering. *Journal of Machine Learning Research*, 2: 125–137.

Berglund, E. and Sitte, J. (2006). The parameterless self-organizing map algorithm. *IEEE Transactions on Neural Networks*, 17(2): 305–316.

Berkhin, P. (2001). Survey of clustering data mining techniques. http://www.accrue.com/products/rp_cluster_review.pdf, cached at http://citeseer.nj.nec.com/berkhin02survey.html.

Beyer, K., Goldstein, J., Ramakrishnan, R., and Shaft, U. (1999). When is nearest neighbor meaningful. In *Proceedings of 7th International Conference on Database Theory*, pp. 217–235.

Bezdek, J. (1974). Cluster validity with fuzzy sets. *Journal of Cybernetics*, 3(3): 58–72.

Bezdek, J. (1975). Mathematical models for systematics and taxonomy. In *Proceedings of 8th International Conference on Numerical Taxonomy*, G. Estabrook, Ed., San Francisco, CA: Freeman, pp. 143–166.

Bezdek, J. (1981). *Pattern recognition with fuzzy objective function algorithms*. New York, NY: Plenum Press.

Bezdek, J., Hall, L., and Clarke, L. (1993). Review of MR image segmentation techniques using pattern recognition. *Medical Physics*, 20(4): 1033–1048.

Bezdek, J. and Hathaway, R. (1992). Numerical convergence and interpretation of the fuzzy *c*-shells clustering algorithms. *IEEE Transactions on Neural Networks*, 3(5): 787–793.

Bezdek, J. and Pal, N. (1995). Two soft relatives of learning vector quantization. *Neural Networks*, 8(5): 729–743.

Bezdek, J. and Pal, N. (1998). Some new indexes of cluster validity. *IEEE Transactions on Systems, Man, and Cybernetics—Part B: Cybernetics*, 28(3): 301–315.

Bishop, C. (1995). *Neural networks for pattern recognition*. New York, NY: Oxford University Press.

Bishop, C., Svensén, M., and Williams, C. (1998). GTM: The generative topographic mapping. *Neural Computation*, 10: 215–234.

Bobrowski, L. and Bezdek, J. (1991). *c*-Means clustering with the l_1 and l_∞; norms. *IEEE Transactions Systems, Man, and Cybernetics*, 21(3): 545–554.

Bock, H. (1996). Probabilistic models in cluster analysis. *Computational Statistics & Data Analysis*, 23: 5–28.

Bolten, E., Sxhliep, A., Schneckener, S., Schomburg, D., and Schrader, R. (2001). Clustering protein sequences—structure prediction by transitive homology. *Bioinformatics*, 17: 935–941.

Borg, I. and Groenen, P. (1997). *Modern multidimensional scaling: Theory and applications*. New York, NY: Springer-Verlag.

Boujemaa, N. (2000). Generalized competitive clustering for image segmentation. In *Proceedings of 19th International Meeting of the North American Fuzzy Information Processing Society—NAFIPS 2000*.

Box, G., Jenkins, G., and Reinsel, G. (1994). *Time series analysis: Forecasting and control, 3rd edition*. Englewood Cliffs, NJ: Prentice Hall.

Bozdogan, H. (1994). Mixture-model cluster analysis using model selection criteria and a new information measure of complexity. In *Proceedings of the 1st US/Japan Conference on the Frontiers of Statistical Modeling: An Informational Approach*, pp. 69–113, Dordrecht, Netherlands: Kluwer Academic Publishers.

Bracco, M., Ridella, S., and Zunino, R. (2003). Digital implementation of hierarchical vector quantization. *IEEE Transactions on Neural Networks*, 14(5): 1072–1084.

Bradley, P., Bennett, K., and Demiriz, A. (2000b). Constrained *K*-means clustering. Microsoft Research Technical Report.

Bradley, P. and Fayyad, U. (1998). Refining initial points for *K*-Means clustering. In *Proceedings of 15th International Conference on Machine Learning*, San Francisco, CA: Morgan Kaufmann, pp. 91–99.

Bradley, P., Fayyad, U., and Reina, C. (1998). Scaling clustering algorithms to large databases. In *Proceedings of 4th International Conference on Knowledge Discovery and Data Mining (KDD98)*, New York, NY: AAAI Press, pp. 9–15.

Bradley, P., Fayyad, U., and Reina, C. (2000a). Clustering very large databases using EM mixture models. In *Proceedings of 15th International Conference on Pattern Recognition*, Vol. 2, pp. 76–80.

Brown, D. and Huntley, C. (1992). A practical application of simulated annealing to clustering. *Pattern Recognition*, 25(4): 401–412.

Butler, D. and Jiang, J. (1996). Distortion equalized fuzzy competitive learning for image data vector quantization. In *Proceedings of IEEE International Conference on Acoustics, Speech, and Signal Processing—ICASSP96*, Vol. 6, pp. 3390–3393.

Burges, C. (1998). A tutorial on support vector machines for pattern recognition. *Data Mining and Knowledge Discovery* 2, pp. 121–167.

Burke, J., Davison, D., and Hide, W. (1999). d2_Cluster: A validated method for clustering EST and full-length cDNA sequences. *Genome Research*, 9: 1135–1142.

Cadez, I., Gaffney, S., and Smyth, P. (2000a). A general probabilistic framework for clustering individuals and objects. In *Proceedings of the 6th ACM SIGKDD International Conference on Knowledge Discovery and Data Mining*, pp. 140–149.

Cadez, I., Heckerman, D., Meek, C., Smyth, P., and White, S. (2000b). Visualization of navigation patterns on a web site using model-based clustering. In *Proceedings of the 6th ACM SIGKDD International Conference on Knowledge Discovery and Data Mining*, pp. 280–284.

Cai, D., He, X., and Han, J. (2005). Document clustering using locality preserving indexing. *IEEE Transactions on Knowledge and Data Engineering*, 17(12): 1624–1637.

Caliński, R. and Harabasz, J. (1974). A dendrite method for cluster analysis. *Communications in Statistics*, 3: 1–27.

Camastra, F. (2003). Data dimensionality estimation methods: A survey. *Pattern Recognition*, 360(1212): 2945–2954.

Cannon, R., Dave, J., and Bezdek, J. (1986). Efficient implementation of the fuzzy *c*-means clustering algorithms. *IEEE Transactions on Pattern Analysis and Machine Intelligence*, PMI-8: 248–255.

Carpenter, G. (2003) Default ARTMAP. In *Proceedings of International Joint Conference on Neural Networks 2003*, Vol. 2, pp. 1396–1401.

Carpenter, G. and Grossberg, S. (1987a). A massively parallel architecture for a self-organizing neural pattern recognition machine. *Computer Vision, Graphics, and Image Processing*, 37: 54–115.

Carpenter, G. and Grossberg, S. (1987b). ART2: Self-organization of stable category recognition codes for analog input patterns, *Applied Optics*, 26(23): 4919–4930.

Carpenter, G. and Grossberg, S. (1988). The ART of adaptive pattern recognition by a self-organizing neural network. *IEEE Computer*, 21(3): 77–88.

Carpenter, G. and Grossberg, S. (1990). ART3: Hierarchical search using chemical transmitters in self-organizing pattern recognition Architectures. *Neural Networks*, 3(23): 129–152.

Carpenter, G., Grossberg, S., Markuzon, N., Reynolds, J., and Rosen, D. (1992). Fuzzy ARTMAP: A neural network architecture for incremental supervised learning of analog multidimensional maps. *IEEE Transactions on Neural Networks*, 3(5): 698–713.

Carpenter, G., Grossberg, S., and Reynolds, J. (1991a). ARTMAP: Supervised real-time learning and classification of nonstationary data by a self-organizing neural network. *Neural Networks*, 4(5): 169–181.

Carpenter, G., Grossberg, S., and Rosen, D. (1991b). Fuzzy ART: Fast stable learning and categorization of analog patterns by an adaptive resonance System. *Neural Networks*, 4: 759–771.

Carpenter, G. and Markuzon, N. (1998). ARTMAP-IC and medical diagnosis: Instance counting and inconsistent cases. *Neural Networks*, 11: 323–336.

Carpenter, G., Milenova, B., and Noeske, B. (1998). Distributed ARTMAP: A neural network for fast distributed supervised learning. *Neural Networks*, 11: 793–813.

Castro, R., Coates, M., and Nowak, R. (2004). Likelihood based hierarchical clustering. *IEEE Transactions on Signal Processing*, 52(8): 2308–2321.

Caudell, T., Smith, S., Escobedo, R., and anderson, M. (1994). NIRS: Large scale ART-1 neural architectures for engineering design retrieval. *Neural Networks*, 7(9): 1339–1350.

Caudell, T., Smith, S., Johnson, G., and Wunsch, D. II (1991). An application of neural networks to group technology. In *Proceedings of SPIE*, vol. 1469, *Applications of Neural Networks II*, pp. 612–621.

Celebi, M., Aslandogan, Y., and Bergstresser, P. (2005). Mining biomedical images with density-based clustering. In *Proceedings of the International Conference on Information Technology: Coding and Computing*, Vol. 1, pp. 163–168.

Celeux, G. and Govaert, G. (1991). Clustering criteria for discrete data and latent class models. *Journal of Classification*, 8: 157–176.

Celeux, G. and Govaert, G. (1992). A classification EM algorithm for clustering and two stochastic versions. *Computational Statistics and Data Analysis*, 14: 315–332.

Chan, K. and Fu, A. (1999). Efficient time series matching by wavelets. In *Proceedings of the 15th IEEE International Conference on Data Engineering*, pp. 126–133.

Charalampidis, D. (2005). A modified *K*-means algorithm for circular invariant clustering. *IEEE Transactions on Pattern Analysis and Machine Intelligence*, 27(12): 1856–1865.

Cheeseman, P. and Stutz, J. (1996). Bayesian classification (AutoClass): Theory and results. In *Advances in Knowledge Discovery and Data Mining*, U. Fayyad, G. Piatetsky-Shapiro, P. Smyth, and R. Uthurusamy Eds., New York, NY: AAAI Press, pp. 153–180.

Chen, L. and Chang, S. (1994). An adaptive conscientious competitive learning algorithm and its applications. *Pattern Recognition*, 27(12): 1787–1813.

Chen, C. and Ye, C. (2004). Particle swarm optimization algorithm and its application to clustering analysis. In *Proceedings of the 2004 IEEE International Conference on Networking, Sensing & Control*, Vol. 2, pp. 789–794.

Cheng, C., Fu, A., and Zhang, Y. (1999). Entropy-based subspace clustering for mining numerical data. In *Proceedings of ACM SIGKDD International Conference on Knowledge Discovery and Data Mining*, pp. 84–93.

Cheng, T., Goldgof, D., and Hall, L. (1998). Fast fuzzy clustering. *Fuzzy Sets Systems*, 93: 49–56.

Cherkassky, V. and Mulier, F. (1998). *Learning from data: concepts, theory, and methods*. New York, NY: John Wiley & Sons.

Cherng, J. and Lo, M. (2001). A hypergraph based clustering algorithm for spatial data sets. In *Proceedings of IEEE International Conference on Data Mining—ICDM 2001*, pp. 83–90.

Cheung, Y. (2005). On rival penalization controlled competitive learning of clustering with automatic cluster number selection. *IEEE Transactions on Knowledge and Data Engineering*, 17(11): 1583–1588.

Chiang, J. and Gader, P. (1997). Recognition of handprinted numerals in VISA card application forms. *Machine Vision and Applications*, 10(3): 144–149.

Chiang, J. and Hao, P. (2003). A new kernel-based fuzzy clustering approach: support vector clustering with cell growing. *IEEE Transactions on Fuzzy Systems*, 11(4): 518–527.

Chiang, J., Yue, S., and Yin, Z. (2004). A new fuzzy cover approach to clustering. *IEEE Transactions on Fuzzy Systems*, 12(2): 199–208.

Chinrungrueng, C. and Séquin, C. (1995). Optimal adaptive *K*-Means algorithm with dynamic adjustment of learning rate. *IEEE Transactions on Neural Networks*, 6(1): 157–169.

Chiu, T., Fang, D., Chen, J., Wang, Y., and Jeris, C. (2001). A robust and scalable clustering algorithm for mixed type attributes in large database environment. In *Proceedings of 7th ACM SIGKDD International Conference on Knowledge Discovery and Data Mining*, pp. 263–268.

Cho, R., Campbell, M., Winzeler, E., Steinmetz, L., Conway, A., Wodicka, L., Wolfsberg, T., Gabrielian, A., Landsman, D., Lockhart, D., and Davis, R. (1998). A genome-wide transcriptional analysis of the mitotic cell cycle. *Molecular Cell*, 2: 65–73.

Choy, C. and Siu, W. (1997). Distortion sensitive competitive learning for vector quantizer. In *Proceedings of IEEE International Conference on Acoustics, Speech, and Signal Processing—ICASSP97*, Vol. 4, pp. 3405–3408.

Chu, S., DeRisi, J., Eisen, M., Mulholland, J., Botstein, D., Brown, P., and Herskowitz, I. (1998). The transcriptional program of sporulation in budding yeast. *Science*, 282: 699–705.

Chu, S. and Roddick, J. (2000). A clustering algorithm using the Tabu search approach with simulated annealing. *Data Mining II—Proceedings of Second International Conference on Data Mining Methods and Databases*, N. Ebecken and C. Brebbia, Eds., Cambridge, UK: WIT Press, pp. 515–523.

Chung, F. and Lee, T. (1994). Fuzzy competitive learning. *Neural Networks*, 3: 539–551.

Clerc, M. and Kennedy, J. (2002). The particle swarm—explosion, stability, and convergence in a multidimensional complex space. *IEEE Transactions on Evolutionary Computation*, 6(1): 58–73.

Coifman, R. and Lafon, S. (2006). Diffusion maps. *Applied and Computational Harmonic Analysis*, 21: 5–30.

Comon, P. (1994). Independent component analysis—A new concept? *Signal Processing*, 36(3): 287–314.

Componation, P. and Byrd, J., Jr., (2000). Utilizing cluster analysis to structure concurrent engineering teams. *IEEE Transactions on Engineering Management*, 47(2): 269–280.

Consortium, I.H.G.S. (2001). Initial sequencing and analysis of the human genome. *Nature*, 409: 860–921.

Corchado, J. and Fyfe, C. (2000). A comparison of kernel methods for instantiating case based reasoning systems. *Computing and Information Systems*, 7: 29–42.

Cosman, P., Oehler, K., Riskin, E., and Gray, R. (1993). Using vector quantization for image processing. *Proceedings of the IEEE*, 81(9): 1326–1341.

Cottrell, M., Hammer, B., Hasenfu, A., and Villmann, T. (2006). Batch and median neural gas. *Neural Networks*, 19(6–7): 762–771.

Cover, T. (1965). Geometrical and statistical properties of systems of linear inequalities with applications in pattern recognition. *IEEE Transactions on Electronic Computers*, EC-14: 326–334.

Cowgill, M., Harvey, R., and Watson, L. (1999). A genetic algorithm approach to cluster analysis. *Computers and Mathematics with Applications*, 37: 99–108.

Cressie, N. (1993). *Statistics for spatial data*. New York, NY: John Wiley & Sons.

Cummings, C. and Relman, D. (2000). Using DNA microarray to study host-microbe interactions. *Genomics*, 6(5): 513–525.

Cunningham, K. and Ogilvie, J. (1972). Evaluation of hierarchical grouping techniques: A preliminary study. *Computer Journal*, 15: 209–213.

Dahlhaus, E. (2000). Parallel algorithms for hierarchical clustering and applications to split decomposition and parity graph recognition. *Journal of Algorithms*, 36(2): 205–240.

Darrell, T. and Pentland, A. (1995). Cooperative robust estimation using layers of support. *IEEE Transactions on Pattern Analysis and Machine Intelligence*, 17(5): 474–487.

Dasey, T. and Micheli-Tzanakou, E. (2000). Detection of multiple sclerosis with visual evoked potentials—an unsupervised computational intelligence system. *IEEE Transactions on Information Technology in Biomedicine*, 4(3): 216–224.

Davé, R. (1990). Fuzzy shell clustering and applications to circle detection in digital images. *International Journal of general systems*, 16: 343–355.

Davé, R. (1991). Characterization and detection of noise in clustering. *Pattern Recognition letter*, 12(11): 657–664.

Davé, R. and Bhaswan, K. (1992). Adaptive fuzzy *c*-Shells clustering and detection of ellipses. *IEEE Transactions on Neural Networks*, 3(5): 643–662.

Davé, R. and Krishnapuram, R. (1997). Robust clustering methods: A unified view. *IEEE Transactions on Fuzzy Systems*, 5(2): 270–293.

Davies, D. and Bouldin, D. (1979). A cluster separation measure. *IEEE Transactions on Pattern Analysis and Machine Intelligence*, 1: 224–227.

Delattre, M. and Hansen, P. (1980). Bicriterion cluster analysis. *IEEE Transactions on Pattern Analysis and Machine Intelligence*, 2: 277–291.

Delcher, A., Kasif, S., Fleischman, R., Peterson, J. White, O., and Salzberg, S. (1999). Alignment of whole genomes. *Nucleic Acids Research*, 27(11): 2369–2376.

Delgado, M., Skármeta, A., and Barberá, H. (1997). A Tabu search approach to the fuzzy clustering problem. In *Proceedings of the 6th IEEE International Conference on Fuzzy Systems*, Vol. 1, pp. 125–130.

Dembélé, D. and Kastner, P. (2003). Fuzzy *c*-means method for clustering microarray data. *Bioinformatics*, 19(8): 973–980.

Dempster, A., Laird, N., and Rubin, D. (1977). Maximum-likelihood from incomplete data via the EM algorithm. *Journal of the Royal Statistical Society, Series B*, 39(1): 1–38.

DeSarbo, W. and Cron, W. (1988). A maximum likelihood methodology for clusterwise linear regression. *Journal of Classification*, 5(1): 249–282.

Desieno, D. (1988). Adding a conscience to competitive learning. In *Proceedings of IEEE International Conference on Neural Networks*, Vol. 1, pp. 117–124.

Dhillon, I. and Modha, D. (2001). Concept decompositions for large sparse text data using clustering. *Machine Learning*, 42(1): 143–175.

Dice, L. (1945). Measures of the amount of ecologic association between species. *Journal of Ecology*, 26: 297–302.

Dixon, J. (1979). Pattern recognition with partly missing data. *IEEE Transactions on Systems, Man and Cybernetics, SMC* 9: 617–621.

Donoho, D. and Grimes, C. (2003). Hessian eigenmaps: New locally linear embedding techniques for high-dimensional data. Technical Report, Stanford University.

Dopazo, J., Zanders, E., Dragoni, I., Amphlett, G., and Falciani, F. (2001). Methods and approaches in the analysis of gene expression data. *Journal of Immunological Methods*, 250: 93–112.

Dubes, R. (1993). Cluster analysis and related issue. In *Handbook of Pattern Recognition and Computer Vision*, C. Chen, L. Pau, and P. Wang, Eds., River Edge, NJ: World Science Publishing Company, pp. 3–32.

Duda, R., Hart, P., and Stork, D. (2001). *Pattern classification, 2nd edition*. New York, NY: John Wiley & Sons.

Dunn, J. (1974a). A fuzzy relative of the ISODATA process and its use in detecting compact well separated clusters. *Journal of Cybernetics*, 3(3): 32–57.

Dunn, J. (1974b). Well separated clusters and optimal fuzzy partitions. *Journal of Cybernetics*, 4: 95–104.

Duran, B. and Odell, P. (1974). *Cluster analysis: A survey*. New York, NY: Springer-Verlag.

Durbin, R., Eddy, S., Krogh, A., and Mitchison, G. (1998). *Biological sequence analysis: Probabilistic models of proteins and nucleic acids*. Cambridge, UK: Cambridge University Press.

D'Urso, P. (2005). Fuzzy clustering for data time arrays with inlier and outlier time trajectories. *IEEE Transactions on Fuzzy Systems*, 13(5): 583–604.

Dy, J. and Brodley, C. (2000). Feature subset selection and order identification for unsupervised learning. In *Proceedings of the 17th International Conference on Machine Learning*, pp. 247–254.

Eisen, M. and Brown, P. (1999). DNA arrays for analysis of gene expression. *Methods Enzymol*, 303: 179–205.

Eisen, M., Spellman, P., Brown, P., and Botstein, D. (1998). Cluster analysis and display of genome-wide expression patterns. *Proceedings of National Academic Science USA 95*, 95(25): 14863–14868.

El-Sonbaty, Y. and Ismail, M. (1998). Fuzzy clustering for symbolic data. *IEEE Transactions on Fuzzy Systems*, 6(2): 195–204.

Eltoft, T. and deFigueiredo, R. (1998). A new neural network for cluster-detection-and-labeling. *IEEE Transactions on Neural Networks*, 9(5): 1021–1035.

Enright, A. and Ouzounis, C. (2000). GeneRAGE: A robust algorithm for sequence clustering and domain detection. *Bioinformatics*, 16: 451–457.

Eschrich, S., Ke, J., Hall, L., and Goldgof, D. (2003). Fast accurate fuzzy clustering through data reduction. *IEEE Transactions on Fuzzy Systems*, 11(2): 262–270.

Ester, M., Kriegel, H., Sander, J., and Xu, X. (1996). A density-based algorithm for discovering clusters in large spatial databases with noise. In *Proceedings of 2nd International Conference on Knowledge Discovery and Data Mining (KDD96)*, New York, NY: AAAI Press.

Estivill-Castro, V. and Lee, I. (1999). AMOEBA: Hierarchical clustering based on spatial proximity using Delaunay diagram. In *9th International Symposium on Spatial Data Handling—SDH99*, pp. 7a.26–7a.41.

Estivill-Castro, V. and Yang, J. (2000). A fast and robust general purpose clustering algorithm. In *Proceedings 6th Pacific Rim International Conference on Artificial Intelligence PRICAI 2000*, R. Mizoguchi and J. Slaney, Eds., *Lecture Notes in Artificial Intelligence 1886*, New York, NY: Springer-Verlag, pp. 208–218.

Everitt, B. (1980). *Cluster analysis, 2nd edition*. London: Social Science Research Council.

Everitt, B., Landau, S., and Leese, M. (2001). *Cluster analysis, 4th edition*. London: Arnold.

Faloutsos, C. and Lin, K. (1995). Fastmap: A fast algorithm for indexing, datamining and visualization of traditional and multimedia databases. In *Proceedings of 1995 ACM SIGMOD International Conference on Management of Data*, pp. 163–174.

Farnstrom, F., Lewis, J., and Elkan, C. (2000). Scalability for clustering algorithms revisited. *SIGKDD Explorations*, 2(1): 51–57.

Fasulo, D. (1999). An analysis of recent work on clustering algorithms. Technical Report #01-03-02, Department of Computer Science & Engineering, University of Washington.

Figueiredo, M. and Jain, A. (2002). Unsupervised learning of finite mixture models. *IEEE Transactions on Pattern Analysis and Machine Intelligence*, 24(3): 381–396.

Fodor, I. (2002). A survey of dimension reduction techniques. Technical Report, UCRL-ID-148492, Center for Applied Scientific Computing, Lawrence Livermore National Laboratory.

Fogel, D. (1994). An introduction to simulated evolutionary optimization. *IEEE Transactions on Neural Networks*, 5(1): 3–14.

Fogel, D. (2005). *Evolutionary computation: Toward a new philosophy of machine intelligence, 3rd edition*. Piscataway, NJ: Wiley-IEEE Press.

Fogel, L., Owens, A., and Walsh, M. (1966). *Artificial intelligence through simulated evolution*. New York, NY: Wiley Publishing.

Forgy, E. (1965). Cluster analysis of multivariate data: efficiency vs. interpretability of classifications. *Biometrics*, 21: 768–780.

Fowler, J., Jr., Adkins, K., Bibyk, S., and Ahalt, S. (1995). Real-time video compression using differential vector quantization. *IEEE Transactions on Circuits and Systems for Video Technology*, 5(1): 14–24.

Fowlkes, E. and Mallows, C. (1983). A method for comparing two hierarchical clustering. *Journal of the American Statistical Association*, 78: 553–569.

Fraley C. and Raftery A. (1998). How many clusters? Which clustering method?—Answers via model-based cluster analysis. *The Computer Journal*, 41: 578–588.

Fraley, C. and Raftery, A. (1999). MCLUST: Software for model-based cluster analysis. *Journal of Classification*, 16: 297–306.

Fraley, C. and Raftery, A. (2002). Model-Based clustering, discriminant analysis, and density estimation. *Journal of the American Statistical Association*, 97: 611–631.

Frank, T., Kraiss, K., and Kuhlen, T. (1998). Comparative analysis of fuzzy ART and ART-2A network clustering performance. *IEEE Transactions on Neural Networks*, 9(3): 544–559.

Frey, B. and Jojic, N. (2003). Transformation-invariant clustering using the EM algorithm. *IEEE Transactions on Pattern Analysis and Machine Intelligence*, 25(1): 1–17.

Friedman, J. (1987). Exploratory projection pursuit. *Journal of the American Statistical Association*, 82: 249–266.

Friedman, J. (1995). An overview of prediction learning and function approximation. In V. Cherkassky, J. Friedman, and H. Wechsler, Eds., *From Statistics to Neural Networks: Theory and Pattern Recognition Applications*. New York, NY: Springer-Verlag.

Frigui, H. and Krishnapuram, R. (1999). A robust competitive clustering algorithm with applications in computer vision. *IEEE Transactions on Pattern Analysis and Machine Intelligence*, 21(5): 450–465.

Fritzke, B. (1994). Growing cells structures—a self-organizing network for unsupervised and supervised learning. *Neural Networks*, 7(9): 1441–1460.

Fritzke, B. (1995). A growing neural gas network learns topologies. In G. Tesauro, D. Touretzky, and T. Leen, Eds., *Advances in Neural Information Processing Systems* 7, Cambridge, MA: MIT Press, pp. 625–632.

Fritzke, B. (1997). Some competitive learning methods. Draft Document, http://www.neuroinformatik.ruhr-uni-bochum.de/ini/VDM/research/gsn/JavaPaper.

Fukushima, K. (1975). Cognitron: A self-organizing multi-layered neural network. *Biological Cybernetics*, 20: 121–136.

Fukuyama, Y. and Sugeno, M. (1993). A new metod of choosing the number of clusters for the fuzzy *c*-means method. In *Proceedings of the 5th Fuzzy System Symposium*, pp. 247–250.

Gabrys, B. and Bargiela, A. (2000). General fuzzy min-max neural network for clustering and classification. *IEEE Transactions on Neural Networks*, 11(3): 769–783.

Gaffney, S. and Smyth, P. (1999). Trajectory clustering using mixtures of regression models. In *Proceedings of the 5th ACM SIGKDD International Conference on Knowledge Discovery and Data Mining*, S. Chaudhuri and D. Madigan Eds., New York, NY: ACM Press, pp. 63–72.

Gaffney, S. and Smyth, P. (2003). Curve clustering with random effects regression mixtures. In *Proceedings of the 9th International Workshop on Artificial Intelligence and Statistics*, C. Bishop and B. Frey, Eds.

Gaffney, S. and Smyth, P. (2004). Joint probabilistic curve clustering and alignment. In *Advances in Neural Information Processing Systems*, Cambridge, MA: MIT Press.

Ganti, V., Ramakrishnan, R., Gehrke, J., Powell, A., and French, J. (1999). Clustering large datasets in arbitrary metric spaces. In *Proceedings of the 15th International Conference on Data Engineering*, pp. 502–511.

Garber, M., Troyanskaya, O., Schluens, K., Petersen, S., Thaesler, Z., Pacyna-Gengelbach, M., van de Rijn, M., Rosen, G., Perou, C., Whyte, R., Altman, R., Brown, P., Botstein, D., and Petersen, I. (2001). Diversity of gene expression in adenocarcinoma of the lung. *Proceedings of National Academy of Sciences*, 98(24): 13784–13789.

Gasch, A. and Eisen, M. (2002). Exploring the conditional coregulation of yeast gene expression through fuzzy *K*-means clustering. *Genome Biology*, 3(11): 1–22.

Gath, I. and Geva, A. (1989). Unsupervised optimal fuzzy clustering. *IEEE Transactions on Pattern Analysis and Machine Intelligence*, 11(7): 773–781.

Gersho, A. and Gray, R. (1992) *Vector quantization and signal compression*. Norwell, MA: Kluwer Academic Publishers.

Geva, A. (1999). Hierarchical unsupervised fuzzy clustering. *IEEE Transactions on Fuzzy Systems*, 7(6): 723–733.

Ghosh, D. and Chinnaiyan, A. (2002). Mixture modeling of gene expression data from microarray experiments. *Bioinformatics*, 18(2): 275–286.

Ghozeil, A. and Fogel, D. (1996). Discovering patterns in spatial data using evolutionary programming. In *Proceedings of 1st Annual Conference on Genetic Programming*, Cambridge, MA: MIT Press, pp. 512–520.

Girolami, M. (2002). Mercer kernel based clustering in feature space. *IEEE Transactions on Neural Networks*, 13: 780–784.

Glover, F. (1989). Tabu search, Part I. *ORSA Journal of Computing*, 1(3): 190–206.

Glover, F. and Laguna, M. (1997). *Tabu search*. Norwell, MA: Kluwer Academic Publishers.

Golub, T., Slonim, D., Tamayo, P., Huard, C., Gaasenbeek, M., Mesirov, J., Coller, H., Loh, M., Downing, J., Caligiuri, M., Bloomfield, C., and Lander, E. (1999). Molecular classification of cancer: Class discovery and class prediction by gene expression monitoring. *Science*, 286: 531–537.

Gonzalez, A., Graña, M., and D'Anjou, A. (1995). An analysis of the GLVQ algorithm. *IEEE Transactions on Neural Networks*, 6(4): 1012–1016.

Goodall, D. (1966). A new similarity index based on probability. *Biometrics*, 22: 882–907.

Goodhill, G. and Barrow, H. (1994). The role of weight normalization in competitive learning. *Neural Computation*, 6: 255–269.

Gorban, A., Pitenko, A., Zinovyev, A., and Wunsch, D. II (2001). Visualization of any data using elastic map method. *Smart Engineering System Design*, vol. 11, pp. 363–368.

Gorban, A., Zinovyev, A., and Wunsch, D. II (2003). Application of the method of elastic maps in analysis of genetic texts. In *Proceedings of International Joint Conference on Neural Networks*, vol. 3, pp. 1826–1831.

Gordon, A. (1998). Cluster validation. In *Data Science, Classification, and Related Methods*, C. Hayashi, N. Ohsumi, K. Yajima, Y. Tanaka, H. Bock, and Y. Bada, Eds., New York, NY: Springer-Verlag, pp. 22–39.

Gordon, A. (1999). *Classification, 2nd edition*. London, UK: Chapman and Hall/CRC Press.

Goutte, C., Toft, P., and Rostrup, E. (1999). On clustering fMRI time series. *Neuroimage*, 9(3): 298–310.

Gower, J. (1967). A comparison of some methods of cluster analysis. *Biometrics*, 23(4): 623–628.

Gower, J. (1971). A general coefficient of similarity and some of its properties. *Biometrics*, 27: 857–872.

Gower, J. and Legendre, P. (1986). Metric and Euclidean properties of dissimilarity coefficients. *Journal of Classification*, 3: 5–48.

Gray, R. (1990). *Entropy and information theory*. New York, NY: Springer-Verlag.

Griffiths, A., Miller, J., Suzuki, D., Lewontin, R., and Gelbart, W. (2000). *An introduction to genetic analysis, 7th ed.*, New York, NY: W. H. Freeman and Company.

Grira, N., Crucianu, M., and Boujemaa, N. (2004). Unsupervised and semi-supervised clustering: A brief survey. In *A Review of Machine Learning Techniques for Processing Multimedia Content*, report of the MUSCLE European Network of Excellence (FP6).

Gröll, L. and Jäkel, J. (2005). A new convergence proof of fuzzy *c*-means. *IEEE Transactions on Fuzzy Systems*, 13(5): 717–720.

Grossberg, S. (1976a). Adaptive pattern classification and universal recoding: I. Parallel development and coding of neural feature detectors. *Biological Cybernetics*, 23: 121–134.

Grossberg, S. (1976b). Adaptive pattern recognition and universal encoding II: Feedback, expectation, olfaction, and illusions. *Biological Cybernetics*, 23: 187–202.

Grossberg, S. (1978). A theory of human memory: Self-organization and performance of sensory-motor codes, maps, and plans. In R. Rosen and F. Snell, Eds., *Progress in Theoretical Biology*, vol. 5, New York, NY: Academic Press.

Grossberg, S. (1987). Competitive learning: From interactive activation to adaptive resonance. *Cognitive Science*, 11: 23–63.

Grünwald, P., Kontkanen, P., Myllymäki, P., Silander, T., and Tirri, H. (1998). Minimum encoding approaches for predictive modeling. In *Proceedings of 14th International Conference on Uncertainty in AI—UAI'98*, G. Cooper and S. Moral, Eds., pp. 183–192.

Guan, X. and Du, L. (1998). Domain identification by clustering sequence alignments. *Bioinformatics*, 14: 783–788.

Guha, S., Meyerson, A., Mishra, N., Motwani, R., and O'Callaghan, L. (2003). Clustering data streams: Theory and practice. *IEEE Transactions on Knowledge and Data Engineering*, 15(3): 515–528.

Guha, S., Rastogi, R., and Shim, K. (1998). CURE: An efficient clustering algorithm for large databases. In *Proceedings of ACM SIGMOD International Conference on Management of Data*, pp. 73–84.

Guha, S., Rastogi, R., and Shim, K. (2000). ROCK: A robust clustering algorithm for categorical attributes. *Information Systems*, 25(5): 345–366.

Guo, P., Chen, C., and Lyu, M. (2002). Cluster number selection for a small set of samples using the Bayesian Ying-Yang model. *IEEE Transactions on Neural Networks*, 13(3): 757–763.

Gupta, S., Rao, K., and Bhatnagar, V. (1999). *K*-means clustering algorithm for categorical attributes. In *Proceedings of 1st International Conference on Data Warehousing and Knowledge Discovery—DaWaK '99*, pp. 203–208.

Guralnik, V. and Karypis, G. (2001). A scalable algorithm for clustering sequential data. In *Proceedings of the 1st IEEE International Conference on Data Mining—ICDM 2001*, pp. 179–186.

Gusfield, D. (1997). *Algorithms on strings, trees, and sequences: Computer science and computational biology*. Cambridge, UK: Cambridge University Press.

Halkidi, M., Batistakis, Y., and Vazirgiannis, M. (2002). Cluster validity methods: Part I & II, *SIGMOD Record*, 31(2&3).

Hall, L., Özyurt, I., and Bezdek, J. (1999). Clustering with a genetically optimized approach. *IEEE Transactions on Evolutionary Computation*, 3(2): 103–112.

Ham, J., Lee, D., Mika, S., and Schölkopf, B. (2003). A kernel view of the dimensionality reduction of manifolds. Technical Report TR-110, Mac Planck Institute for Biological Cybernetics.

Hammah, R. and Curran, J. (2000). Validity measures for the fuzzy cluster analysis of orientations. *IEEE Transactions on Pattern Analysis and Machine Intelligence*, 22(12): 1467–1472.

Hammer, B., Micheli, A., Sperduti, A., and Strickert, M. (2004). Recursive self-organizing network models. *Neural Networks*, 17(8–9): 1061–1085.

Hammouda, K. and Kamel, M. (2004). Efficient phrase-based document indexing for web document clustering. *IEEE Transactions on Knowledge and Data Engineering*, 16(10): 1279–1296.

Han, J. and Kim, H. (2006). Optimization of requantization codebook for vector quantization. *IEEE Transactions on Image Processing*, 15(5): 1057–1061.

Hansen, P. and Jaumard, B. (1997). Cluster analysis and mathematical programming. *Mathematical Programming*, 79: 191–215.

Hansen, P. and Mladenović, N. (2001). J-Means: A new local search heuristic for minimum sum of squares clustering. *Pattern Recognition*, 34: 405–413.

Harary, F. (1969). *Graph theory*. Reading, MA: Addison-Wesley.

Hartigan, J. (1975). *Clustering algorithms*. New York, NY: John Wiley & Sons.

Hartuv, E. and Shamir, R. (2000). A clustering algorithm based on graph connectivity. *Information Processing Letters*, 76: 175–181.

Hastie, T. and Stuetzle, W. (1989). Principal curves. *Journal of the American Statistical Association*, 84(406): 502–516.

Hathaway, R. and Bezdek, J. (2001). Fuzzy *c*-Means clustering of incomplete data. *IEEE Transactions on System, Man, and Cybernetics*, 31(5): 735–744.

Hathaway, R., Bezdek, J., and Hu, Y. (2000). Generalized fuzzy *c*-Means clustering strategies using L_p norm distances. *IEEE Transactions on Fuzzy Systems*, 8(5): 576–582.

Hay, B., Wets, G., and Vanhoof, K. (2001). Clustering navigation patterns on a website using a sequence alignment method. In *Intelligent Techniques for Web Personalization: IJCAI 2001 17th International Joint Conference on Artificial Intelligence*, s.1., pp. 1–6.

Haykin, S. (1999). *Neural networks: A comprehensive foundation, 2nd edition*. Upper Saddle River, NJ: Prentice Hall.

He, Q. (1999). A review of clustering algorithms as applied to IR. Technical Report UIUCLIS-1999/6+IRG, University of Illinois at Urbana-Champaign.

Healy, M., Caudell, T., and Smith, S. (1993). A neural architecture for pattern sequence verification through inferencing. *IEEE Transactions on Neural Networks*, 4(1): 9–20.

Hebb, D. (1949). *The organization of behavior*. New York, NY: Wiley.

Hecht-Nielsen, R. (1987). Counterpropagation networks. *Applied Optics*, 26(23): 4979–4984.

Heger, A. and Holm, L. (2000). Towards a covering set of protein family profiles. *Progress in Biophysics & Molecular Biology*, 73: 321–337.

Heinke, D. and Hamker, F. (1998). Comparing neural networks: A benchmark on growing neural gas, growing cell structures, and fuzzy ARTMAP. *IEEE Transactions on Neural Networks*, 9(6): 1279–1291.

Herwig, R., Poustka, A., Müller, C., Bull, C., Lehrach, H., and O'Brien, J. (1999). Large-scale clustering of cDNA-fingerprinting data. *Genome Research*, 9(11): 1093–1105.

Hinneburg, A. and Keim, D. (1998). An efficient approach to clustering in large multimedia databases with noise. In *Proceedings of 4th International Conference on Knowledge Discovery and Data Mining—KDD98*, New York, NY: AAAI Press, pp. 58–65.

Hinneburg, A. and Keim, D. (1999). Optimal grid-clustering: Towards breaking the curse of dimensionality in high-dimensional clustering. In *Proceedings of the 25th VLDB Conference*, pp. 506–517.

Hoeppner, F. (1997). Fuzzy shell clustering algorithms in image processing: Fuzzy *c*-rectangular and 2-rectangular shells. *IEEE Transactions on Fuzzy Systems*, 5(4): 599–613.

Hoey, J. (2002). Clustering contextual facial display sequences. In *Proceedings of the 5th IEEE International Conference on Automatic Face and Gesture Recognition—FGR'02*, pp. 354–359.

Hofmann, T. and Buhmann, J. (1997). Pairwise data clustering by deterministic annealing. *IEEE Transactions on Pattern Analysis and Machine Intelligence*, 19(1): 1–14.

Hogg, R. and Tanis, E. (2005). *Probability And Statistical Inference*, 7 edition. Upper Saddle River, NJ: Prentice Hall.

Holland, J. (1975). *Adaption in natural and artificial systems*. Ann Arbor, MI: University of Michigan Press.

Holm, L. and Sander, C. (1998). Removing near-neighbor redundancy from large protein sequence collections. *Bioinformatics*, 14: 423–429.

Honda, K. and Ichihashi, H. (2005). Regularized linear fuzzy clustering and probabilistic PCA mixture models. *IEEE Transactions on Fuzzy Systems*, 13(4): 508–516.

Honkela, T., Kaski, S., Lagus, K., and Kohonen, T. (1997). WEBSOM–self-organizing maps of document collections. In *Proceedings of Workshop on Self-Organizing Maps—WSOM'97, Helsinki University of Technology, Neural Networks Research Center, Espoo, Finland*, pp. 310–315.

Höppner, F. and Klawonn, F. (2003). A contribution to convergence theory of fuzzy *c*-means and derivatives. *IEEE Transactions on Fuzzy Systems*, 11(5): 682–694.

Höppner, F., Klawonn, F., and Kruse, R. (1999). *Fuzzy cluster analysis: Methods for classification, data analysis and image recognition*. New York, NY: Wiley.

Horng, Y., Chen, S., Chang, Y., and Lee, C. (2005) A new method for fuzzy information retrieval based on fuzzy hierarchical clustering and fuzzy inference techniques. *IEEE Transactions on Fuzzy Systems*, 13(2): 216–228.

Huang, J., Georgiopoulos, M., and Heileman, G. (1995). Fuzzy ART properties. *Neural Networks*, 8(2): 203–213.

Huang, Z. (1998). Extensions to the *K*-Means algorithm for clustering large data sets with categorical values. *Data Mining and Knowledge Discovery* 2, pp. 283–304.

Huang, Z., Ng., M., Rong, H., and Li, Z. (2005). Automated variable weighting in *k*-means type clustering. *IEEE Transactions on Pattern Analysis and Machine Intelligence*, 27(5): 657–668.

Huber, P. (1985). Projection pursuit. The *Annals of Statistics*, 13(2): 435–475.

Hubert, L. (1974). Approximate evaluation techniques for the single-link and complete-link hierarchical clustering procedures. *Journal of the American Statistical Association*, 69: 698–704.

Hubert, L. and Arabie, P. (1985). Comparing partitions. *Journal of Classification*, 2: 193–218.

Hughey, R., Karplus, K., and Krogh, A. (2003). SAM: Sequence alignment and modeling software system. Technical Report UCSC-CRL-99-11, University of California, Santa Cruz, CA.

Hughey, R. and Krogh, A. (1996). Hidden Markov models for sequence analysis: Extension and analysis of the basic method. *CABIOS*, 12(2): 95–107.

Hull, J. (1994). A database for handwritten text recognition research. *IEEE Transactions on Pattern Analysis and Machine Intelligence*, 16(5): 550–554.

Hung, M. and Yang, D. (2001). An efficient fuzzy *c*-Means clustering algorithm. In *Proceedings of IEEE International Conference on Data Mining*, pp. 225–232.

Hunt, L. and Jorgensen, J. (1999). Mixture model clustering using the MULTIMIX program. *Australia and New Zealand J Statistics*, 41: 153–171.

Hwang, J., Vlontzos, J., and Kung, S. (1989). A systolic neural network architecture for hidden Markov models. *IEEE Transactions on Acoustics, Speech, and Signal Processing*, 37(12): 1967–1979.

Hyvärinen, A. (1999). Survey of independent component analysis. *Neural Computing Surveys*, 2: 94–128.

Iyer, V., Eisen, M., Ross, D., Schuler, G., Moore, T., Lee, J., Trent, J., Staudt, L., Hudson, J., Boguski, M., Lashkari, D., Shalon, D., Botstein, D., and Brown, P. (1999). The transcriptional program in the response of human fibroblasts to serum. *Science*, 283: 83–87.

Jaccard, P. (1908). Nouvelles recherches sur la distribution florale. *Bulletin de la Société Vaudoise de Sciences Naturelles*, 44: 223–370.

Jain, A. and Dubes, R. (1988). *Algorithms for clustering data*. Englewood Cliffs, NJ: Prentice Hall.

Jain, A., Duin, R., and Mao, J. (2000). Statistical pattern recognition: A review. *IEEE Transactions on Pattern Analysis and Machine Intelligence*, 22(1): 4–37.

Jain, A., Murty, M., and Flynn, P. (1999). Data clustering: A review. *ACM Computing Surveys*, 31(3): 264–323.

Jenssen, R. and Eltoft, T. (2006). An information theoretic perspective to kernel *K*-means. In *Proceedings of IEEE International Workshop on Machine Learning for Signal Processing—MLSP2006*, pp. 161–166.

Jenssen, R., Eltoft, T., Erdogmus, D., and Principe, J. (2006a). Some equivalences between kernel methods and information theoretic methods. *Journal of VLSI Signal Processing*, 45: 49–65.

Jenssen, R., Eltoft, T., Girolami, M., and Erdogmus, D. (2006b). Kernel maximum entropy data transformation and an enhanced spectral clustering algorithm. In *Advances in Neural Information Processing Systems* 18.

Jenssen, R., Erdogmus, D., Hild K., Principe, J., and Eltoft, T. (2007). Information cut for clustering using a gradient descent approach. *Pattern Recognition*, 40: 796–806.

Jenssen, R., Erdogmus, D., Principe, J., and Eltoft, T. (2005). The Laplacian PDF distance: A cost function for clustering in a kernel feature space. In *Advances in Neural Information Processing Systems 17*, pp. 625–632.

Jenssen, R., Principe, J., and Eltoft, T. (2003). Information cut and information forces for clustering. In *Proceedings of IEEE International Workshop on Neural Networks for Signal Processing—NNSP2003*, pp. 459–468.

Jiang, D., Tang, C., and Zhang, A. (2004). Cluster analysis for gene expression data: A survey. *IEEE Transactions on Knowledge and Data Engineering*, 16(11): 1370–1386.

Jin, H., Wong, M., and Leung, K. (2005). Scalable model-based clustering for large databases based on data summarization. *IEEE Transactions on Pattern Analysis and Machine Intelligence*, 27(11): 1710–1719.

Johnson, E., Mehrotra, A., and Nemhauser, G. (1993). Min-cut clustering. *Mathematical Programming*, 62: 133–151.

Johnson, R. and Wichern, D. (2002). *Applied multivariate statistical analysis*. Upper Saddle River, NJ: Prentice Hall.

Johnson, S. (1967). Hierarchical clustering schemes. *Psychometrika*, 32(3): 241–254.

Jolion, J., Meer, P., and Bataouche, S. (1991). Robust clustering with applications in computer vision. *IEEE Transactions on Pattern Analysis and Machine Intelligence*, 13(8): 791–802.

Jollife, I. (1986). *Principal component analysis*. New York, NY: Springer-Verlag.

Jutten, C. and Herault, J. (1991). Blind separation of sources, Part I: An adaptive algorithm based on neuromimetic architecture. *Signal Processing*, 24(1): 1–10.

Kannan, R., Vampala, S., and Vetta, A. (2000). On clustering—good, bad and spectral. In *Foundations of Computer Science 2000*, pp. 367–378.

Kanungo, T., Mount, D., Netanyahu, N., Piatko, C., Silverman, R., and Wu, A. (2000). An efficient *K*-Means clustering algorithm: Analysis and implementation. *IEEE Transactions on Pattern Analysis and Machine Intelligence*, 24(7): 881–892.

Kaplan, N., Sasson, O., Inbar, U., Friedlich, M., Fromer, M., Fleischer, H., Portugaly, E., Linial, N., and Linial, M. (2005). ProtoNet 4.0: A hierarchical classification of one million protein sequences. *Nucleic Acids Research*, 33: D216–D218.

Karayiannis, N. (1997). A methodology for construction fuzzy algorithms for learning vector quantization. *IEEE Transactions on Neural Networks*, 8(3): 505–518.

Karayiannis, N. (2000). Soft learning vector quantization and clustering algorithms based on ordered weighted aggregation operators. *IEEE Transactions on Neural Networks*, 11(5): 1093–1105.

Karayiannis, N. and Bezdek, J. (1997). An integrated approach to fuzzy learning vector quantization and fuzzy *c*-means clustering. *IEEE Transactions on Fuzzy Systems*, 5(4): 622–628.

Karayiannis, N., Bezdek, J., Pal, N., Hathaway, R., and Pai, P. (1996). Repairs to GLVQ: A new family of competitive learning schemes. *IEEE Transactions on Neural Networks*, 7(5): 1062–1071.

Karayiannis, N. and Pai, P. (1996). Fuzzy algorithms for learning vector quantization. *IEEE Transactions on Neural Networks*, 7(5): 1196–1211.

Karayiannis, N. and Pai, P. (1999). Segmentation of magnetic resonance images using fuzzy algorithms for learning vector quantization. *IEEE Transactions on Medical Imaging*, 18(2): 172–180.

Karhunen, J., Oja, E., Wang, L., Vigário, R., and Joutsensalo, J. (1997). A class of neural networks for independent component analysis. *IEEE Transactions on Neural Networks*, 8(3): 486–504.

Karplus, K., Barrett, C., and Hughey, R. (1998). Hidden Markov models for detecting remote protein homologies. *Bioinformatics*, 14(10): 846–856.

Karypis, G., Han, E., and Kumar, V. (1999). Chameleon: Hierarchical clustering using dynamic modeling. *IEEE Computer*, 32(8): 68–75.

Kasif, S. (1999). Datascope: Mining biological sequences. *IEEE Intelligent Systems*, 14(6): 38–43.

Kaski, S., Honkela, T., Lagus, K., and Kohonen, T. (1998). WEBSOM—self-organizing maps of document collections. *Neurocomputing*, 21: 101–117.

Kaufman, L. and Rousseeuw, P. (1990). Finding groups in data: An introduction to cluster analysis. New York, NY: John Wiley & Sons.

Keller, A. (2000). Fuzzy clustering with outliers. In *Proceedings of 19th International Conference of the North American Fuzzy Information Processing Society—NAFIPS 2000*, pp. 143–147.

Kennedy, J., Eberhart, R., and Shi, Y. (2001). *Swarm intelligence*. San Francisco, CA: Morgan Kaufmann Publishers.

Kennedy, J. and Mendes, R. (2002). Population structure and particle swarm performance. In *Proceedings of the 2002 Congress on Evolutionary Computation*, Vol. 2, pp. 1671–1676.

Kent, W. and Zahler, A. (2000). Conservation, regulation, synteny, and introns in a large-scale C. Briggsae—C. elegans genomic alignment. *Genome Research*, 10: 1115–1125.

Keogh, E., Lin, J., and Truppel, W. (2003). Clustering of time series subsequences is meaningless: Implications for previous and future research. In *Proceedings of the 3rd IEEE International Conference on Data Mining—ICDM'03*, pp. 115–122.

Kersten, P. (1997). Implementation issues in the fuzzy *c*-Medians clustering algorithm. In *Proceedings of the 6th IEEE International Conference on Fuzzy Systems*, Vol. 2, pp. 957–962.

Kettenring, J. (2006). The practice of cluster analysis. *Journal of Classification*, 23: 3–30.

Khan, J., Wei, J., Ringnér, M., Saal, L., Ladanyi, M., Westermann, F., Berthold, F., Schwab, M., Antonescu, C., Peterson, C., and Meltzer, P. (2001). Classification and diagnostic prediction of cancers using gene expression profiling and artificial neural networks. *Nature Medicine*, 7(6): 673–679.

Kim, Y., Boyd, A., Athey, B., and Patel, J. (2005). miBLAST: Scalable evaluation of a batch of nucleotide sequence queries with BLAST. *Nucleic Acids Research*, 33(13): 4335–4344.

Kim, N., Shin, S., and Lee, S. (2005). ECgene: Genome-based EST clustering and gene modeling for alternative splicing. *Genome Research*, 15(4): 566–576.

Kim, Y., Street, W., and Menczer, F. (2000). Feature selection for unsupervised learning via evolutionary search. In *Proceedings of the 6th ACM SIGKDD International Conference on Knowledge Discovery and Data Mining*, pp. 365–369.

Kirkpatrick, S., Gelatt, C., and Vecchi, M. (1983). Optimization by simulated annealing. *Science*, 220(4598): 671–680.

Klein, R. and Dubes, R. (1989). Experiments in projection and clustering by simulated annealing. *Pattern Recognition*, 22: 213–220.

Kleinberg, J. (2002). An impossibility theorem for clustering. In *Proceedings of the 2002 Conference on Advances in Neural Information Processing Systems* 15, pp. 463–470.

Kohavi, R. (1995). A study of cross-validation and bootstrap for accuracy estimation and model selection. In *Proceedings of the 14th International Joint Conference on Artificial Intelligence*, pp. 338–345.

Kohonen, T. (1989). *Self-organization and associative memory, 3rd edition*, Berlin: Springer Verlag.

Kohonen, T. (1990). The self-organizing map. *Proceedings of the IEEE*, 78(9): 1464–1480.

Kohonen, T. (2001). *Self-organizing maps, 3rd edition*, Berlin, Hiedelberg: Springer.

Kohonen, T. (2006). Self-organizing neural projections. *Neural Networks*, 19(6–7): 723–733.

Kohonen, T., Kaski, S., Lagus, K., Salojärvi, J., Honkela, J., Paatero, V., and Saarela, A. (2000). Self organization of a massive document collection. *IEEE Transactions on Neural Networks*, 11(3): 574–585.

Kolatch, E. (2001). Clustering algorithms for spatial databases: A survey. cached at http://citeseer.nj.nec.com/436843.html.

Kolen, J. and Hutcheson, T. (2002). surnameucing the time complexity of the fuzzy *c*-Means algorithm. *IEEE Transactions on Fuzzy Systems*, 10(2): 263–267.

Kollios, G., Gunopulos, D., Koudas, N., and Berchtold, S. (2003). Efficient biased sampling for approximate clustering and outlier detection in large data sets. *IEEE Transactions on Knowledge and Data Engineering*, 13(5): 1170–1187.

Kothari, R. and Pitts, D. (1999). On finding the number of clusters. *Pattern Recognition Letters*, 20: 405–416.

Koza, J. (1992). *Genetic programming: On the programming of computers by means of natural selection*. Cambridge, MA: MIT Press.

Koza, J. (1994). *Genetic programming II: Automatic discovery of reusable programs*. Cambridge, MA: MIT Press.

Kramer, M. (1991). Nonlinear principal component analysis using autoassociative neural networks. *AIChe Journal*, 37(2); 233–243.

Krishna, K. and Murty, M. (1999). Genetic *K*-Means algorithm. *IEEE Transactions on Systems, Man, and Cybernetics—Part B: Cybernetics*, 29(3): 433–439.

Krishnapuram, R., Frigui, H., and Nasraoui, O. (1995). Fuzzy and possiblistic shell clustering algorithms and their application to boundary detection and surface approximation—Part I and II. *IEEE Transactions on Fuzzy Systems*, 3(1): 29–60.

Krishnapuram, R. and Keller, J. (1993). A possibilistic approach to clustering. *IEEE Transactions on Fuzzy Systems*, 1(2): 98–110.

Krishnapuram, R. and Keller, J. (1996). The possibilistic *c*-means algorithm: Insights and recommendations. *IEEE Transactions on Fuzzy Systems*, 4(3): 385–393.

Krishnapuram, R., Nasraoui, O., and Frigui, H. (1992). The fuzzy *c* spherical shells algorithm: A new approach. *IEEE Transactions on Neural Networks*, 3(5): 663–671.

Krogh, A., Brown, M., Mian, I., Sjölander, K., and Haussler, D. (1994). Hidden Markov models in computational biology: Applications to protein modeling. *Journal of Molecular Biology*, 235: 1501–1531.

Kumar, M., Stoll, R., and Stoll, N. (2006) A min-max approach to fuzzy clustering, estimation, and identification. *IEEE Transactions on Fuzzy Systems*, 14(2): 248–262.

Lafon, S. and Lee, A. (2006). Diffusion maps and coarse-graining: A unified framework for dimensionality reduction, graph partitioning, and data set parameterization. *IEEE Transactions on Pattern Analysis and Machine Intelligence*, 28(9): 1393–1403.

Lagus, K., Kaski, S., and Kohonen, T. (2004). Mining massive document collections by the WEBSOM method. *Information Sciences*, 163(1–3): 135–156.

Laiho, J., Raivio, K., Lehtimäki, P., Hätönen, K., and Simula, O. (2005). Advanced analysis methods for 3G cellular networks. *IEEE Transactions on Wireless Communications*, 4(3): 930–942.

Lance, G. and Williams, W. (1967). A general theory of classification sorting strategies: 1. Hierarchical systems. *Computer Journal*, 9: 373–380.

Laszlo, M. and Mukherjee, S. (2006) A genetic algorithm using hyper-quadtrees for low-dimensional K-means clustering. *IEEE Transactions on Pattern Analysis and Machine Intelligence*, 28(4): 533–543.

Law, M., Figueiredo, M., and Jain, A. (2004). Simultaneous feature selection and clustering using mixture models. *IEEE Transactions on Pattern Analysis and Machine Intelligence*, 26(9): 1154–1166.

Law, M. and Kwok, J. (2000). Rival penalized competitive learning for model-based sequence clustering. In *Proceedings of the 15th International Conference on Pattern Recognition*, Vol. 2, pp. 195–198.

LeCun, Y., Jackel, L., Bottou, L., Brunot, A., Cortes, C., Denker, J., Drucker, H., Guyon, I., Muller, U., Sackinger, E., Simard, P., and Vapnik, V. (1995). Comparison of learning algorithms for handwritten digit recognition. In *Proceedings of International Conference on Artificial Neural Networks*, pp. 53–60.

Lee, J. and Lee, D. (2005). An improved cluster labeling method for support vector clustering. *IEEE Transactions on Pattern Analysis and Machine intelligence*, 27(3): 461–464.

Lee, M. (2004). *Analysis of microarray gene expression data.* Norwell, MA: Kluwer Academic Publishers.

Lee, W., Yeung, C., and Tsang, C. (2005). Hierarchical clustering based on ordinal consistency. *Pattern Recognition*, 38: 1913–1925.

Lenk, P. and DeSarbo, W. (2000). Bayesian inference for finite mixtures of generalized linear models with random effects. *Psychometrika*, 65(1): 93–119.

Lerner, B., Guterman, H., Aladjem, M., and Dinstein, I. (1999). A comparative study of neural network based feature extraction paradigms. *Pattern Recognition Letters*, 20(1): 7–14.

Leski, J. (2003). Generalized weighted conditional fuzzy clustering. *IEEE Transactions on Fuzzy Systems*, 11(6): 709–715.

Leung, Y., Zhang, J., and Xu, Z. (2000). Clustering by scale-space filtering. *IEEE Transactions on Pattern Analysis and Machine Intelligence*, 22(12): 1396–1410.

Li, C. and Biswas, G. (1999). Temporal pattern generation using hidden Markov model based unsupervised classification. In *Advances in Intelligent Data Analysis, Lecture Notes in Computer Science 1642*, D. Hand, K. Kok, and M. Berthold, eds., New York, NY: Springer.

Li, C. and Biswas, G. (2002). Unsupervised learning with mixed numeric and nominal data. *IEEE Transactions on Knowledge and Data Engineering*, 14(4): 673–690.

Li, C., Chang, E., Garcia-Molina, H., and Wiederhold, G. (2002). Clustering for approximate similarity search in high-dimensional spaces. *IEEE Transactions on Knowledge and Data Engineering*, 14(4): 792–808.

Li, W., Jaroszewski, L., and Godzik, A. (2001). Clustering of highly homologous sequences to reduce the size of large protein databases. *Bioinformatics*, 17: 282–283.

Li, W., Jaroszewski, L., and Godzik, A. (2002). Tolerating some redundancy significantly speeds up clustering of large protein databases. *Bioinformatics*, 18: 77–82.

Li, X. (1990). Parallel algorithms for hierarchical clustering and cluster validity. *IEEE Transactions on Pattern Analysis and Machine Intelligence*, 12(11): 1088–1092.

Liao, T. (2005). Clustering of time series data—a survey. *Pattern Recognition*, 38: 1857–1874.

Lian, W., Cheung, D., Mamoulis, N., and Yiu, S. (2004). An efficient and scalable algorithm for clustering XML documents by structure. *IEEE Transactions on Knowledge and Data Engineering*, 16(1): 82–96.

Liew, W., Yan, H., and Yang, M. (2005). Pattern recognition techniques for the emerging field of bioinformatics: A review. *Pattern Recognition*, 38(11): 2055–2073.

Likas, A., Vlassis, N., and Verbeek, J. (2003). The global *K*-Means clustering algorithm. *Pattern Recognition*, 36(2): 451–461.

Lin, C. and Chen, M. (2002). On the optimal clustering of sequential data. In *Proceedings of the 2nd SIAM International Conference on Data Mining*, pp. 141–157.

Lin, S. and Kernighan, B. (1973). An effective heuristic algorithm for the traveling salesman problem. *Operations Research*, 21: 498–516.

Linares-Barranco, B., Serrano-Gotarredona, M., and andreaou, A. (1998). *Adaptive resonance theory microchips: Circuit design techniques*. Norwell, MA: Kluwer Academic Publisher.

Linde, Y., Buzo, A., and Gray, R. (1980). An algorithm for vector quantizer design. *IEEE Transactions on Communications*. COM-28(1): 84–95.

Lingras, P. and West, C. (2004). Interval set clustering of web users with rough *k*-means. *Journal of Intelligent Information Systems*, 23: 5–16.

Lipshutz, R., Fodor, S., Gingeras, T., and Lockhart, D. (1999). High density synthetic Oligonucleotide arrays. *Nature Genetics*, 21: 20–24.

Little, R. and Rubin, D. (1987). *Statistical analysis with missing data*. New York, NY: Wiley.

Liu, G. (1968). *Introduction to combinatorial mathematics*. New York, NY: McGraw Hill.

Liu, H. and Yu, L. (2005). Toward integrating feature selection algorithms for classification and clustering. *IEEE Transactions on Knowledge and Data Engineering*, 17(4): 491–502.

Liu, J. and Rost, B. (2002). Target space for structural genomics revisited. *Bioinformatics*, 18(7): 922–933.

Liu, X., Wang, W., and Chai, T. (2005). The fuzzy clustering analysis based on AFS theory. *IEEE Transactions on Systems, Man, and Cybernetics—Part B: Cybernetics*, 35(5): 1013–1027.

Lloyd, S. (1982). Least squared quantization in PCM. *IEEE Transactions on Information Theory*. IT-28(2): 129–137.

Lockhart, D., Dong, H., Byrne, M., Follettie, M., Gallo, M., Chee, M., Mittmann, M., Wang, C., Kobayashi, M., Horton, H., and Brown, E. (1996). Expression monitoring by hybridization to high-density oligonucleotide arrays. *Nature Biotechnology*, 14: 1675–1680.

Lo Conte, L., Ailey, B., Hubbard, T., Brenner, S., Murzin, A., and Chothia, C. (2000). SCOP: A structural classification of proteins database. *Nucleic Acids Research*, 28: 257–259.

Lozano, J. and Larrañaga, P. (1999). Applying genetic algorithms to search for the best hierarchical clustering of a dataset. *Pattern Recognition Letters*, 20: 911–918.

Lukashin, A. and Fuchs, R. (2001). Analysis of temporal gene expression profiles: Clustering by simulated annealing and determining the optimal number of clusters. *Bioinformatics*, 17(5): 405–414.

Ma, C., Chan, C., Yao, X., and Chiu, K. (2006). An evolutionary clustering algorithm for gene expression microarray data analysis. *IEEE Transactions on Evolutionary Computation*, 10(3): 296–314.

Ma, J. and Wang, T. (2006). A cost-function approach to rival penalized competitive learning (RPCL). *IEEE Transactions on Systems, Man, and Cybernetics—Part B*: Cybernetics, 36(4): 722–737.

Macnaughton-Smith, P., Williams, W., Dale, M., and Mockett, L. (1964). Dissimilarity analysis: A new technique of hierarchical sub-division. *Nature*, 202: 1034–1035.

MacQueen, J. (1967). Some methods for classification and analysis of multivariate observations. In *Proceedings of the Fifth Berkeley Symposium*, Vol. 1, pp. 281–297.

Madeira, S. and Oliveira, A. (2004). Biclustering algorithms for biological data analysis: a survey. *IEEE/ACM Transactions on Computational Biology and Bioinformatics*, 1(1): 24–45.

Maharaj, E. (2000). Clusters of time series. *Journal of Classification*, 17: 297–314.

Malthouse, E. (1998). Limitations of nonlinear PCA as performed with generic neural networks. *IEEE Transactions on Neural Networks*, 9(1): 165–173.

Man, Y. and Gath, I. (1994). Detection and separation of ring-shaped clusters using fuzzy clustering. *IEEE Transactions on Pattern Analysis and Machine Intelligence*, 16(8): 855–861.

Mandelbrot, M. (1983). *The Fractal Geometry of Nature*. New York, NY: W.H. Freeman.

Mao, J. and Jain, A. (1996). A self-organizing network for hyperellipsoidal clustering (HEC). *IEEE Transactions on Neural Networks*, 7(1): 16–29.

Marcotte, E., Pellegrini, M., Thompson, M., Yeates, T., and Eisenberg, D. (1999). A combined algorithm for genome-wide prediction of protein function. *Nature*, 402: 83–86.

Marsland, S., Shapiro, J., and Nehmzow, U. (2002). A self-organizing network that grows when required. *Neural Networks*, 15(8–9): 1041–1058.

Martinetz, T., Berkovich, S., and Schulten, K. (1993). "Neural-Gas" network for vector quantization and its application to time-series prediction. *IEEE Transactions on Neural Networks*, 4(4): 558–569.

Martinetz, T. and Schulten, K. (1994). Topology representing networks. *Neural Networks*, 7(3): 507–522.

Martinez, A. and Vitrià, J. (2001). Clustering in image space for place recognition and visual annotations for human-robot interaction. *IEEE Transactions on Systems, Man, And Cybernetics—Part B: Cybernetics*, 31(5): 669–682.

Maulik, U. and Bandyopadhyay, S. (2000). Genetic algorithm-based clustering technique. *Pattern Recognition*, 33: 1455–1465.

McLachlan, G., Do, K., and Ambroise, C. (2004). *Analyzing microarray gene expression data*. Hoboken, NJ: John Wiley & Sons.

McLachlan, G. and Krishnan, T. (1997). *The EM algorithm and extensions*. New York, NY: Wiley.

McLachlan, G. and Peel, D. (1998). Robust cluster analysis via mixtures of multivariate t-distributions. In *Lecture Notes in Computer Science*, Vol. 1451, Amin, A., Dori, D., Pudil, P., and Freeman, H., eds., Berlin: Springer-Verlag, pp. 658–666.

McLachlan, G. and Peel, D. (2000). *Finite mixture models*. New York, NY: John Wiley & Sons.

McLachlan, G., Peel, D., Basford, K., and Adams, P. (1999). The EMMIX software for the fitting of mixtures of normal and t-components. *Journal of Statistical Software*, 4(2): 1–14.

McQuitty, L. (1966). Similarity analysis by reciprocal pairs for discrete and continuous data. *Educational and Psychological Measurement*, 27: 21–46.

Mercer, J. (1909). Functions of positive and negative type, and their connection with the theory of integral equations. Transactions *of the London Philosophical Society (A)*, 209: 415–446.

Merwe, D. and Engelbrecht, A. (2003) Data clustering using particle swarm optimization. In *Proceedings of the 2003 Congress on Evolutionary Computation*, Vol. 1, pp. 215–220.

Metropolis, N., Rosenbluth, A., Rosenbluth, M., Teller, A., and Teller, E. (1953) Equations of state calculations by fast computing machines. *Journal of Chemical Physics*, 21: 1087–1092.

Miller, C., Gurd, J., and Brass, A. (1999). A RAPID algorithm for sequence database comparisons: Application to the identification of vector contamination in the EMBL databases. *Bioinformatics*, 15: 111–121.

Miller, R., Christoffels, A., Gopalakrishnan, C., Burke, J., Ptitsyn, A., Broveak, T., and Hide, W. (1999). A comprehensive approach to clustering of expressed human gene sequence: The sequence tag alignment and consensus knowledge base. *Genome Research*, 9(11): 1143–1155.

Miller, W. (2001). Comparison of genomic DNA sequences: Solved and unsolved problems. *Bioinformatics*, 17: 391–397.

Milligan, G. (1981). A Monte Carlo study of 30 internal criterion measures for cluster analysis. *Psychometrika*, 46: 187–195.

Milligan, G. and Cooper, M. (1985). An examination of procedures for determining the number of clusters in a data set. *Psychometrika*, 50: 159–179.

Mirkin, B. (2005). *Clustering for data mining: A data recovery approach*. Boca Raton, FL: Chapman and Hall / CRC.

Mirkin, B. (2007). The iterative extraction approach to clustering. In *Principal Manifolds for Data Visualization and Dimension surnameuction, Lecture Notes in Computational Science and Engineering*, Vol. 58, A. Gorban, B. Kégl, D. Wunsch, and A. Zinovyev, eds., Berlin/Heidelberg: Springer, pp. 151–177.

Mitra, P., Murthy, C., and Pal, S. (2002). Unsupervised feature selection using feature similarity. *IEEE Transactions on Pattern Analysis and Machine Intelligence*, 24(3): 301–312.

Moghaddam, B. (2002). Principal manifolds and probabilistic subspaces for visual recognition. *IEEE Transactions on Pattern Analysis and Machine Intelligence*, 24(6): 780–788.

Mollineda, R. and Vidal, E. (2000). A relative approach to hierarchical clustering. In *Pattern Recognition and Applications, Frontiers in Artificial Intelligence and Applications*, M. Torres and A. Sanfeliu, eds., Amsterdam: IOS Press, Vol. 56, pp. 19–28.

Moore, B. (1989). ART1 and pattern clustering. In *Proceedings of the 1988 Connectionist Models Summer School*, pp. 174–185.

Moore, S. (2001). Making chips to probe genes. *IEEE Spectrum*, 38: 54–60.

Moreau, Y., Smet, F., Thijs, G., Marchal, K., and Moor, B. (2002). Functional bioinformatics of microarray data: From expression to regulation. *Proceedings of the IEEE*, 90(11): 1722–1743.

Morgenstern, B., Frech, K., Dress, A., and Werner, T. (1998). DIALIGN: Finding local similarities by multiple sequence alignment. *Bioinformatics*, 14: 290–294.

Morzy, T., Wojciechowski, M., and Zakrzewicz, M. (1999). Pattern-oriented hierarchical clustering. In *Proceedings of the 3rd East European Conference on Advances in Databases and Information Systems, LNCS 1691*, New York, NY: Springer–Verlag.

Motwani, R. and Raghavan, P. (1995). *Randomized algorithms*. Cambridge, UK: Cambridge University Press.

Mulder, S. and Wunsch, D. (2003). Million city traveling salesman problem solution by divide and conquer clustering with adaptive resonance neural networks. *Neural Networks*, 16: 827–832.

Müller, K., Mika, S., Rätsch, G., Tsuda, K., and Schölkopf, B. (2001). An introduction to kernel-based learning algorithms. *IEEE Transactions on Neural Networks*, 12(2): 181–201.

Murtagh, F. (1983). A survey of recent advances in hierarchical clustering algorithms. Computer *Journal*, 26(4): 354–359.

Murtagh, F., Starck, J., and Berry, M. (2000). Overcoming the curse of dimensionality in clustering by means of the wavelet transform. Computer *Journal*, 43(2): 107–120.

Nagesh, H., Goil, S., and Choudhary, A. (2001). Adaptive grids for clustering massive data sets. In *Proceedings of 1st SIAM International Conference on Data Mining*.

Nascimento, S., Mirkin, B., and Moura-Pires, F. (2003). Modeling proportional membership in fuzzy clustering. *IEEE Transactions on Fuzzy Systems*, 11(2): 173–186.

Nath, J. and Shevade, S. (2006). An efficient clustering scheme using support vector methods. *Pattern Recognition*, 39: 1473–1480.

Needleman, S. and Wunsch, C. (1970). A general method applicable to the search for similarities in the amino acid sequence of two proteins. *Journal of Molecular Biology*, 48: 443–453.

Ng, E., Fu, A., and Wong, R. (2005). Projective clustering by histograms. *IEEE Transactions on Knowledge and Data Engineering*, 17(3): 369–383.

Ng, R. and Han, J. (2002). CLARANS: A method for clustering objects for spatial data mining. *IEEE Transactions on Knowledge and Data Engineering*, 14(5): 1003–1016.

Nguyen, H. (2005). A soft decoding scheme for vector quantization over a CDMA channel. *IEEE Transactions on Communications*, 53(10): 1603–1608.

Noseworthy, J., Lucchinetti, C., Rodriguez, M., and Weinshenker, B. (2000). Multiple sclerosis. *The New England Journal of Medicine*, 343(13): 938–952.

Oates, T., Firoiu, L., and Cohen, P. (2000). Using dynamic time warping to bootstrap HMM-based clustering of time series. In *Sequence Learning, LNAI 1828*, R. Sun and C. Giles, Eds., Berlin, Heidelberg: Springer-Verlag, pp. 35–52.

Ohashi, Y. (1984). Fuzzy clustering and robust estimation. In *9th Meeting of SAS Users Group International*, Hollywood Beach, FL.

Oja, E. (1982). A simplified neuron model as principal component analyzer. *Journal of Mathematical biology*, 15: 267–273.

Oja, E. (1992). Principal components, minor components, and linear neural networks. *Neural Networks*, 5: 927–935.

Okabe, A., Boots, B., Sugihara, K., and Chiu, S. (2000) *Spatial tessellations: Concepts and applications of Voronoi diagrams, 2nd edition*. West Sussex: John Wiley & Sons.

Oliver, J., Baxter, R., and Wallace, C. (1996). Unsupervised learning using MML. In *Proceedings of the 13th International Conference on Machine Learning—ICML 96*, San Francisco, CA: Morgan Kaufmann Publishers, pp. 364–372.

Olson, C. (1995). Parallel algorithms for hierarchical clustering. *Parallel Computing*, 21: 1313–1325.

Ordonez, C. (2003). Clustering binary data streams with K-means. In *Proceedings of the 8th ACM SIGMOD Workshop on Research Issues in Data Mining and Knowledge Discovery*, pp. 12–19.

Ordonez, C. (2006). Integrating K-means clustering with a relational DBMS using SQL. *IEEE Transactions on Knowledge and Data Engineering*, 18(2): 188–201.

Ordonez, C. and Omiecinski, E. (2004). Efficient disk-based K-means clustering for relational databases. *IEEE Transactions on Knowledge and Data Engineering*, 16(8): 909–921.

Osiński, S. and Weiss, D. (2005) A concept-driven algorithm for clustering search results. *IEEE Intelligent Systems*, 20(3): 48–54.

Owsley, L., Atlas, L., and Bernard, G. (1997). Self-organizing feature maps and hidden Markov models for machine-tool monitoring. *IEEE Transactions on Signal Processing*, 45(11): 2787–2798.

Pal, N. and Bezdek, J. (1995). On cluster validity for the fuzzy *c*-Means model. *IEEE Transactions on Fuzzy Systems*, 3(3): 370–379.

Pal, N., Bezdek, J., and Tsao, E. (1993). Generalized clustering networks and Kohonen's self-organizing scheme. *IEEE Transactions on Neural Networks*, 4(4): 549–557.

Pal, N., Pal, K., Keller, J., and Bezdek, J. (2005). A possibilistic fuzzy c-means clustering algorithm. *IEEE Transactions on Fuzzy Systems*, 13(4): 517–530.

Pan, J. (1997). Extension of two-stage vector quantization-lattice vector quantization. *IEEE Transactions on Communications*, 45(12): 1538–1547.

Patanè, G. and Russo, M. (2001). The enhanced-LBG algorithm. *Neural Networks*, 14(9): 1219–1237.

Patanè, G. and Russo, M. (2002). Fully automatic clustering system. *IEEE Transactions on Neural Networks*, 13(6): 1285–1298.

Pawlak, Z. (1992). *Rough sets: Theoretical aspects of reasoning about data*. Boston, MA: Kluwer Academic Publishers.

Pearson, W. and Lipman, D. (1988). Improved tools for biological sequence comparison. *Proceedings of the National Academy of Science, USA 85*, 85: 2444–2448.

Pedrycz, W. and Waletzky, J. (1997). Fuzzy clustering with partial supervision. *IEEE Transactions on Systems, Man, and Cybernetics—Part B: Cybernetics*, 27(5): 787–795.

Peel, D. and McLachlan, G. (2000). Robust mixture modeling using the t-distribution. *Statistics and Computing*, 10: 339–348.

Pelleg, D. and Moore, A. (1999). Accelerating exact *K*-means algorithms with geometric reasoning. In *Proceedings of ACM SIGKDD International Conference on Knowledge Discovery and Data Mining*, pp. 277–281.

Pelleg, D. and Moore, A. (2000). X-means: Extending *K*-means with efficient estimation of the number of clusters. In *Proceedings of the 17th International Conference on Machine Learning—ICML2000*.

Peña, J., Lozano, J., and Larrañaga, P. (1999). An empirical comparison of four initialization methods for the *K*-Means algorithm. *Pattern Recognition Letters*, 20: 1027–1040.

Peters, G. (2006). Some refinements of rough *k*-means clustering. *Pattern Recognition*, 39: 1481–1491.

Pham, L., Xu, C., and Prince, J. (2000). Current methods in medical image segmentation. *Annual Review of Biomedical Engineering*, 2: 315–337.

Pizzuti, C. and Talia, D. (2003). P-AutoClass: scalable parallel clustering for mining large data sets. *IEEE Transactions on Knowledge and Data Engineering*, 15(3): 629–641.

Plumbley, M. (2003). Algorithms for nonnegative independent component analysis. *IEEE Transactions on Neural Networks*, 14(3): 534–543.

Policker, S. and Geva, A. (2000). Nonstationary time series analysis by temporal clustering. *IEEE Transactions on Systems, Man, and Cybernetics—Part B: Cybernetics*, 30(2): 339–343.

Porrmann, M., Witkowski, U., and Ruckert, U. (2003). A massively parallel architecture for self-organizing feature maps. *IEEE Transactions on Neural Networks*, 14(5): 1110–1121.

Procopiuc, C., Jones, M., Agarwal, P., and Murali, T. (2002). A Monte Carlo algorithm for fast projective clustering. In *Proceedings of ACM SIGMOD International Conference on Management of Data*.

Qin, A. and Suganthan, P. (2004). Robust growing neural gas algorithm with application in cluster analysis. *Neural Networks*, 17(8–9): 1135–1148.

Quinlan, J. (1993). *C4.5: Programs for machine learning*. San Mateo, CA: Morgan Kaufmann Publishers.

Rabiner, L. (1989). A tutorial on hidden Markov models and selected applications in speech recognition. *Proceedings of the IEEE*, 77: 257–286.

Rajasekaran, S. (2005). Efficient parallel hierarchical clustering algorithms. *IEEE Transactions on Parallel and Distributed Systems*, 16(6): 497–502.

Ramoni, M., Sebastiani, P., and Cohen, P. (2002). Bayesian clustering by dynamics. *Machine Learning*, 47(1): 91–121.

Rand, W. (1971). Objective criteria for the evaluation of clustering methods. *Journal of the American Statistical Association*, 66: 846–850.

Rauber, A., Paralic, J., and Pampalk, E. (2000). Empirical evaluation of clustering algorithms. *Journal of Information and Organizational Sciences (JIOS)*, 24(2): 195–209.

Rechenberg, I. (1973). *Evolutions strategie: Optimierung technischer systeme nach prinzipien der biologischen evolution*. Frommann-Holzboog, Stuttgart.

Richardson, S. and Green, P. (1997). On Bayesian analysis of mixtures with an unknown number of components. *Journal of the Royal Statistical Society, Series B*, 59(4): 731- 758.

Ridella, S., Rovetta, S., and Zunino, R. (1998). Plastic algorithm for adaptive vector quantization. *Neural Computing and Applications*, 7: 37–51.

Rissanen, J. (1978). Modeling by shortest data description. *Automatica*, 14: 465–471.

Rissanen, J. (1996). Fisher information and stochastic complexity. *IEEE Transactions on Information Theory*, 42(1): 40–47.

Roberts, S. (1997). Parametric and non-parametric unsupervised cluster analysis. *Pattern Recognition*, 30: 327–345.

Roberts, S., Husmeier, D., Rezek, I., and Penny, W. (1998). Bayesian approaches to Gaussian mixture modeling. *IEEE Transactions on Pattern Analysis and Machine Intelligence*, 20(11): 1133–1142.

Rogers, D. and Tanimoto, T. (1960). A computer program for classifying plants. *Science*, 132: 1115–1118.

Rohlf, F. (1970). Adaptive hierarchical clustering schemes. *Systematic Zoology*, 19: 58–82.

Rohlf, F. and Fisher, D. (1968). Tests for hierarchical structure in random data sets. *Systematic Zoology*, 17: 407–412.

Romdhani, S., Gong, S., and Psarrou, A. (1999). A multiview nonlinear active shape model using kernel PCA. In *Proceedings of the 10th British Machine Vision Conference*, pp. 483–492.

Romesburg, C. (1984). *Cluster analysis for researchers*. London: Wadsworth.

Rose, K. (1998). Deterministic annealing for clustering, compression, classification, regression, and related optimization problems. *Proceedings of The IEEE*, 86(11): 2210–2239.

Rose, K., Gurewitz, E., and Fox, G. (1992). Vector quantization by deterministic annealing. *IEEE Transactions on Information Theory*, 38(4): 1249–1257.

Rose, K., Gurewitz, E., and Fox, G. (1993). Constrained clustering as an optimization method. *IEEE Transactions on Pattern Analysis and Machine Intelligence*, 15(8): 785–794.

Rosenblatt, F. (1962). *Principles of neurodynamics*. New York, NY: Spartan.

Rosenwald, A., Wright, G., Chan, W., Connors, J., Campo, C., Fisher, R., Gascoyne, R., Muller-Hermelink, H., Smeland, E., and Staudt, L. (2002). The use of molecular profiling to predict survival after chemotherapy for diffuse large-B-cell lymphoma. *The New England Journal of Medicine*, 346(25): 1937–1947.

Rosipal, R., Girolami, M., and Trejo, L. (2000). Kernel PCA feature extraction of event-related potentials for human signal detection performance. In *Perspectives in Neural Computation. Proceedings of ANNIMAB-1 Conference o Artificial Neural Networks in Medicine and Biology*, Malmgren, H., Borga, M., and Niklasson, L., eds., London, UK: Springer-Verlag, pp. 321–326.

Roth, F., Hughes, J., Estep, P., and Church, G. (1998). Finding DNA-regulatory motifs within unaligned noncoding sequences clustered by whole-genome mRNA quantitation. *Nature Biotechnology*, 16(10): 939–945.

Roth, V. and Lange, T. (2004). Feature selection in clustering problems. In *Advances in Neural Information Processing Systems 16*, Cambridge, MA: MIT Press.

Roweis, S. and Ghahramani, Z. (1999). A unifying review of linear Gaussian models. *Neural Computation*, 11(2): 305–345.

Roweis, S. and Saul, L. (2000). Nonlinear dimensionality reduction by locally linear embedding. *Science*, 90(5500): 2323–2326.

Rumelhart, D. and Zipser, D. (1985). Feature discovery by competitive learning. Cognitive *Science*, 9: 75–112.

Sæbø, P., Andersen, S., Myrseth, J., Laerdahl, J., and Rognes, T. (2005). PARALIGN: Rapid and sensitive sequence similarity searches powered by parallel computing technology. *Nucleic Acids Research*, 23: W535–W539.

Sammon, J. (1969). A nonlinear mapping for data structure analysis. *IEEE Transactions on Computers*, 18: 401–409.

Sander, J., Ester, M., Kriegel, H., and Xu, X. (1998). Density-based clustering in spatial data bases: The algorithm GDBSCAN and its applications. *Data Mining and Knowledge Discovery*, 2(2): 169–194.

Sankoff, D. and Kruskal, J. (1999). *Time warps, string edits, and macromolecules: The theory and practice of sequence comparison*. Stanford, CA: CSLI publications.

Sasson, O., Linial, N. and Linial, M. (2002). The metric space of proteins—comparative study of clustering algorithms. *Bioinformatics*, 18: s14–s21.

Schena, M., Shalon, D., Davis, R., and Brown, P. (1995). Quantitative monitoring of gene expression patterns with a complementary DNA microarray. *Science*, 270(5235): 467–470.

Scherf, U., Ross, D., Waltham, M., Smith, L., Lee, J., Tanabe, L., Kohn, K., Reinhold, W., Myers, T., Andrews, D., Scudiero, D., Eisen, M., Sausville, E., Pommier, Y., Botstein, D., Brown, P., and Weinstein, J. (2000). A gene expression database for the molecular pharmacology of cancer. *Nature Genetics*, 24(3): 236–244.

Scheunders, P. (1997). A comparison of clustering algorithms applied to color image quantization. *Pattern Recognition Letter*, 18: 1379–1384.

Schölkopf, B., Burges, C., and Smola, A. (1999). *Advances in kernel methods: support vector learning*. Cambridge, MA: The MIT Press.

Schölkopf, B., Mika, S., Burges, C., Knirsch, P., Müller, K., Rätsch, G., and Smola, A. (1999). Input space versus feature space in kernel-based methods. *IEEE Transactions on Neural Networks*, 10(5): 1000–1017.

Schölkopf, B. and Smola, A. (2002). *Learning with kernels: Support vector machines, regularization, optimization, and beyond*. Cambridge, MA: The MIT Press.

Schölkopf, B., Smola, A., and Müller, K. (1998). Nonlinear component analysis as a kernel eigenvalue problem. *Neural Computation*, 10(5): 1299–1319.

Schwarz, G. (1978). Estimating the dimension of a model. *Annals of Statistics*, 6(2): 461–464.

Schwartz, S., Zhang, Z., Frazer, K., Smit, A., Riemer, C., Bouck, J., Gibbs, R., Hardison, R., and Miller, W. (2000). PipMaker—a web server for aligning two genomic DNA sequences. *Genome Research*, 10(4): 577–586.

Scott, G., Clark, D., and Pham, T. (2001). A genetic clustering algorithm guided by a descent algorithm. In *Proceedings of the 2001 Congress on Evolutionary Computation*, Vol. 2, pp. 734–740.

Sebastiani, P., Ramoni, M., and Cohen, P. (2000). Sequence learning via Bayesian clustering by dynamics. In *Sequence Learning, LNAI 1828*, R. Sun and C. Giles, Eds., Berlin, Heidelberg: Springer-Verlag, pp. 11–34.

Selim, S. and Alsultan, K. (1991). A simulated annealing algorithm for the clustering problems. *Pattern Recognition*, 24(10): 1003–1008.

Selim, S. and Ismail, M. (1984). K-means-type algorithms: A generalized convergence theorem and characterization of local optimality. *IEEE Transactions on Pattern Analysis and Machine Intelligence*, 6(1): 81–87.

Seo, S. and Obermayer, K. (2004). Self-organizing maps and clustering methods for matrix data. *Neural Networks*, 17(8–9): 1211–1229.

Shamir, R. and Sharan, R. (2002). Algorithmic approaches to clustering gene expression data. In *Current Topics in Computational Molecular Biology*, T. Jiang, T. Smith, Y. Xu, and M. Zhang, Eds., Cambridge, MA: MIT Press, pp. 269–300.

Sharan, R. and Shamir, R. (2000). CLICK: A clustering algorithm with applications to gene expression analysis. In *Proceedings of the 8th International Conference on Intelligent Systems for Molecular Biology*, pp. 307–316.

Sheikholeslami, G., Chatterjee, S., and Zhang, A. (1998). WaveCluster: A multi-resolution clustering approach for very large spatial databases. In *Proceedings of the 24th VLDB conference*, pp. 428–439.

Sheng, W., Swift, S., Zhang, L., and Liu, X. (2005) A weighted sum validity function for clustering with a hybrid niching genetic algorithm. *IEEE Transactions on Systems, Man, and Cybernetics—Part B: Cybernetics*, 35(6): 1156–1167.

Shi, J. and Malik, J. (2000). Normalized cuts and image segmentation. *IEEE Transactions on Pattern Analysis and Machine Intelligence*, 22(8): 888–905.

Shi, Y. and Eberhart, R. (1998). Parameter selection in particle swarm optimization, In *Proceedings of the 7th Annual Conference on Evolutionary Programming*, pp. 591–601.

Shi, Y., Song, Y., and Zhang, A. (2005). A shrinking-based clustering approach for multidimensional data. *IEEE Transactions on Knowledge and Data Engineering*, 17(10): 1389–1403.

Simon, G., Lee, J., and Verleysen, M. (2006). Unfolding preprocessing for meaningful time series clustering. *Neural Networks*, 19: 877–888.

Simpson, P. (1993). Fuzzy min-max neural networks—Part 2: *Clustering. IEEE Transactions on Fuzzy Systems*, 1(1): 32–45.

Sklansky, J. and Siedlecki, W. (1993). Large-scale feature selection. In *Handbook of Pattern Recognition and Computer Vision*, C. Chen, L. Pau, and P. Wang, Eds., Singapore: World Science Publishing Company, pp. 61–124.

Smith, T. and Waterman, M. (1980). New stratigraphic correlation techniques. *Journal of Geology*, 88: 451–457.

Smyth, P. (1996). Clustering using Monte Carlo cross-Validation. In *Proceedings of the 2nd International Conference on Knowledge Discovery and Data Mining*, New York, NY: AAAI Press, pp. 126–133.

Smyth, P. (1997). Clustering sequences with hidden Markov models. In *Advances in Neural Information Processing 9*, M. Mozer, M. Jordan and T. Petsche, Eds., Cambridge, MA: MIT Press, pp. 648–654.

Smyth, P. (1998). Model selection for probabilistic clustering using cross validated likelihood. *Statistics and computing*, 10: 63–72.

Smyth, P. (1999). Probabilistic model-based clustering of multivariate and sequential data. In *Proceedings of the 7th International Workshop on Artificial Intelligence and Statistics*, San Francisco, CA: Morgan Kaufmann, pp. 299–304.

Sneath, P. (1957). The application of computers to taxonomy. *Journal of General Microbiology*, 17: 201–226.

Sneath, P. and Sokal, R. (1973). *Numerical Taxonomy*. San Francisco, CA: Freeman.

Sokal, R. and Michener, C. (1958). A statistical method for evaluating systematic relationships. *University of Kansas Science Bulletin*, 38: 1409–1438.

Sokal, R. and Sneath, P. (1963). *Principles of Numerical Taxonomy*. San Francisco: Freeman.

Somervuo, P. and Kohonen, T. (2000). Clustering and visualization of large protein sequence databases by means of an extension of the self-organizing map. In *Proceedings of the 3rd International Conference on Discovery Science, Lecture Notes in Artificial Intelligence 1967*, pp. 76–85.

Sorensen, T. (1948). A method of establishing groups of equal amplitude in plant sociology based on similarity of species content and its application to analyses of the vegetation on Danish commons. *Biologiske Skrifter*, 5: 1–34.

Späth, H. (1980). *Cluster analysis algorithms*. Chichester, UK: Ellis Horwood.

Spellman, P., Sherlock, G., Ma, M., Iyer, V., Anders, K., Eisen, M., Brown, P., Botstein, D., and Futcher, B. (1998). Comprehensive identification of cell cycle-regulated genes of the Yeast Saccharomyces Cerevisiae by microarray hybridization. *Molecular Biology of the Cell*, 9: 3273–3297.

Stahl, H. (1986). Cluster analysis of large data sets. In *Classification as a Tool of Research*, W. Gaul and M. Schader, eds., New York, NY: *Elsevier North-Holland*, pp. 423–430.

Steinbach, M., Karypis, G., and Kumar, V., (2000). A comparison of document clustering techniques. In *KDD Workshop on Text Mining*.

Stekel, D. (2003). *Microarray bioinformatics*. Cambridge, UK: Cambridge University Press.

Stoffel, K. and Belkoniene, A. (1999). Parallel *K*-Means clustering for large data sets. *EuroPar'99 Parallel Processing, Lecture Notes in Computer Science 1685*, Berlin, Heidelberg: Springer-Verlag.

Su, M. and Chang, H. (2000). Fast self-organizing feature map algorithm. *IEEE Transactions on Neural Networks*, 11(3): 721–733.

Su, M. and Chou, C. (2001). A modified version of the *K*-Means algorithm with a distance based on cluster symmetry. *IEEE Transactions on Pattern Analysis and Machine Intelligence*, 23(6): 674–680.

Suh, I., Kim, J., and Rhee, F. (1999). Convex-set-based fuzzy clustering. *IEEE Transactions on Fuzzy Systems*, 7(3): 271–285.

Sun, R. (2000). Introduction to sequence learning. In *Sequence Learning, LNAI 1828*, R. Sun and C. Giles, eds., Berlin, Heidelberg: Springer-Verlag, pp. 1–10.

Sun, R. and Giles, C. (2000). *Sequence learning: Paradigms, algorithms, and applications, LNAI 1828*. Berlin, Heidelberg: Springer-Verlag.

Sung, C. and Jin, H. (2000). A Tabu-search-based heuristic for clustering. *Pattern Recognition*, 33: 849–858.

Talavera, L. (2000). Dependency-based feature selection for clustering symbolic data. *Intelligent Data Analysis*, 4: 19–28.

Tamayo, P., Slonim, D., Mesirov, J., Zhu, Q., Kitareewan, S., Dmitrovsky, E., Lander, E., and Golub, T. (1999). Interpreting patterns of gene expression with self-organizing maps: Methods and application to Hematopoietic differentiation. *Proceedings of the National Academy of Science, USA 96*, 96: 2907–2912.

Tavazoie, S., Hughes, J., Campbell, M., Cho, R., and Church, G. (1999). Systematic determination of genetic network architecture. *Nature Genetics*, 22: 281–285.

Teicher, H. (1961). Identifiability of mixtures. *Annals of Mathematical Statistics*, 32(1): 244–248.

Tenenbaum, J., Silva, V., and Langford, J. (2000). A global geometric framework for nonlinear dimensionality reduction. *Science*, 290: 2319–2323.

Theodoridis, S. and Koutroumbas, K. (2006). *Pattern recognition, 3rd*. San Diego, CA: Academic Press.

Tibshirani, R., Hastie, T., Eisen, M., Ross, D., Botstein, D., and Brown, P. (1999). Clustering methods for the analysis of DNA microarray data. Technical report, Department of Statistics, Stanford University.

Tibshirani, R. and Knight, K. (1999). The covariance inflation criterion for adaptive model selection. *Journal of the Royal Statistical Society: Series B*, 61: 529–546.

Tipping, M. (2000). Sparse kernel principal component analysis. In *Advances in Neural Information Processing Systems* 13, pp. 633–639.

Titterington, D., Smith, A., and Makov, U. (1985). *Statistical analysis of finite mixture distributions*. Chichester, U.K.: John Wiley & Sons.

Tomida, S., Hanai, T., Honda, H., and Kobayashi, T. (2002). Analysis of expression profile using fuzzy adaptive resonance theory. *Bioinformatics*, 18(8): 1073–1083.

Törönen, P., Kolehnainen, M., Wong, G., and Castrén, E. (1999). Analysis of gene expression data using self-organizing maps. *FEBS (Federation of European Biochemical Societies) Letters*, 451: 142–146.

Trauwaert, E. (1987). L_1 in fuzzy clustering. In *Statistical Data Analysis Based on the L_1-Norm and Related Methods*, Dodge, Y., ed., Amsterdam: Elsevier Science Publishers, pp. 417–426.

Tsao, E., Bezdek, J. and Pal, N. (1994). Fuzzy Kohonen clustering network. *Pattern Recognition*, 27(5): 757–764.

Tseng, L. and Yang, S. (2001). A genetic approach to the automatic clustering problem. *Pattern Recognition*, 34: 415–424.

Twining, C. and Taylor, C. (2001). Kernel principal component analysis and the construction of non-linear active shape models. In *Proceedings of the British Machine Vision Conference*, pp. 23–32.

Ueda, N., Nakano, R., Ghahramani, Z., and Hinton, G. (2000). SMEM algorithm for mixture models. *Neural Computation*, 12(9): 2109–2128.

Urquhart, R. (1982). Graph theoretical clustering based on limited neighborhood sets. *Pattern Recognition*, 15: 173–187.

Vapnik, V. (1998). *Statistical learning theory*. New York, NY: John Wiley & Sons.

Vapnik, V. (2000). *The nature of statistical learning theory, 2nd edition*. New York, NY: Springer-Verlag.

Venter, J., Adams, M., Myers, E., Li, P., Mural, R., Sutton, G., Smith, H., Yandell, M., Evans, C., Holt, R., Gocayne, J., Amanatides, P., Ballew, R., Huson, D., Wortman, J., Zhang, Q., Kodira, C., Zheng, X., Chen, L., Skupski, M., Subramanian, G., Thomas, P., Zhang, J., Miklos, G., Nelson, C., Broder, S., Clark, A., Nadeau, J., McKusick, V., Zinder, N., Levine, A., Roberts, R., Simon, M., Slayman, C., Hunkapiller, M., Bolanos, R., Delcher, A., Dew, I., Fasulo, D., Flanigan, M., Florea, L., Halpern, A., Hannenhalli, S., Kravitz, S., Levy, S., Mobarry, C., Reinert, K., Remington, K., Abu-Threideh, J., Beasley, E., Biddick, K., Bonazzi, V., Brandon, R., Cargill, M., Chandramouliswaran, I., Charlab, R., Chaturvedi, K., Deng, Z., Di Francesco, V., Dunn, P., Eilbeck, K., Evangelista, C., Gabrielian, A., Gan, W., Ge, W., Gong, F., Gu, Z., Guan, P., Heiman, T., Higgins, M., Ji, R., Ke, Z., Ketchum, K., Lai, Z., Lei, Y., Li, Z., Li, J., Liang, Y., Lin, X., Lu, F., Merkulov, G., Milshina, N., Moore, H., Naik, A., Narayan, V., Neelam, B., Nusskern, D., Rusch, D., Salzberg, S., Shao, W., Shue, B., Sun, J., Wang, Z., Wang, A., Wang, X., Wang, J., Wei, M., Wides, R., Xiao, C., Yan, C., Yao, A., Ye, J., Zhan, M., Zhang, W., Zhang, H., Zhao, Q., Zheng, L., Zhong, F., Zhong, W., Zhu, S., Zhao, S., Gilbert, D., Baumhueter, S., Spier, G., Carter, C., Cravchik, A., Woodage, T., Ali, F., An, H., Awe, A., Baldwin, D., Baden, H., Barnstead, M., Barrow, I., Beeson, K., Busam, D., Carver, A., Center, A., Cheng, M., Curry, L., Danaher, S., Davenport, L., Desilets, R., Dietz, S., Dodson, K., Doup, L., Ferriera, S., Garg, N., Gluecksmann, A., Hart, B., Haynes, J., Haynes, C., Heiner, C., Hladun, S., Hostin, D., Houck, J., Howland, T., Ibegwam, C., Johnson, J., Kalush, F., Kline, L., Koduru, S., Love, A., Mann, F., May, D., McCawley, S., McIntosh, T., McMullen, I., Moy, M., Moy, L., Murphy, B., Nelson, K., Pfannkoch, C., Pratts, E., Puri, V., Qureshi, H., Reardon, M., Rodriguez, R., Rogers, Y., Romblad, D., Ruhfel, B., Scott, R., Sitter, C., Smallwood, M., Stewart, E., Strong, R., Suh, E., Thomas, R., Tint, N., Tse, S., Vech, C., Wang, G., Wetter, J., Williams, S., Williams, M., Windsor, S., Winn-Deen, E., Wolfe, K., Zaveri, J., Zaveri, K., Abril, J., Guigó, R., Campbell, M., SjolAnder, K., Karlak, B., Kejariwal, A., Mi, H., Lazareva, B., Hatton, T., Narechania, A., Diemer, K., Muruganujan, A., Guo, N., Sato, S., Bafna, V., Istrail, S., Lippert, R., Schwartz, R., Walenz, B., Yooseph, S., Allen, D., Basu, A., Baxendale, J., Blick, L., Caminha, M., Carnes-Stine, J., Caulk, P., Chiang, Y., Coyne, M., Dahlke, C., Mays, A., Dombroski, M., Donnelly, M., Ely, D., Esparham, S., Fosler, C., Gire, H., Glanowski, S., Glasser, K., Glodek, A., Gorokhov, M., Graham, K., Gropman, B., Harris, M., Heil, J., Henderson, S., Hoover, J., Jennings, D., Jordan, C., Jordan, J., Kasha, J., Kagan, L., Kraft, C., Levitsky, A., Lewis, M., Liu, X., Lopez, J., Ma, D., Majoros, W., McDaniel, J., Murphy, S., Newman, M., Nguyen, T., Nguyen, N., Nodell, M., Pan, S., Peck, J., Peterson, M., Rowe, W., SAnders, R., Scott, J., Simpson, M., Smith, T., Sprague, A.,

Stockwell, T., Turner, R., Venter, E., Wang, M., Wen, M., Wu, D., Wu, M., Xia, A., Zandieh, A., and Zhu, X. (2001). The sequence of the human genome. *Science*, 291: 1304–1351.

Vesanto, J. and Alhoniemi, E. (2000). Clustering of the self-organizing map. *IEEE Transactions on Neural Networks*, 11(3): 586–600.

Viterbi, A. (1967). Error bounds for convolutional codes and an asymptotically optimal decoding algorithm. *IEEE Transactions on Information Theory*, IT-13: 260–269.

Vlachos, M., Lin, J., Keogh, E., and Gunopulos, D. (2003). A wavelet-based anytime algorithm for K-means clustering of time series. In *Proceedings of the 3rd SIAM International Conference on Data Mining, San Francisco, CA*.

von der Malsburg, C. (1973). Self-organization of orientation sensitive cells in the striate cortex. *Kybernetik*, 14: 85–100.

Wagstaff, K., Cardie, C., Rogers, S., and Schroedl, S. (2001). Constrained *K*-means clustering with background knowledge. In *Proceedings of the 8th International Conference on Machine Learning*, pp. 577–584.

Wallace, C. and Boulton, D. (1968). An information measure for classification. Computer *Journal*, 11: 185–194.

Wallace, C. and Dowe, D. (1994). Intrinsic classification by MML—the SNOB program. In *Proceedings of the 7th Australian Joint Conference on Artificial Intelligence*, pp. 37–44.

Wang, J., Lindsay, B., Leebens-Mack, J., Cui, L., Wall, K., Miller, W., and dePamphilis, C. (2004). EST clustering error evaluation and correction. *Bioinformatics*, 20(17): 2973–2984.

Wang, H., Wang, W., Yang, J., and Yu, P. (2002). Clustering by pattern similarity in large data sets. In *Proceedings of the ACM SIGMOD International Conference on Management of Data*, pp. 394–405.

Wang, W., Yang, J., and Muntz, R. (1997). STING: A statistical information grid approach to spatial data mining. In *Proceedings of the 23rd International Conference on Very Large Data Bases*, pp. 186–195.

Wang, W., Yang, J., and Muntz, R. (1999). STING+: A approach to active spatial data mining. In *Proceedings of 15th International Conference on Data Engineering, Sydney, Australia*, pp. 116–125.

Ward, J. (1963). Hierarchical groupings to optimize an objective function. *Journal of the American statistical association*, 58: 236–244.

Wei, C., Lee, Y., and Hsu, C. (2000). Empirical comparison of fast clustering algorithms for large data sets. In *Proceedings of 33rd Hawaii International Conference on System Sciences*.

Weingessel, A. and Hornik, K. (2000). Local PCA algorithms. *IEEE Transactions on Neural Networks*, 11(6): 1242–1250.

West, D. (2001). *Introduction to graph theory, 2nd edition*. Upper Saddle River, NJ: Prentice-Hall.

Whaite, P. and Ferrie, F. (1991). From uncertainty to visual exploration. *IEEE Transactions on Pattern Analysis and Machine Intelligence*, 13(10): 1038–1049.

Williamson, J. (1996). Gaussian ARTMAP: A neural network for fast incremental learning of noisy multidimensional maps. *Neural Networks*, 9(5): 881–897.

Williamson, J. (1997). A constructive, incremental-learning network for mixture modeling and classification. *Neural Computation*, 9: 1517–1543.

Windham, M. and Culter, A. (1992). Information ratios for validating mixture analysis. *Journal of the American Statistical Association*, 87: 1188–1192.

Wong, C., Chen, C., and Su, M. (2001). A novel algorithm for data clustering. *Pattern Recognition*, 34: 425–442.

Wu, Z. and Leahy, R. (1993). An optimal graph theoretic approach to data clustering: Theory and its application to image segmentation. *IEEE Transactions on Pattern Analysis and Machine Intelligence*, 15(11): 1101–1113.

Wunsch, D. II (1991). An optoelectronic learning machine: Invention, experimentation, analysis of first hardware implementation of the ART1 neural network. Ph.D. dissertation, University of Washington.

Wunsch, D. II, Caudell, T., Capps, C., Marks, R., and Falk, R. (1993). An optoelectronic implementation of the adaptive resonance neural network. *IEEE Transactions on Neural Networks*, 4(4): 673–684.

Wunsch, D. II and Mulder, S. (2004). Evolutionary algorithms, Markov decision processes, adaptive critic designs, and clustering: commonalities, hybridization, and performance. In *Proceedings of IEEE International Conference on Intelligent Sensing and Information Processing*.

Xie, X. and Beni, G. (1991). A validity measure for fuzzy clustering. *IEEE Transactions on Pattern Analysis and Machine Intelligence*, 13(8): 841–847.

Xiong, Y. and Yeung, D. (2004). Time series clustering with ARMA mixtures. *Pattern Recognition*, 37: 1675–1689.

Xu, J., Erdogmus, D., Jenssen, R., and Principe, J. (2005). An information-theoretic perspective to kernel independent components analysis. In *Proceedings of IEEE International Conference on Acoustics, Speech, and Signal Processing—ICASS2005*, Vol. 5, pp. 249–252.

Xu, L. (1997). Bayesian Ying-Yang machine, clustering and number of clusters. *Pattern Recognition Letters*, 18: 1167–1178.

Xu, L. (1998). Rival penalized competitive learning, finite mixture, and multisets clustering. In *Proceedings of IEEE International Joint Conference on Neural Networks 1998*, Vol. 3, pp. 2525–2530.

Xu, L. (2001). Best harmony, unified RPCL and automated model Selection for unsupervised and supervised learning on Gaussian mixtures, ME-RBF models and three-layer nets. *International Journal of Neural Systems*, 11(1): 3–69.

Xu, L. (2002). BYY harmony learning, structural RPCL, and topological self-organizing on mixture models. *Neural Networks*, 15(8/9): 1125–1151.

Xu, R., Anagnostopoulos, G., and Wunsch, D. II (2002). Tissue classification through analysis of gene expression data using a new family of ART architectures. In *Proceedings of International Joint Conference on Neural Networks 2002*, Vol. 1, pp. 300–304.

Xu, R., Anagnostopoulos, G., and Wunsch, D. II (2007a). Multi-class cancer classification using semi-supervised ellipsoid ARTMAP and particle swarm optimization with gene expression data. *IEEE/ACM Transactions on Computational Biology and Bioinformatics*, 4(1): 65–77.

Xu, X., Ester, M., Kriegel, H., and Sander, J. (1998). A distribution-based clustering algorithm for mining in large spatial databases. In *Proceedings of 14th International Conference on Data Engineering, Orlando, FL*, pp. 324–331.

Xu, L., Krzyżak, A., and Oja, E. (1993). Rival penalized competitive learning for clustering analysis, RBF net, and curve detection. *IEEE Transactions on Neural Networks*, 4(4): 636–649.

Xu, Y., Olman, V., and Xu, D. (2002). Clustering gene expression data using graph-theoretic approach: An application of minimum spanning trees. *Bioinformatics*, 18(4): 536–545.

Xu, R. and Wunsch, D. II (2005). Survey of clustering algorithms. *IEEE Transactions on Neural Networks*, 16(3): 645–678.

Xu, R., Wunsch D. II, and Frank, R. (2007b). Inference of genetic regulatory networks with recurrent neural network models using particle swarm optimization. *IEEE/ACM Transactions on Computational Biology and Bioinformatics*, to appear.

Xu, H., Wang, H., and Li, C. (2002). Fuzzy tabu search method for the clustering problem. In *Proceedings of the 2002 International Conference on Machine Learning and Cybernetics*, Vol. 2, pp. 876–880.

Yager, R. (2000). Intelligent control of the hierarchical agglomerative clustering process. *IEEE Transactions on Systems, Man, and Cybernetics*, 30(6): 835–845.

Yager, R. and Filev, D. (1994). Approximate clustering via the mountain method. *IEEE Transactions on Systems, Man, and Cybernetics*, 24(8): 1279–1284.

Yair, E., Zeger, K., and Gersho, A. (1992). Competitive learning and soft competition for vector quantizer design. *IEEE Transactions on Signal Processing*, 40(2): 294–309.

Yang, J., Estivill-Castro, V., and Chalup, S. (2002). Support vector clustering through proximity graph modeling. In *Proceedings of the 9th International Conference on Neural Information Processing*, Vol. 2, pp. 898–903.

Yang, M. and Wu, K. (2006). Unsupervised possibilistic clustering. *Pattern Recognition*, 39: 5–21.

Yeung, K., Haynor, D., and Ruzzo, W. (2001). Validating clustering for gene expression data. *Bioinformatics*, 17(4): 309–318.

Yiu, M. and Mamoulis, N. (2005). Iterative projected clustering by subspace mining. *IEEE Transactions on Knowledge and Data Engineering*, 17(2): 176–189.

Yip, K., Cheung, D., and Ng, M. (2004). HARP: A practical projected clustering algorithm. *IEEE Transactions on Knowledge and Data Engineering*, 16(11): 1387–1397.

Yona, G., Linial, N., and Linial, M. (1999). ProtoMap: Automatic classification of protein sequences, a hierarchy of protein families, and local maps of the protein space. *Proteins: Structure, Function and Genetics*, 37: 360–378.

Young, F. and Hamer, R. (1987). *Multidimensional scaling: History, theory, and applications*. Hillsdale, NJ: Lawrence Erlbaum Associates.

Yu, D., Chatterjee, S., and Zhang, A. (1999). Efficiently detecting arbitrary shaped clusters in image databases. In *Proceedings of 11th IEEE International Conference on Tools with Artificial Intelligence*, pp. 187–194.

Zadeh, L. (1965). Fuzzy sets. *Information and Control*, 8: 338–353.

Zahn, C. (1971). Graph-theoretical methods for detecting and describing Gestalt clusters. *IEEE Transactions on Computers*, C20: 68–86.

Zha, H., He, X., Ding, C., Simon, H., and Gu, M. (2001). Spectral relaxation for K-means clustering. In *Proceedings of Neural Information Processing Systems*, Vol. 14, pp. 1057–1064.

Zhang, J. and Leung, Y. (2004). Improved possibilistic *c*-means clustering algorithms. *IEEE Transactions on Fuzzy Systems*, 12(2): 209–217.

Zhang, Y. and Liu, Z. (2002). Self-splitting competitive learning: A new on-line clustering paradigm. *IEEE Transactions on Neural Networks*, 13(2): 369–380.

Zhang, T., Ramakrishnan, R., and Livny, M. (1996) BIRCH: An efficient data clustering method for very large databases. In *Proceedings of the ACM SIGMOD Conference on Management of Data*, pp. 103–114.

Zhang, R. and Rudnicky, A. (2002). A large scale clustering scheme for kernel K-means. In *Proceedings of the 16th International Conference on Pattern Recognition*, Vol. 4, pp. 289–292.

Zhang, Z., Schwartz, S., Wagner, L., and Miller, W. (2000). A greedy algorithm for aligning DNA sequences. *Journal of Computational Biology*, 7: 203–214.

Zhao, Y. and Karypis, G. (2005a). Hierarchical clustering algorithms for document datasets. *Data Mining and Knowledge Discovery*, 10: 141–168.

Zhao, Y. and Karypis, G. (2005b). Data clustering in life sciences. *Molecular Biotechnology*, 31(1): 55–80.

Zhong, W., Altun, G., Harrison, R., Tai, P., and Pan, Y. (2005). Improved K-means clustering algorithm for exploring local protein sequence motifs representing common structural property. *IEEE Transactions on Nanobioscience*, 4(3): 255–265.

Zhuang, X., Huang, Y., Palaniappan, K., and Zhao, Y. (1996). Gaussian mixture density modeling, decomposition, and applications. *IEEE Transactions on Image Processing*, 5(9): 1293–1302.

Zubin, T. (1938). A technique for measuring like-mindedness. *Journal of Abnormal and Social Psychology*, 33: 508–516.

AUTHOR INDEX

Y

Z

SUBJECT INDEX

Clustering, by Rui Xu and Donald C. Wunsch, II

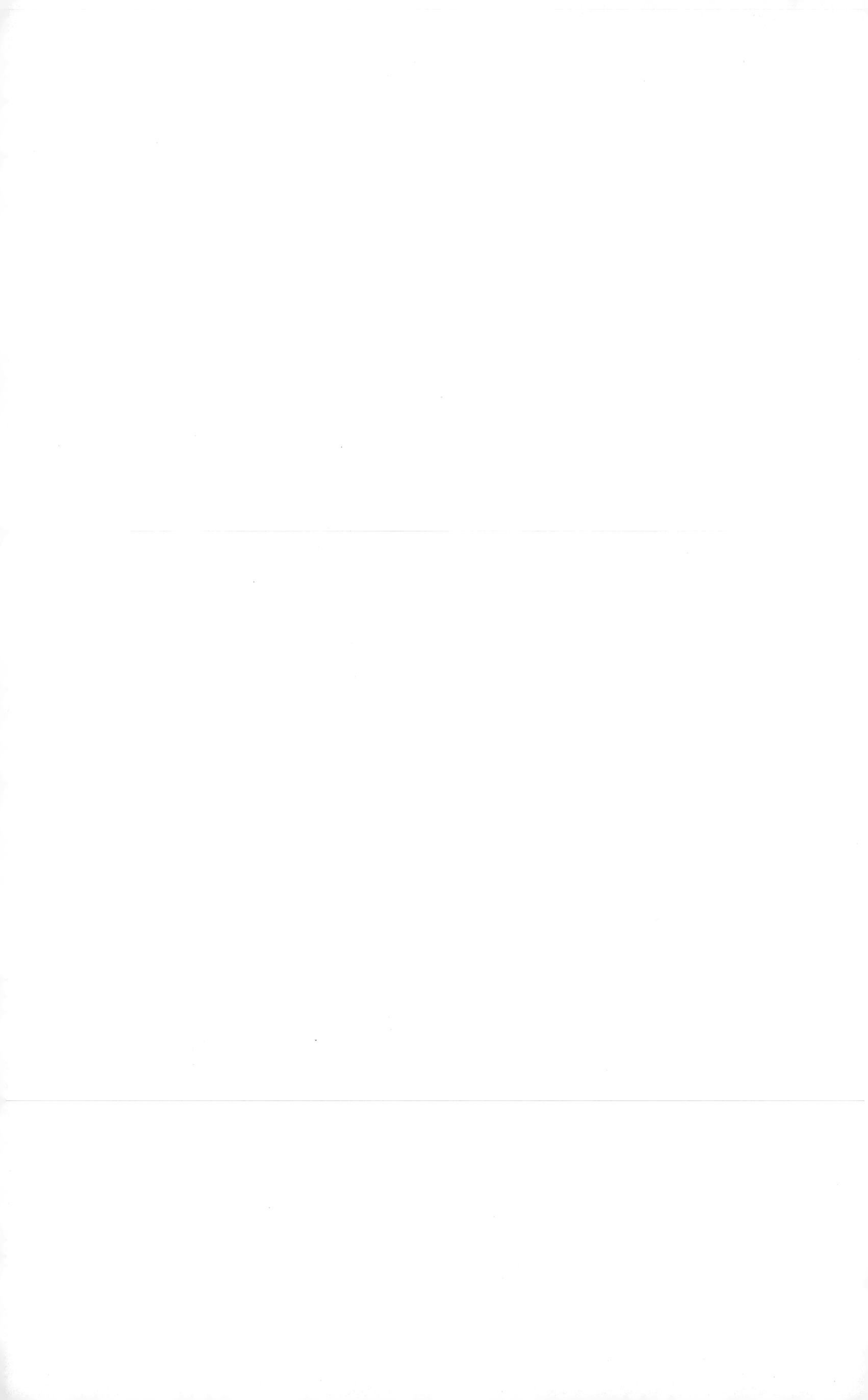